Theories of Population Variation
in Genes and Genomes

Theories of Population Variation in Genes and Genomes

FREDDY BUGGE CHRISTIANSEN

PRINCETON UNIVERSITY PRESS
Princeton and Oxford
2008

Published by Princeton University Press
41 William Street, Princeton, New Jersey 08540

In the United Kingdom: Princeton University Press
3 Market Place, Woodstock, Oxfordshire OX20 1SY

Library of Congress Control Number: 2007939437

ISBN: 978-0-691-13367-6 (acid-free paper)

British Library Cataloging-in-Publication Data is available

This book has been composed in LaTeX

The publisher would like to acknowledge the author
of this volume for providing the camera-ready copy
from which this book was printed.

Printed on acid-free paper. ∞

press.princeton.edu

Printed in Malaysia

10 9 8 7 6 5 4 3 2 1

Contents

Preface and Acknowledgments

This book evolved from lecture notes written for the course "Molecular Population Genetics" that I teach with Mikkel Heide Schierup at the Department of Biology, University of Aarhus. A second text in the course is Hein, Schierup, and Wiuf's (2005) book on coalescent analysis. Students of biology and bioinformatics have followed this course. They were recruited from computer science, mathematics, and statistics, in addition to biology, molecular biology, and other sciences where a biological view is fundamental.

Teaching, and therefore also lecture notes, are subject to a barrage of critiques and suggestions from students. I see this as a blessing for which I am grateful, and I truly appreciate the contributions of the students and teaching assistants of the course. My colleagues at the Bioinformatics Research Center, University of Aarhus—Tomas Bataillon, Ole Christensen, Asger Hobolt, Leif Schauser, Mikkel Schierup, and Carsten Wiuf—have read parts or all of the ancestral notes or the book manuscript. I gratefully acknowledge their comments to the text and their suggestions of good and illustrative examples. Thoughtful remarks and suggestions in a similar vein were offered by Bernt Guldbrandtsen and Dave Parker. As the manuscript took form I spent a pleasant week at Lund University discussing it with Bengt Olle Bengtsson, Torbjörn Säll, and their students. Their comments were critical in kick-starting a book format and leave the lecture notes behind. Tomas Bataillon provided the idea behind Figure 9.15, Ivar Heuch updated me on the *Acraea* story, and Rikke Bakker Jørgensen provided up-to-date information for Table 10.1.

I have been accused of not including correct spelling and grammar in my list of priorities concerning writing. That is not correct, and I value being repeatedly rescued from making a fool of myself by my wife Else Løvdal Nielsen. Most of all, however, I am grateful for her suggestions on phrasing of descriptions and presentation of arguments.

Introduction

Genomes and genomic variation entered into the study of genetic variation in natural populations in this century. The human genome sequencing projects led to increasingly affordable procedures for studying sequence variation, and by 2001[1] population genetic studies of genes were already dominated by analyses of sequence variation. During the work on the human sequence more than a million places in the genome were discovered to vary among 24 individuals representing the ethnic variation in the world.[2] This corresponds to one single nucleotide polymorphism for every 2000 base pairs in the DNA of the human genome, corresponding to a recombination distance of less than one crossover in 10,000 meioses—quite a dense map, and even denser ones are now available because more genomes are sequenced, and hence more single nucleotide polymorphisms are revealed as differences among more people.[3] The genome sequence of the common chimpanzee has also been determined, allowing the recent evolution of the human genome sequence to be addressed. Scores of genomes in other animals, fungi, and plants have been sequenced, and even more are currently under way.

Such a brief account can only provide a superficial impression of the amount of population data currently available, presently accumulating at an immense rate, and expected to keep increasing in the foreseeable future. This offers ample opportunity for investigations into the history and dynamics of gene and genome variation in natural populations. Fundamental questions of evolution may be asked and new ones formulated. To profit from this wealth of data, however, great challenges have to be overcome. Population genetics, as most branches of genetics, is thus in the most exciting of circumstances for a scientific field. For scientists the wish "may you live in interesting times" is surely not a curse—given, of course, that scientific matters catch the public awareness.

Population genetics has been around for about a century, emerging right at the dawn of genetics. The field can be defined as the study of the distribution of hereditary variation across time and space in species and populations. A human population is in this context a biologically reasonable assemblage of people. Individuals usually find their mates within their own population, and it commonly comprises all humans inhabiting a more or less well-defined area.

---

[1] The draft human genome was published by Lander et al. (2001) and Venter et al. (2001).
[2] Sachidanandam et al. (2001).
[3] Every person adds new genetic variation (Levy et al. 2007)

1

All of humanity may in some circumstances be considered a population, but usually a more restricted definition is used, for instance a city and its environs, an island or peninsula, or a continent. Denmark, Faroe Islands, and Greenland are three distinct populations in the Kingdom of Denmark, even though each is of mixed origin and regularly receives immigrants from the others, and from the rest of the world, for that matter. Each of the three may be further separated into local populations on islands—real islands in Denmark and the Faroes, and islands of habitable areas in Greenland.

The description of population variation has two foci: the general understanding of biological evolution and the application of genetic variation for human welfare. Mendelian genetics supplied Darwin's theory of evolution with crucial elements, many of which were developed and matured by population geneticists in the decades before the so-called neo-Darwinian synthesis around 1930. Medical applications are centered on the understanding of the prevalence of rare genetic and hereditary diseases, and on the basis of the hereditary aspects of many common diseases.[4] This has driven recent developments in human genetics, not least the sequencing of the human genome. Genetic investigations of human diseases necessarily require population genetic studies—inheritance and segregation are observed in existing families. The methods used in these areas have benefited the development of other aspects of population genetics, in particular applications to animal and plant breeding.

The field of population genetics builds on experiences from observations and experiments and is supported by a well-developed framework of theory founded early in the twentieth century.[5] The basic laws of genetic transmission are probability laws. Genetics and, in particular, population genetics has thus always relied on observations interpreted through statistical analysis, and many developments in statistics have their origin in genetic applications. Population genetic theory entertains the whole spectrum characteristic of population sciences from statistical modeling to descriptive dynamic theory. Through time, the weight of the statistical and descriptive aspects of theory have changed—as well as the interest in theoretical or empirical developments of the field in general. Surprisingly, these two historical oscillations seem largely uncorrelated, but at present we are in a statistical and empirical era due to important breakthroughs within both of these approaches in the recent past.

The analysis and interpretation of data on molecular genetic variation relies on population genetic theory, coalescent theory in particular, and the implementation and execution of such analysis requires skills in computer science and statistics. Much of this activity occurs within the field of bioinformatics. Accordingly, the writing of this volume was largely carried out at the Bioinformatics Research Center (BiRC) at the University of Aarhus. One aim is to communicate some of the experiences from a century of population genetics, and relate them to contemporary developments for the benefit of students and

---

[4]Everyone is expected to carry genetic determinants of rare diseases and susceptility genes for common diseases. This expectation is typified in genome sequences of the individual (Levy et al. 2007)

[5]See Provine (1971).

colleagues in bioinformatics. A second aim is to participate in the process of incorporating some of the recent developments of population genetics into biology teaching.

The two aims seem contradictory, and the only reason I attempt to combine them is that my students taught me that it is possible. For some years, students of bioinformatics have attended the biology course, which developed into the present book, and biology students frequented courses in the elements of bioinformatics. Population genetics is a field of study where formal and quantitative theory play an integral part. The language of such theory includes mathematical formulations and reasoning, and throughout its history the field has attracted attention from students with a background in mathematics and its applications. Population genetic theory, however, resides firmly within biology, because its issues arise from biological phenomena, and its results refer to those phenomena. Any biology student of population genetics considers theory as a background for his or her activities—variation exists only in the level of theory deemed necessary.

The structure of the book tries to accommodate the broad range of potential readers. Short introductions to genetic subjects and concepts are given at appropriate places to avoid the overwhelming task of studying an introductory text in genetics. In a course setting, students with a biological background naturally offer assistance—and welcome the recap of basic genetics. The mathematical requirements correspond to the introductory mathematic courses given in undergraduate biology teaching. They include few formal requirements beyond elementary algebra and calculus, but assume that fear of adding and multiplying letters has been alleviated. Students of biology, many of whom are decidedly not theoretically inclined, have used the ancestral lecture notes with success, but they needed to realize that the biological content is the focus. The level of statistical background seems to vary considerably more among biology students than does their baground in mathematics, so the necessary statistical concepts are briefly explained in Appendix A. Still, with a reasonable consensus on minimal requirements, a textbook should meet and challenge the students on their home ground, while keeping the material palatable for all readers. In the running text calculations and mathematical arguments are relegated to text boxes, and if their contents require difficult or lengthy arguments, they are marked by a * in front of the box title. Most are, however, fairly uncomplicated and do not require a high level of mathematical abilities—many of the arguments may look more complicated for biologists than they really are. Their presence is intended to tickle the curiosity of readers with a more mathematical background. On a similar note, some of the scattered exercises and footnotes have the warning star. The answers to the exercises should often vary with the background of the student. Solutions suggested in Appendix B tend to be short and incomplete, to leave room for developments along lines of personal interests.

The focus of the book is on the theoretical background of contemporary population genetics, while acknowledging that population genetics is a subject that grew and continues to grow in the close interplay between empiricism and theory. Theoretical results refer to the material world, and observations rarely make sense without reference to the theoretical background of the field. This interplay is acknowledged in the discussion of empirical investigations that range the history of population genetics. The coverage of observations and experiments is in no way intended to be broad or representative of the field. Rather, the empirical references are chosen mainly for their qualities as illustrations of the development of thoughts within the science of population genetics.

Part I

Genetic Variation

The information in biological inheritance is carried by genes. The genes of an individual human being are copies of genes that his or her parents transmitted through the egg cell and the sperm cell that united to form its original cell. A stretch of DNA sequence on a chromosome in the nucleus of one of our cells traces its origin to a stretch of DNA in either the egg or the sperm that formed us, and from there the ancestral sequences form an unbroken line back through the history of life. The genetic variation in a population of humans therefore originates in the genes transmitted to them from the population of their parents, and we may study the inheritance of the total of the genes carried by the population. Population genetics thus describes the genetic variation in a population and its transmission between generations. The rules of transmission originate in the fundamental laws of inheritance described by Gregor Mendel in 1866. Each individual carries two versions of a particular gene, one from each parent, and they are transmitted according to Mendel's first law of inheritance, which states that an offspring is equally likely to receive either of the two genes carried by a given parent. An offspring thus receives one version of the gene from each parent, and the gene transmitted by a parent is picked as a random copy of the two versions available. In terms of the life cycle of the individual, as it forms, it receives a version of the gene from each parent, and when it subsequently reproduces, it transmits the gene of maternal origin on average in half of the cases and the paternal gene in the other half.

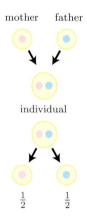

Mendel's law lifted to the level of a population of individuals posits that the parental genes each have the same probability of being transmitted to the offspring population. This transmission rule implies conservation of population variation—exactly the property of hereditary transmission that Darwin's theory of evolution was in need of. Darwin's theory assumed direct transmission of traits of a character from parents to offspring, but he lacked a mechanism for the maintenance of population variation. Mendel's law defines biological inheritance as indirect, that is, genes determining a trait and not the trait itself are inherited, and conservation of the heterogeneity within the individual immediately produces conservation of the genetic variability among individuals in a population.

Population genetics, and for that matter genetics, would be topics void of issue without genetic variation, and a basic introduction to the study of genetic variation is offered in Chapter 1 (a short introduction to Mendelian genetics is also offered and a few basic genetic concepts are introduced in Chapter 1—further genetic prerequisites are introduced as needed). Variation abounds and its emergence and decay are main themes in this volume, as is the consequence of genomic structure for the distribution and dynamics of genetic variation. The study and description of variation is therefore an integral part of population genetics, and it commences in Chapter 2.

Chapter 1

Genetics

Mendel based his description of heredity on experiments with the edible pea, *Pisum sativum*. Offspring of a cross between plants from a true-breeding line[1] of yellow peas (first parental line P_1) and one of green peas (P_2) gave yellow offspring (first filial generation F_1) peas in the pods. Selfing plants grown from F_1 peas gave most pods with both yellow and green peas, the F_2 generation. These qualitative results were known among plant breeders, but Mendel contributed a quantitative analysis. The 258 plants grown from the F_1 peas yielded 8023 pea seeds (the F_2 generation), of which 6022 were yellow and 2001 green, and he noted that the observed frequencies were close to $\frac{3}{4}$ and $\frac{1}{4}$, respectively. He made similar observations on six other characters (seed shape, flower color, plant height, ...), and having been educated as a mathematician, he placed much emphasis on these simple proportions and devised a model to explain their occurrence.

The F_1 peas in the experiment showed the yellow trait, which is then called the *dominant* trait. The green trait was obviously transmitted through the F_1 peas because it reappeared in F_2; as it failed to appear in F_1, it is called the *recessive* trait. Mendel hypothesized that the hybrid plants (the F_1 plants) carried determinants for both traits, and that plants from the true-breeding lines only carried determinants (or factors) for one of the traits. The factors are, say, A and a, with the capital letter corresponding to the dominant trait. Further, he suggested that hybrid plants transmitted one determinant through pollen and one through the egg cell, and the two determinants of hybrids were transmitted equally often through each type of gamete—egg cell or pollen. The determinants acted like discrete particles. With the additional assumption that gametes unite at random with respect to the determinant carried, the segregation of traits among F_2 peas becomes $\frac{3}{4}$ and $\frac{1}{4}$, the proportions suggested by the experiments.

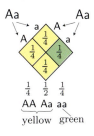

---

[1]A variety of plants that look the same generation after generation when crossed among themselves.

The key hypothesis of equal segregation from the hybrids was thoroughly tested by Mendel in a series of additional experiments, which are still part of the Mendelian analysis of inheritance. The *backcross* performs a direct check in that F_1 plants are pollinated by individuals from the green recessive line. The seed set is then expected to segregate evenly in the two colors:

$$\boxed{F_1} \times \boxed{P_2} \rightarrow \boxed{\tfrac{1}{2}} \ \boxed{\tfrac{1}{2}}$$

The segregation in F_2 may be checked by observing the seed set by letting plants grown from F_2 peas self-fertilize to form F_3. All plants from green peas are true breeding, and the expected two types of yellow plants emerge as $\frac{1}{3}$ true-breeding yellow and $\frac{2}{3}$ showing the segregation characteristic of selfed F_1 plants.[2] Alternatively, the F_2 individuals may be crossed back to the green parental line. Then the offspring of plants grown from yellow peas either segregate like a backcross or have entirely yellow peas:

$$\boxed{F_2} \times \boxed{P_2} \rightarrow \begin{cases} \tfrac{1}{3} & \boxed{1} \\[2mm] \tfrac{2}{3} & \boxed{\tfrac{1}{2}} \ \boxed{\tfrac{1}{2}} \end{cases}$$

Mendel's observations bore out his hypotheses. He distributed his description of inheritance broadly among his contemporaries, but it was not accepted as generally applicable. The missing element was probably corroborating evidence for material objects that segregate like Mendelian factors.

General acceptance came about shortly after three investigators, de Vries, Correns, and von Tschermak, independently "rediscovered" Mendel's law in the year 1900 (see Stern and Sherwood 1966). These events came shortly after the description (in the last decade of the nineteenth century) of the meiotic cell division, where chromosomes behave much like Mendelian determinants. By then it was known that cells of most higher organisms, peas in particular, are diploid, carrying a pair of each morphological chromosome, and that their gametes are haploid, with only a single complement of chromosomes.[3] Diploid cells are formed by fertilization—the fusion of gametes to form a diploid zygote, which is the fertilized egg and thus the first cell in the diploid phase. The zygote in turn proliferates by mitotic cell divisions, which maintains the genetic constitution of the zygote in the cells of the body in multicellular

Life Cycle

female   male
gamete   gamete

zygote

gametes

---

[2] Unless a large number of peas are counted per plant, a deviation of the segregation from 1:2 is expected, because with probability $(\frac{3}{4})^n$, an Aa plant will segregate like an AA plant.

[3] The drawing is simplified and shows only one chromosome present as a homologous pair in the zygote. The number of chromosomes varies among species. Flies and mosquitos have few, humans have 46, and ferns have hundreds.

organisms. To complete the life cycle, meiosis reduces the diploid cell to haploid gametes (details of this process are given in Chapter 6).

Soon experiments established the generality of Mendel's law of inheritance. The determinants were given the name *genes* Johannsen (1905, 1909), and the distinguishable types of homologous genes were named *alleles* (Bateson and Saunders 1902). The genes determining pea color carried by the two lines in Mendel's experiments are of allele type A and a, respectively. A is designated the dominant allele and a the recessive. The genetic constitution of individuals is their *genotype*. With respect to the gene for pea color, the genotypes are AA, Aa, and aa. The pure lines are *homozygotes* AA or aa, and the hybrid type Aa is a *heterozygote* (Bateson and Saunders 1902).

1.1   Genetic Variation

A human population varies a lot. Usually we can easily identify people from their appearance, and a fair proportion of such characteristics are hereditary—we can recognize family resemblance. In addition, different human populations differ from each other, at least in their bulk appearance, allowing for very similar individuals even in rather different populations. The subject of population genetics is such population variation, but a population is delimited as a biological entity, not a demographic, social, or administrative unit. A *population* is thus a collection of interbreeding individuals.

Mitosis

The two homologous or *allelic* genes brought together in the formation of a diploid zygote are transmitted unaltered when at sexual maturity the individual produces gametes. Rare exceptions from this constancy occur (mutation, considered in Chapter 4). This conservative transmission of genes from parents to offspring causes conservation of the population frequencies of the various variants, or alleles, when Mendelian segregation prevails and genotypes show equal fertility and survival.

Classical population genetics was developed for the study of variation in Mendelian characters in natural and experimental populations. The scenario is a gene that exists in two allelic forms in the population, and the presence of the two alleles, say, A and a, is described by the gene frequencies p of allele A and q of allele a, where $p + q = 1$ (see Box 1).[4] Only one of the gene frequencies need be known (the other is easily calculated), and the model of population variation is therefore one dimensional. In addition, the two alleles are symmetric, and results show symmetry in the variables p and q. This simple modeling framework anchored in the laws of inheritance has made all workers in the fields of

---

[4]The way allele names are written in Box 1 follows the convention of writing them in *italic* font. The present use of a **special** font was adopted to avoid confusion with mathematical symbols written in an *italic* font.

Box 1: Description of a population

In Kalø Cove, Denmark, a sample of 12,607 adult eelpouts (*Zoarces viviparus*, a teleostean fish) was investigated for their genotype with respect to variation in the gene that codes for the enzyme esterase III (Christiansen et al. 1977). The variation was known to be caused by two alleles *EstIII*^{1} and *EstIII*^{2}.

Genotypes in *Zoarces viviparus*

genotype	11	12	22	Σ
number	1701	5676	5230	12,607

In this sample we observed $2 \times 1701 + 5676 = 9078$ genes of allele *EstIII*^{1} and $5676 + 2 \times 5230 = 16,136$ of allele *EstIII*^{2} among the 25,214 genes observed. The observed gene frequencies of the two alleles are thus $p_1 = 0.640$ and $p_2 = 0.360$.

---

population and evolutionary genetics view their fields as theoretically based.

Variation in a population described by a few discrete traits is called a *polymorphism*, and if the various traits are reflections of allelic variation, it is called a genetic polymorphism. An example of a two-allele polymorphism is given in Box 1, and a sample is represented in Table 1.1. The genetic polymorphism is

Table 1.1: Genotypes of n individuals

genotype	AA	Aa	aa	Σ
number	n_{AA}	n_{Aa}	n_{aa}	n

summarized in the gene frequencies

$$p = \frac{2n_{AA} + n_{Aa}}{2n} \quad \text{and} \quad q = \frac{n_{Aa} + 2n_{aa}}{2n}.$$

The opposite of polymorphism is *monomorphism*. Multiallelic polymorphisms were found early in the twentieth century, for instance the ABO blood groups in humans determined by the three alleles I_A, I_B, and I_O. Alleles I_A and I_B are dominant to I_O, and I_A and I_B are *codominant* in that the heterozygote $I_A I_B$ displays the traits of both homozygotes. Four ABO blood types thus exist: A ($I_A I_A$ or $I_A I_O$), B ($I_B I_B$ or $I_B I_O$), AB ($I_A I_B$), and O ($I_O I_O$). Genetic polymorphism was considered quite rare, and multiallelic polymorphism even rarer, because most known genetic polymorphisms could be understood as two-allele polymorphisms. But alas, by 1966 this simple description of natural variation was superseded when Harris (1966) in humans and Lewontin and Hubby (1966) in fruit flies (of the species *Drosophila pseudoobscura*) showed immense protein polymorphism in natural populations, usually segregating multiple alleles. Protein variation was investigated by the method of electrophoresis (Box 2).

Genetic variation related to protein function, especially that of enzymes, was well known at that time. In 1902, the physician Archibald Garrod sug-

Box 2: Electrophoretically defined genetic polymorphisms

Electrophoresis is a biochemical procedure for analyzing charged macro-molecules. The *EstIII* polymorphism in Box 1 is revealed by protein electrophoresis. A tissue sample (brain) is taken from each individual and an extract is placed on an electrically conducting starch gel. Voltage is applied across the gel, causing proteins to migrate from the application slot (marked 0). After a while the voltage is turned off, and the distance a given protein has migrated depends on its mobility, which is a function of the ionic charge of the molecule in the given buffer, and of the resistance to movement inflicted by the gel. Variation in these properties is mainly caused by variation in the amino acid

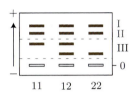

sequence of the protein. Now, staining the gel for proteins would just show a smear, but soaking the gel in a solution that contains a substrate of the enzyme of interest will reveal the presence of the enzyme when the product of the enzymatic reaction is stained.

The figure shows the three esterase III phenotypes corresponding to the geno-types (see Box 1) as they appear on a gel. The esterase III variation is seen between the thin dashed lines. Allele $EstIII^1$ produces the fastest moving protein corresponding to the band farthest from the origin; $EstIII^2$ produces the slowest one closer to the origin. The proteins in a homozygote thus congregate in only one band, whereas the heterozygote shows both bands—the alleles are codominant. Mendelian analysis confirms this interpretation of the bands.

The two bands in front (above) of the two esterase III bands are the enzymes esterase I and II. The single esterase II band is always present, so the population is monomorphic for this character. About 90 percent of individuals have the esterase I band, while the remainder show no band in that region of the gel. This may be interpreted as segregation of an allele $EstI^+$ that makes the enzyme, and an allele $EstI^-$ that does not produce a functional enzyme, often called a null allele. Null alleles are usually recessive because the difference between the amount of enzyme in the bands on the gel corresponding to the genotypes $EstI^+ EstI^+$ and $EstI^+ EstI^-$ is hard to detect, even if the difference corresponds to a factor of two.

The *EstIII* polymorphism is characteristic of a simple protein made up of a single contiguous polypeptide—a so-called monomeric protein. Many functional proteins are dimers or polymers made up of two or several subunits. The regular human hemoglobin, for instance, is made up of four protein subunits, two α and two β subunits coded for by two different genes. Simpler dimers are made up of subunits coded by the same gene, and the three genotypes of a two-allele polymorphism are two single-banded homozygotes and a heterozygote with three bands—two like those of the homozygotes, called homodimeric bands, and a band consisting of heterodimers. If the subunits combine at random, the amount of protein in the three bands is found in the ratio of 1:2:1.

gested that the recessive disease alkaptonuria is caused by a defective enzyme in the metabolism of phenylalanine and tyrosine (amino acids). He described such diseases as "inborn errors of metabolism." He thus discovered one of the fundamental functions of genes, namely to produce enzymes—formulated in the 1940s as the "one gene one enzyme" hypothesis by George Beadle and Edward Tatum and based on work on the biochemical genetics of *Neurospora* fungi. This simple description of the physiological function of genes is very applicable, and in general, genes control the production of proteins, including enzymes and structural proteins, even though other kinds of functional genes exist.

Alkaptonuria is but one of scores of rare inborn errors of metabolism caused by usually recessive alleles of the gene that codes for a crucial enzyme. The disease allele produces a defective enzyme or no enzyme at all, thus causing the malfunction. Attention to these alleles is caused by the familial aggregation of the disease, that is, by the phenotype and its aggregation in sibships.[5]

The link between genes and proteins was firmly established by the mid twentieth century. Investigations in genetics and molecular biology led to the development of very sensitive methods to analyze even small differences between proteins, and among these, electrophoresis had matured into a population genetic tool by the early 1960s. The method allowed geneticists to probe into hitherto unseen biochemical traits (Box 2). For instance, hemoglobin may easily be isolated from an individual, purified in large amounts, and rendered visible on a gel by the red color of the protein. In many organisms this disclosed variation in mobility among bands. Using enzyme electrophoresis, however, proteins present in very low concentrations in a tissue could be "stained" by way of their specific catalytic capacity (Box 2). Given access to a stain for the specific metabolites and accepting that enzymes are the primary products of gene action, enzyme electrophoresis provided a way of probing the amount of variation at a level close to the gene.

Armed with a series of recipes for the electrophoretic investigation of enzymes, Harris (1966) in humans and Hubby and Lewontin (1966) in *Drosophila pseudoobscura* evaluated the amount of genetic variation in natural populations. What is key is not the quantities they found, but rather that they were a lot higher than expected. The real achievement was that they established population genetics as founded in Mendel's indirect inheritance. Thereafter genetic variation can be addressed without referring to its function. This was revolutionary, and it threw the field into a turmoil for several decades—even today more or less implicit references to those debates crop up in the literature. The quarrel is known as *the neutralist–selectionist controversy*. The resolution, however, is straightforward and originates in the neo-Darwinian thesis of indirect inheritance: *Heritable variation at the phenotypic level requires variation at the genotypic level, but genetic variation need not produce phenotypic variation—* a statement that could be dubbed the *central dogma* of population genetics.

---

[5]Rare recessive diseases are not hereditary in the ordinary sense. The reason is the low probability that a gamete from an individual with the disease is united with a gamete carrying the defective allele. In addition, many inborn errors of metabolism are genetically lethal, because the afflicted person dies before the age of sexual maturity.

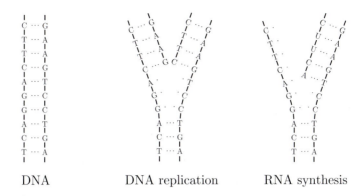

DNA   DNA replication   RNA synthesis

Figure 1.1: The principles of replication and transcription.

It's a one-way street. Natural selection occurs at the phenotypic level, so gene variation may well exist that does not participate in the current processes of Darwinian evolution. Notice, however, that nothing in Darwin's theory of evolution requires phenotypic variation to cause natural selection. On the other hand, for evolution by natural selection to occur, heritable phenotypic variation must be present.

1.1.1   Gene structure and function

The description of the basic function of the gene was formalized after the discovery in 1944 by Avery, MacLeod, and McCarthy that hereditary information is stored in chromosomal DNA (deoxyribonucleic acid). The double-helix structure of this molecule found by Watson and Crick in 1953 suggested a mechanism for the stable storage of information in cells and transmission of it from cell to cell—and thus ultimately from parents to offspring. The DNA molecule is a polymer made up of *nucleotides* consisting of the sugar deoxyribose with a phosphate group and a base attached. Four nucleotides designated T, C, A, and G exist (named after the DNA bases thymine, cytosine, adenine, and guanine). The polymer is formed by linking the phosphate group on one sugar to another sugar, forming a linear DNA strand. DNA occurs in the chromosomes as a double-stranded helix, where the two strands are intertwined and linked by characteristic pairing of the bases such that T pairs with A and C pairs with G. Thymine and cytosine are relatively small molecules called pyrimidines, while adenine and guanine are larger purines, and the base pairing in double-stranded DNA is always between a purine and a pyrimidine.

    Replication of the DNA occurs by opening the helix and synthesizing two double-stranded helices, one on each of the strands of the old molecule (Figure 1.1). This process ensures a highly reliable synthesis of two copies of the original DNA molecule, and thus exact copying as required in Mendel's law (rare errors, mutations, occur, and we return to those in Chapter 4). The DNA of a chromosome exists as one double-helical molecule, and each string stays intact during replication (as shown by Meselson and Stahl in 1958). The synthesis of DNA occurs in the cell interphase well before the cell divides.

The information in the DNA of a gene is made available to the cell by a process called *transcription*, where an RNA (ribonucleic acid) copy of one of the strands is synthesized in the same way as in replication (Figure 1.1). The RNA molecule is very similar to the DNA molecule, except that the backbone of the molecule contains ribose instead of deoxyribose, and U (uracil, a pyrimidine) is exchanged for T. The RNA copy of the gene is called messenger RNA or mRNA. This molecule is transported from the nucleus of the cell to the cytoplasm, where the ribosomes *translate* the messenger into the amino acid sequence of a protein. Only pieces of DNA are transcribed into RNA, and for transcripts that contain information about protein, these pieces carry the information for at least one gene. The translated part of the mRNA is described as *an open reading frame* (ORF) of the DNA; ORFs occupy only a few percent of the DNA in genomes of eukaryotes (animals, fungi, plants, and protists).

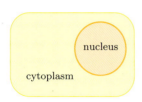

The translation is based on the *genetic code*. It specifies amino acids in terms of nonoverlapping triplets of RNA bases—the so-called *codons*. All triplets are interpretable by the ribosome. The translation always starts at the codon AUG, which also codes for methionine. Three signals cause the ribosome to stop the protein synthesis: the stop codons UAA, UAG, and UGA. The four bases define 64 triplets, and three are stop codons, leaving 61 codons to code for 20 amino acids (Table 1.2). The empirical evidence for this description was based on a series of ingenious experiments. In 1961 Crick and coworkers established the triplet code in a study of mutations in bacterial vira. In the following years the genetic code was established in bacteria by a major effort of the community of molecular geneticists with major contributions by Holley, Khorana, and Nirenberg. In 1967 Crick argued on the basis of known mutants of hemoglobin that the bacterial code was applicable to the translation in human cells. He simply showed that the known changes in amino acids could be viewed as single-base substitutions by assuming the bacterial code.

The code is *degenerate* because some amino acids are coded for by more than one codon. For instance, leucine (Leu) is specified by six codons, proline (Pro) by four, and histidine (His) by two. The second codon position is never degenerate in the sense that changing the second base always leads to a change in the coded amino acid (or to a stop codon). The first codon position is rarely degenerate, an example being some of the leucine codons. Many codons do not need the third position to specify the amino acid; for instance, CU and any base specifies leucine. Many others only need specification of the type of nucleotide base in the third position. CA plus pyrimidine codes for histidine (His) and CA plus purine codes for glutamine (Gln). Tryptophan (Trp) and methionine (Met) are specified by unique codons, and isoleucine (Ile) is coded by AU+pyrimidine and AUA. Codons specifying the same amino acid are referred to as *synonymous* codons.

Table 1.2: Codon table

first	U	C	A	G	last
U	Phe	Ser	Tyr	Cys	**U**
	Phe	Ser	Tyr	Cys	**C**
	Leu	Ser	STOP	STOP	**A**
	Leu	Ser	STOP	Trp	**G**
C	Leu	Pro	His	Arg	**U**
	Leu	Pro	His	Arg	**C**
	Leu	Pro	Gln	Arg	**A**
	Leu	Pro	Gln	Arg	**G**
A	Ile	Thr	Asn	Ser	**U**
	Ile	Thr	Asn	Ser	**C**
	Ile	Thr	Lys	Arg	**A**
	Met	Thr	Lys	Arg	**G**
G	Val	Ala	Asp	Gly	**U**
	Val	Ala	Asp	Gly	**C**
	Val	Ala	Glu	Gly	**A**
	Val	Ala	Glu	Gly	**G**

In eukaryotes the mRNA that reaches the ribosomes is usually very different from the primary RNA transcript. In the genomic DNA of the nucleus, the protein code is often contained in a noncontiguous subset of the sequence. In terms of the primary transcript, this structure may be depicted as a linear piece of RNA read from left to right (often shown as the direction from the 5′ end to

the chemically different 3′ end of the molecule, see Box 3). The open reading frame is shown in color. The protein coding parts are shown in red, and the blue pieces are excised during the maturation of the transcript for translation so that the red pieces are joined into one translated stretch. These blue pieces are called *introns* and the remaining pieces are called *exons*. The black regions are the untranslated regions (UTRs). After intron excision the transcript is reduced to an mRNA composed of the protein code and its flanking regions. The flanking regions are finally modified (e.g., a poly-A string is added to the

end of the transcript, the 3′ end), and then the mRNA is mature to be translated into protein at the ribosomes in the cytoplasm of the cell. The coding sequence (red) starts with an AUG codon in the 5′ end.

This describes the mechanisms behind the *central dogma* of molecular biology: information flows from DNA to RNA to proteins.[6] Open reading frames

---

[6]This general principle is broken only by retrovira.

Box 3: DNA polarity

A nucleotide consists of a deoxyribose molecule with a base (B is either of G, A, T, and C) attached to carbon atom number 1, and with a phosphate group (P) attached at carbon number 5. In a DNA strand the nucleotides are joined at the phosphate group, in that the phosphate attaches to carbon number 3 in the neighbor nucleotide. A DNA strand therefore has a characteristic direction.

The two strands in the DNA molecule have opposite directions, and the molecule therefore does not define a direction. DNA synthesis, however, always progresses in the direction from the 5' to the 3' end of the template DNA. DNA replication, as described in Figure 1.1 on page 13, thus proceeds naturally on one strand only, whereas the other strand should be synthesized in the direction away from the point. This is indeed what happens. The "unnatural" strand is replicated in small pieces which are subsequently joined to form a continuous string.

---

need not code for proteins, as many functions in the cell directly involve RNA molecules. For example, the ribosome is formed by two large and one small RNA molecule, the rRNAs, in addition to proteins. These are referred to by their size, and the eukaryote ribosome is formed by a 28S rRNA, an 18S rRNA, and a 5S rRNA (S refers to a unit measuring sedimentation, or weight in a centrifuge). Another large family consists of transfer RNAs, or tRNAs, that participate in the protein synthesis by bringing the amino acids to the ribosome. For each amino acid there exists at least one tRNA that binds it and recognizes one or more codons on an mRNA bound to a ribosome. The tRNA recognizes the mRNA codon by an exposed anticodon that allows base pairing between the two RNAs. For example, the Trp-tRNA (tryptophan tRNA) exhibits the anticodon CCA, which is reverse complementary to the messenger codon in Table 1.2. Apart from these key RNA molecules, numerous active RNAs have been and are being discovered. RNA enzymes exist, and many enzymes consist of both protein and RNA subunits, for instance the spliceosome that catalyzes the intron excision in mRNAs. In addition, scores of very small modified RNA transcripts are involved in various aspects of regulation of gene expression.

Single-stranded RNA is rather unstable in the cellular environment. However, RNA folds easily, and if possible, RNA molecules form loops stabilized by double-stranded structures of paired bases from different regions of the molecule. tRNAs are dominated by four such structures with only the loops and short sequences flanking or separating the double-stranded regions. The characteristic anticodon is part of a loop. The considerably larger rRNA molecules form a rather complicated structure of loops and double- and single-stranded segments.

The rRNAs are coded for by a large number of similar genes, and their transcripts are modified before the final form is found. Their transcription and processing occur in the nucleolus, the most spectacular structure in the nucleus of an active eukaryotic cell. It is associated with the nucleolar organizer, which is the chromosome segment that contains the rRNA genes. The most humble RNAs, the microRNAs or miRNA are the focus of much current attention because they seem to be important regulators of gene function that supplement the score of proteins that interact with the processes of transcription and translation. The multitude of roles being unveiled for RNA molecules adds credibility to the hypothesis that RNA predates DNA as the information carrier in the evolution of life during a stage known as the RNA world (see, e.g., Fenchel 2001, 2002).

1.1.2   Molecular genetic variation

Electrophoretic mobility is determined by the physical and chemical properties of the protein that, in turn, are functions of the amino acid sequence and the three-dimensional structure of the molecule determined by the underlying DNA sequence. Population variation in the DNA sequence is primary molecular genetic variation, and the study of such variation is the quintessence of population genetics.

Molecular variation is simple. In DNA the information is written in a four-letter alphabet and, if part of a gene coding for a protein, read in triplets and interpreted as a sequence of twenty amino acids. Simple basic rules may, however, define a complex game such as Go. We will return to some of the intricacies of sequence analysis later, and for now restrict attention to a few ways in which variation in the genetic sequence may be probed.

Analysis of sequences depends on the ability to isolate characteristic fragments of DNA. They are obtained by using molecular scissors: restriction enzymes (restriction endonucleases) that are defense mechanisms from bacteria that degrade the DNA of infecting vira. Such enzymes recognize a specific motif only a few bases long in the DNA and cleave it. Digesting DNA with one or more such enzymes produces a lot of fragments. These may be separated by electrophoresis. Staining the DNA in the gel would just produce a smear, and as in enzyme electrophoresis, we need a specific dye to study a specific piece of DNA. Such a dye could be a short marked piece of DNA, a *DNA probe*. Opening the double helix of the DNA in the gel (called melting) and hybridizing with the visible probe highlights the place on the gel where the fragments of interest are situated. These fragments may then be isolated and subjected to further study.

The simplest application of this procedure is to look for variation in the specific motif of a restriction enzyme. Using a probe to mark a piece of DNA, we may look for restriction sites in the neighborhood. Variation at those positions will then reveal the presence or absence of restriction sites in different individuals. A simple example is the existence of two neighborhood configurations, where restriction sites are shown in green, and the probed sequence

is shown in red. All individuals share two restriction sites, one on each side of the probed sequence shown, and some individuals carry an additional restriction site between those two. The segment recognized by the probe is thus shorter in the upper than in the lower sequence. Electrophoresis of DNA fragments isolated from a sample of individuals in a population can therefore exhibit three phenotypes, corresponding to the two homozygotes, long and short, and a heterozygote with both long and short segments. Viewed as Mendelian genes, the two restriction-site configurations are thus codominant alleles.

Electrophoresis of DNA fragments is simpler than protein electrophoresis because DNA is a simpler molecule, and the mobility of a fragment in a polyacrylamide gel simply decreases as its length increases. The "short" homozygote therefore shows a band of higher mobility than that of the "long" homozygote. This kind of polymorphism is therefore dubbed *restriction fragment length polymorphism*, or RFLP. This method may be used to get an impression of sequence variation in natural populations. In a study of *Drosophila melanogaster*, Langley and Aquadro (1987) resolved restriction fragment length variation into changes at the restriction sites and changes caused by insertion or deletion of pieces of DNA in between restriction sites.

The separation of DNA fragments on a polyacrylamide gel is as accurate as desired. If the electrophoresis is run for a sufficient length of time (on a

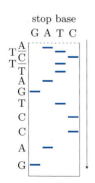

sufficiently long gel), differences as small as one base pair may be resolved. This is the basis of DNA sequencing techniques— the only outstanding problem is to make suitable fragments that differ by only one base, and to identifythat base. Sanger et al. (1977) devised a simple way to do this. The piece of DNA to be sequenced is multiplied to many copies (by polymerase chain reaction, PCR, using a pair of DNA probes called primers) that are made single-stranded, and only one strand is kept (by again using the primers). The single-stranded DNA is used as a template for making copies in a soup containing the radioactively marked nucleotides for later recognition of the copies. This soup is contaminated by a nucleotide made with one of the four bases, but modified in the deoxyribose part (which is dideoxyribose), and when this base is used in the copying process the synthesis of the copy halts. The sequencing procedure is then defined by noting that the synthesis always starts at the 5′ end of the template DNA. The copies thus have a characteristic 3′ end and a variable 5′ end. Repeating the procedure for all four bases allows the displayed gel to be read as beginning with the base sequence GACCTGATTCT....

A section of the DNA in the genome of an organism may be defined by a probe and delimited by restriction sites, and we may then study variation in such a section. Suppose we have a sample of twelve homologous DNA pieces with sequences determined (shown at left). These sequences differ only in positions 3, 4, and 8. Focusing on these differences, the sample contains five alleles:

1:	GACCTGATTCT···
2:	GACATGATTCT···
3:	GACCTGATCCT···
4:	GACCTGATCCT···
5:	GACGTGATTCT···
6:	GACACGATTCT···
7:	GACCTGATTCT···
8:	GACCTGATTCT···
9:	GACATGATTCT···
10:	GACCTGATTCT···
11:	GACCTGATTCT···
12:	GACCTGATCCT···

A_1 :   GACCTGATTCT···
A_2 :   GACATGATTCT···
A_3 :   GACCTGATCCT···
A_4 :   GACGTGATTCT···
A_5 :   GACACGATTCT···

and we observe five genes of allele A_1, two of A_2, three of A_3, and one each of alleles A_4 and A_5. The sequences are referred to as genes, even though they need not correspond to functional genes. We may, however, study their inheritance and confirm that their transmission follows Mendel's law of inheritance. Sequence variation may therefore be considered as variation among Mendelian genes. In the development of population genetic theory, the Mendelian gene concept is more convenient than the functional gene concept. Much of the genetic variation applied in contemporary population genetic investigations is within noncoding DNA. Sequences of introns or DNA between genes commonly show a level of variation higher than that in exons.

Trans-exon DNA includes motifs with little variation. For instance, the miRNA genes seem highly conserved even on an evolutionary timescale joining insects and mammals. The interest in these was excited by the discovery of RNA interference (Fire et al. 1998), and soon miRNA was realized to play an important role in the regulation of gene expression (He and Hannon 2004). Initially the search for miRNA and the corresponding specific recognition sites in the genome was based on evolutionary conservation, but as their structure and function became known, bioinformatic searches in genomes uncovered many additional miRNA that were conserved on considerably shorter timescales (Lindow and Krogh 2005). Related short DNA motifs, the so-called pyknons, are ubiquitous in the genome, and seem to exert their biological function in terms of numbers and kinds in functional genes (Rigoutsos et al. 2006). In total, the DNA sequences in any two of our chromosomes differ in about one out of a thousand bases, which may not seem much. However, the human genome comprises three billion bases, so the contributed maternal and paternal genomes differ at about three million of them.

In the study of functional genes focus is on the transcribed strings of the

DNA molecule. The sequences read from sequence gels correspond to the strings specified by the chosen DNA probes, and they may therefore need to be lifted to the transcribed strings before the coding sequences can be identified. The physical and chemical properties of proteins are determined by their sequences of amino acids, which in turn are determined by the translated nucleotide sequence. The number of protein coding genes in humans is of the order of tens of thousands, but the number of proteins produced in the cells of the human body is considerably higher. This inflation has two causes. Genes may produce more than one kind of transcript, giving variation in the corresponding mRNAs. This variation may manifest itself in several ways, be it by variation among tissues or through the development of the individual. In humans the number of transcripts was found to be higher than the number of genes (Lander et al. 2001). In addition, the proteins may undergo a variety of posttranslational modifications where pieces of the amino acid sequence may be excised or sugar molecules added to individual amino acids.

Chapter 2

Conservation of Variation

Mendel's first law is fundamental to any discussion of inheritance in diploid individuals. Even in a well-executed study a deviation from the expected segregation does not warrant a rejection of the law, but rather suggests that assumptions about the experiment or the considered variation may be unfounded. Upon further analysis the unexpected result may lead to the discovery of a new phenomenon (see, e.g., Sturtevant and Morgan 1923). The corresponding rules of population genetics are in the same way unquestionable, and form the basis for interpretation of observations in nature or in the laboratory.

Observations in Mendel's experiments on plant height[1] presuppose that peas of the various genotypes have the same probability of germination, survival, and growth. In the same way, the conservation law on the population level is based on the premise that the alleles and genotypes do not differ with respect to survival and reproduction. On the other hand, if the genotypes differ in their probability of survival, then deviations from the expected segregation in F_2 is anticipated in a Mendelian cross. In the same way, if the genotypes differ in viability or in their average number of offspring, then conservation on the population level breaks down—we say that *natural selection* on the variation occurs (we consider this in Chapter 8). Genetic variation that does not cause natural selection is said to be *neutral*; for instance, mutually neutral alleles cause variation with no impact on survival and reproduction. The sequences shown in Table 2.1 are expressions of neutral variation. The two alleles differ in the second codon after the translation initiation codon ATG, where CAT and CAC are found, both coding for histidine (See Table 1.2 on page 15).

Table 2.1: Two translated sequences of the β-hemoglobin gene

A	ATGGTGCATCTGACTCCTGAGGAGAAGTCTGCCGT...
a	ATGGTGCACCTGACTCCTGAGGAGAAGTCTGCCGT...

Made available by Fullerton et al. (2000), see Table 4.3.

---

[1]The allele for tall plants is dominant to that producing short plants.

To view this in more detail, we consider the basics of Mendelian genetics in a population polymorphic for two alleles. In some circumstances two-allele models are quite sufficient to describe natural variation, but for now just consider the assumption of two alleles as a simplification. When, for a given gene, the various genotypes in a population have the same probability of survival from zygote to mature adult, the same opportunity for breeding, and the same expected number of offspring, then Mendelian segregation ensures that the various alleles in the population have an equal probability of being transmitted between generations. Their frequency in the population is therefore expected to be conserved. This is the law of *gene frequency conservation*. This principle originally emerged in the study of Mendelian genetic variation and is therefore formulated in terms of the frequency of genes of the various allele types—the gene frequencies of the alleles. Its key, however, is the transmission to offspring of copies of parental DNA, and it is therefore equally valid in terms of sequence variation. Gene frequency conservation therefore describes the transmission of any studied sequence that is well defined within a genome, and as it is transmitted according to Mendel's law, we may consider such a sequence a Mendelian gene. Variants may thus be referred to as alleles. This formal parallel between gene and well defined DNA sequence should obviously only be used when construction as the functional gene is unlikely.

Embedded in Mendel's law is that the gene transmitted to an offspring is an exact copy of one of the parental genes. This is the basic assumption to obtain gene frequency conservation, and it allows the ancestry of a gene to be traced back from generation to generation. Eventually all allelic genes in the population may be traced back to one common ancestor—one piece of DNA of which they are all copies. In reality hereditary transmission is not perfect, but errors in the copying process seldom occur. We return to the effect of that in Chapter 4 and for now assume that transmission is perfect. Genetic recombination (Chapter 6) may also cause deviations from conservative transmission—an effect that becomes increasingly likely with the length of the studied sequence.

Mendelian segregation ensures that the two genes are considered as equal in transmission from a diploid individual—in much the same way, as the further assumptions, that individuals survive and reproduce without influence of the considered genotypes. Accordingly, deviation from Mendelian segregation is ascribed to gametic selection or segregation distortion. The copying process mediated by the DNA replication (see Figure 1.1 on page 13) is therefore the key to gene frequency conservation, and we can thus study its consequences in this and some of the following chapters by simply studying the transmission of genes or DNA sequences between generations. Consequences of sexual reproduction and diploidy will be discussed in Chapter 3.

The law of gene frequency conservation formulates an expectation, neglecting the haphazard happenings of everyday life. The law is not violated in a human population if someone accidentally gets run over by a car. Arguments considering probabilities and expectations correspond to assuming a very large population, effectively with infinitely many genes. These are deterministic arguments in deterministic models. Realizing the finiteness of any population, we need to ponder the influence of random events, thus considering stochastic models.

2.1 Stochastic Fluctuations

Segregation in Mendel's experiments was very close to the expectations that he formulated. The results of his two largest experiments (those on seed shape and albumen coloration) are shown in Table 2.2. The fit of the observed counts to

Table 2.2: Segregation in F_2 in Mendel's (1866) experiments

trait			counts		
dominant	recessive		dominant	recessive	sum
round	wrinkled	observed	5474	1850	7324
		expected	*5493.00*	*1831.00*	7324
yellow	green	observed	6022	2001	8023
		expected	*6017.25*	*2005.75*	8023

those expected based on Mendel's law of inheritance is very good: the Mendelian expectation is a ratio of 3:1 and the observed ratios are 2.96:1 and 3.01:1. The fit is indeed astonishingly good, given that the law is formulated as the probability that a given F_2 individual shows the dominant or the recessive trait.[2] The difference between observed and expected results occurs because of the element of randomness in acquiring a sample of F_2 individuals. Thus, the observed deviation from the strict Mendelian law is trivial in the sense that larger experiments are expected to yield ratios closer to 3:1.

Similar stochastic fluctuations of the frequency of an allele in the offspring population away from the frequency among parents is expected in any population (Table 2.3). These fluctuations, however, are not trivial statistical devia-

Table 2.3: Allele segregation in Table 2.2

trait		gene frequency	
dominant	recessive	dominant	recessive
round	wrinkled	0.498	0.502
yellow	green	0.500	0.500

tions, because we cannot refer to the results in a larger "experiment." Once the parents are dead, the population consists of the offspring and the gene frequencies in the offspring population will now be conserved. The expected constancy of the frequency of a given allele coupled with the inevitable random changes lead us to the conclusion that gene frequencies in real populations change over

---

[2]The experiment on seed shape could have hit bull's-eye on the Mendelian expectation. The probability of this event in an experiment counting 7324 offspring is about 0.01.

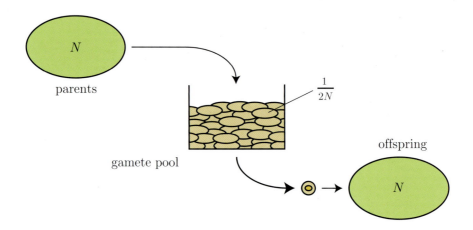

Figure 2.1: The gamete pool model.

time. Such indeterminate stochastic changes in gene frequencies are called *random genetic drift*.

The simplest model of random genetic drift is the Wright–Fisher model (Fisher 1930b, Wright 1931), which is also referred to as the gamete pool model originally formulated for a population of a diploid species. The individuals in the population of parents shed their gametes into a bucket, the gamete pool (Figure 2.1). Each individual contributes a large but equal number of gametes, and the offspring are formed by gametes drawn randomly from this pool. An offspring generation of N diploid individuals is therefore formed by drawing $2N$ gametes and pairing them to produce zygotes. The Wright–Fisher model assumes that the parental population dies after reproduction. In other words, the generations do not overlap. A further simplification is that the population is without any substructure (spacial structure is considered in Chapter 5 and deviations from random gamete pairing in Chapter 3). We shall see that the general process of random genetic drift is well represented by this simple model.

With Mendelian segregation the Wright–Fisher model is recognized as a haploid model for studying the transmission of genes from generation to generation. Following tradition we consider the number of diploid individuals to be N and the number of genes to be $2N$. Now consider an allele, say, A, whose frequency in the parental population is p, which is conserved. The frequency of this allele is also p in the gamete pool. Each gamete we draw from the pool has the probability p of carrying allele A. The number X of As drawn is therefore a stochastic variable that is binomially distributed $b(2N, p)$, where

$$\text{Prob}(X = j) = \binom{2N}{j} p^j (1 - p)^{2N-j} \qquad (2.1)$$

(Figure 2.2). The mean and variance of X are $\mathrm{E}\,X = 2Np$ and $\mathrm{Var}(X) = 2Np(1-p)$ (Appendix A.1.3 on page 351). The large number of gametes in the

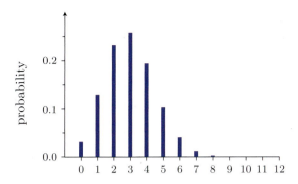

Figure 2.2: The binomial distribution $b(12, \frac{1}{4})$ shown as a histogram.

pool is a metaphor for the assumption that each gamete drawn carries allele A with probability p. In statistics the corresponding metaphor is "sampling with replacement" which sounds a bit odd during an act of breeding. The principle of gene frequency conservation can now be formulated as an expectation because the offspring gene frequency is given by the stochastic variable $\frac{X}{2N}$ with the mean $\mathrm{E}\frac{X}{2N} = p$. The gene frequency in the parental population determines the gene frequency among offspring as specified by formula (2.1). The parental gene frequency was determined in the same way from the gene frequency in the grandparental generation. The above description therefore expresses that, given that the gene frequency in the parental generation is p, the number of genes of allele type A in the offspring population is binomially distributed with probability parameter p. The gene frequency among parental genes is determined by the number X_P of genes of the A allele in the population, and if the number of such genes in the parental population is $2N_P$, we may write the gene frequency conservation as the conditional probability distribution

$$P_{i \to j} = \mathrm{Prob}(X_O = j \mid X_P = i) = \binom{2N_O}{j} \left(\frac{i}{2N_P}\right)^j \left(\frac{2N_P - i}{2N_P}\right)^{2N_O - j}, \quad (2.2)$$

in which we used X_O and $2N_O$ for the number of As and the number of genes in the offspring population. This formula expresses more immediately that the parental gene frequency determines the offspring gene frequency. The change in the population is in general described by the transition probabilities $P_{i \to j}$ for $i = 0, 1, 2, \ldots 2N_P$ and $j = 0, 1, 2, \ldots 2N_O$ with $P_{i \to 0} + P_{i \to 1} + P_{i \to 2} + \cdots + P_{i \to 2N_O} = 1$. In the Wright–Fisher model these are binomially distributed:

$$P_{i \to j} \sim \mathrm{b}\left(2N_O, \frac{i}{2N_P}\right).$$

The conditional expectation of the offspring gene frequency given the parental gene frequency is then

$$\mathrm{E}\left(\frac{X_O}{2N_O} \,\Big|\, \frac{X_P}{2N_P}\right) = \frac{X_P}{2N_P}, \quad (2.3)$$

because we view the population sizes as constants independent of genetics. This expectation expresses the conservation of the gene frequency.

2.2   Wright–Fisher Process

The simplest version of the Wright–Fisher model assumes a population with discrete nonoverlapping generations, each with the same number of individuals N. This is a reference model for the discussion of the effects of random genetic drift on the evolution of a population. Illustrations of the changes in the gene frequency due to random genetic drift in this model are obtained from the "genetic drift" simulator in Holsinger's (2006) simulation package.

Suppose the population in generation 0 segregates A with the frequency $p_0 = p$. We shall study the gene frequencies

$$p_t = \frac{X_t}{2N}, \quad t = 1, 2, 3, \ldots,$$

which are stochastic variables given in terms of the number X_t of A genes in generation t. The first generation is the offspring of the initial population, and it is therefore expected to have the gene frequency $\mathrm{E}\, p_1 = p$ and the number $\mathrm{E}\, X_1 = 2Np$ of A genes. From gene frequency conservation we get

$$\mathrm{E}(p_t|p_{t-1}) = p_{t-1},$$

and therefore $\mathrm{E}\, p_t = \mathrm{E}\,\mathrm{E}(p_t|p_{t-1}) = \mathrm{E}\, p_{t-1}$ (see equation (A.7) on page 349). By continued use of this argument we get $\mathrm{E}\, p_t = p$ as expected from gene frequency conservation.

In generation 1 the number of A genes in the population is binomially distributed with probability parameter p. This means that the number of A genes is chosen like the rolling of dice. The population has a gene frequency of, say, p_1, which will then be conserved in the future, and the number of A genes in generation 2 is binomially distributed with probability parameter p_1. However, the probability distribution of the number of A genes in generation 2 is not binomial when viewed from generation 0. Rather, the distribution in generation 1 is carried to generation 2 by the formula

$$\mathrm{Prob}(X_2 = k \,|\, X_0 = 2Np) =$$
$$\sum_{j=0}^{2N} \mathrm{Prob}(X_2 = k \,|\, X_1 = j)\mathrm{Prob}(X_1 = j \,|\, X_0 = 2Np),$$

which is an application of the law of total probability $\big($see equation (A.6)$\big)$.

Repeating this procedure allows us to calculate the distribution in any future generation.[3] Figure 2.3 shows the probability distribution of the number of A genes in a population with $2N = 10$ genes and initial $p = \frac{1}{2}$ through ten generations of random genetic drift. The binomial distribution appears in

---

[3] * The Wright–Fisher process is a discrete-time Markov chain—a type of stochastic process discussed in Appendix A.2 on page 355.

generation 1, and as time passes the distribution becomes wider and wider and bears less and less resemblance to the binomial distribution—after generation 4 it has three modes, one at 5 and two new ones at 0 and 10. The two new modes correspond to the A allele being lost or fixed in the population, respectively. The probabilities of these two events increase every generation and they are rather high by generation 10. The reason is that a population fixed for allele A $(p_P = 1)$ can only transmit that allele to the next generation, and therefore $p_O = 1$.[4] In the same way a Wright–Fisher population that lost allele A $(p = 0)$ would only contain a in all future generations.

The figure obviously expresses important properties of random genetic drift, but as biologists we need to interpret such theoretical descriptions in a way that allows them to contribute to our biological intuition about the phenomena they describe. This process of interpretation and consumption is an important aspect of theoretical biology, and it requires that biologists participate actively. The interpretation often involves the construction of a biological metaphor to visualize the formal results.

To visualize the probability distributions used in the formal description of random genetic drift, consider a very large number of populations, all with $2N$ genes, distributed on islands such that no migration occurs among the islands. That is, we assume that the populations evolve independently. This population model is often referred to as *Wright's island model* (Figure 2.4). The very large number of populations means that random genetic drift in the total population is negligible compared to the drift on each island. We thus effectively assume that the number of islands is infinite so that the gene frequency conservation is absolute in the total population. Assume that

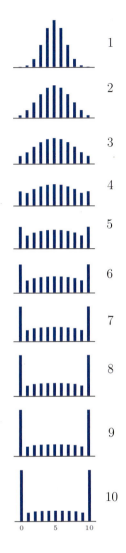

Figure 2.3: Genetic drift in the Wright–Fisher model with $N = 5$ and $p = \frac{1}{2}$. The abscissae show the population number of A genes.

initially all populations have the gene frequency p of allele A. Then the number of A genes in the populations in generation 1 is independent and binomially distributed $b(2N, p)$. For $2N = 10$ and $p = \frac{1}{2}$, the gene frequencies on the

---

[4] * The states $p = 0$ $(X = 0)$ and $p = 1$ $(X = 2N)$ are absorbing states in the Markov chain (page 357).

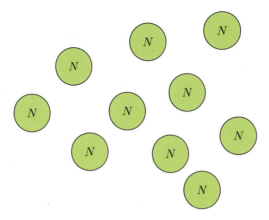

Figure 2.4: Wright's island model.

islands in the following generations are as given by Figure 2.3. The metaphysical distribution of the gene frequency in one population has thus been replaced by the distribution of gene frequencies among physical populations, and various properties of the distribution, for instance the variance, become much more descriptive. A simple argument given in Box 4 shows that the change in the variance of the gene frequency from generation to generation is given by the recurrence equation

$$\mathrm{Var}(p_t) = \frac{p(1-p)}{2N} + \left(1 - \frac{1}{2N}\right)\mathrm{Var}(p_{t-1}). \tag{2.4}$$

In Box 4 this equation is iterated to give the change in variance through time as

$$\mathrm{Var}(p_t) = p(1-p)\left(1 - \left(1 - \frac{1}{2N}\right)^t\right), \quad t = 0, 1, 2, \ldots, \tag{2.5}$$

when starting from a population with gene frequency p at time $t = 0$.

The variance shows a steady increase, and it will eventually be very close to the value $\hat{V} = p(1-p)$, converging to this value as time passes. The gene frequency varies in the interval between 0 and 1 and has the mean p. For such stochastic variables a variance of $p(1-p)$ is known to be the highest possible, and it is only attained for the distribution where the probability of attaining the value 1 is p and that of attaining 0 is $1 - p$ (Figure 2.5). Thus, the analysis of the variance in gene frequencies leads to the conclusion that given enough time, allele A either has the frequency 1 or the frequency 0; that is, it is either fixed in the population or lost. The conservation of the gene frequency p is manifested in the distribution by the properties that the probability of fixing allele A is p and the probability of losing the allele is $1 - p$ (Figure 2.5), and that the mean of the distribution is p. *In a finite population variation is eventually lost, but the probability that an allele is fixed equals its initial gene frequency.* In this sense gene frequencies are conserved.

Box 4: Development of the variance in gene frequency

The variance of the gene frequency in the Wright–Fisher model may be evaluated by using the partitioning of the variance in equation (A.8) on page 350, and we then get

$$\text{Var}(p_t) = \text{E}\big(\text{Var}(p_t|p_{t-1})\big) + \text{Var}\big(\text{E}(p_t|p_{t-1})\big).$$

The last term in this equation is simply $\text{Var}(p_{t-1})$, because $\text{E}(p_t|p_{t-1}) = p_{t-1}$. From the gamete pool model and the ensuing assumption of a binomial distribution we have

$$\text{Var}(p_t|p_{t-1}) = \frac{1}{2N}\, p_{t-1}(1 - p_{t-1}),$$

and therefore

$$\text{Var}(p_t) = \frac{1}{2N}\, \text{E}\big(p_{t-1}(1 - p_{t-1})\big) + \text{Var}(p_{t-1}).$$

The first term is evaluated as

$$
\begin{aligned}
\text{E}\big(p_{t-1}(1 - p_{t-1})\big) &= \text{E}\,p_{t-1} - \text{E}\big(p_{t-1}^2\big) \\
&= \text{E}\,p_{t-1} - \Big(\text{Var}(p_{t-1}) + \big(\text{E}\,p_{t-1}\big)^2\Big) \\
&= p(1 - p) - \text{Var}(p_{t-1})
\end{aligned}
$$

using $\text{E}\,p_{t-1} = p$ and the variance formula in equation (A.3) on page 347. We now get the recurrence equation (2.4), where the variance in the offspring generation is given as a function of the variance in the parent generation and the parameters of the model.

The solution to this *linear recurrence equation* is obtained by initially finding a constant solution where $\text{Var}(p_t) = \text{Var}(p_{t-1})$, $t = 1, 2, \ldots$. This is $\hat{V} = p(1 - p)$. We can then form a *homogeneous* version of equation (2.4) by considering the deviation of the variances from $\hat{V}$. Thus we get

$$\text{Var}(p_t) - p(1 - p) = \left(1 - \frac{1}{2N}\right)\big(\text{Var}(p_{t-1}) - p(1 - p)\big).$$

Iterating this equation is easy, and we get

$$\text{Var}(p_t) - p(1 - p) = \left(1 - \frac{1}{2N}\right)^t \big(\text{Var}(p_0) - p(1 - p)\big).$$

The population starts with a gene frequency of p, so $\text{Var}(p_0) = 0$ (all populations in Wright's island model have the same gene frequency), and therefore the solution to the recurrence equation (2.4) is given by equation (2.5).

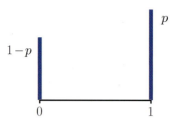

Figure 2.5: Limit distribution for random genetic drift.

2.2.1   Multiple alleles

The description of random genetic drift extends readily to multiple-allele polymorphisms (see, for instance, Crow and Kimura 1970), in that the counts of the various alleles in the offspring populations are multinomially distributed (see Appendix A.1.3 on page 351). The two-allele model, however, is well suited for describing the dynamics of the frequency of any given allele. We just focus on one allele, say, A, and collect the remaining alleles under the label a. Allele a is thus a *collective allele*, but as no differences exist in transmission of the various alleles from generation to generation, the description of the consequences of conservation of the gene frequency p corresponds to that in our two-allele model.

One aspect of multiple-allele variation requires attention, however, namely description of the covariation in the gene frequencies of the various alleles. For two alleles this is almost trivial: when p goes up then q goes down by the same amount, because $q = 1 - p$. The correlation between the gene frequencies is therefore -1, and the covariance becomes

$$\mathrm{Cov}(p_t, q_t) = -pq\left(1 - \left(1 - \frac{1}{2N}\right)^t\right)$$

from equation (2.5). For multiple alleles, say, A_1, A_2, ..., A_k, the covariance between the frequencies of alleles A_1 and A_2 in generation t is

$$\mathrm{Cov}(p_{A_1 t}, p_{A_2 t}) = -p_{A_1 0} p_{A_2 0}\left(1 - \left(1 - \frac{1}{2N}\right)^t\right). \tag{2.6}$$

This is shown by an argument similar to that used in Box 4. Equation (2.5) in the same notation produces

$$\mathrm{Var}(p_{A_1 t}) = p_{A_1 0}(1 - p_{A_1 0})\left(1 - \left(1 - \frac{1}{2N}\right)^t\right). \tag{2.7}$$

The difference between the variance and covariance is thus reminiscent of that for the multinomial distribution (Appendix A.1.3).

2.3   Gene Identity in a Finite Population

Random genetic drift can be described in terms of a process of transmission of genes rather than one of describing the frequencies of alleles in a population. The gamete pool representation of the Wright–Fisher model is well suited for this purpose (Figure 2.1 on page 24). Copies of the $2N$ allelic genes in the parental population are represented equally in the gamete pool, each with a frequency of $1/(2N)$. When sampling $2N$ gametes to form the offspring population, we are likely to sample copies of the same gene twice, necessitating that another parental gene not be transmitted to the offspring population. Two copies of the same gene are said to be *identical by descent*.

This way of looking at the stochastic element of random genetic drift was developed by Gustave Malécot in 1948 using his description of consanguinity (Box 5) to monitor the evolution of the population.[5] Repeated sampling of copies of the same parental gene creates an assembly of identical genes in the offspring population, and because this is done every generation, random genetic drift causes some of these assemblies to expand and the level of gene identity to increase.

The state of a population may be described by the probability of identity by descent of two genes chosen at random among those carried by the individuals in generation t, and we designate this probability F_t. F_t is thus the probability that two genes drawn at random from the gamete pool produced by the individuals in generation $t-1$ are identical, and we will refer to it as the *population coefficient of identity by descent* or just the *identity coefficient*. Two genes are copies of the same gene in generation $t-1$ with probability $\frac{1}{2N}$,[6] and they are copies of two different genes with probability $1 - \frac{1}{2N}$. The identity coefficient in generation t is therefore given by

$$F_t = \frac{1}{2N} + \left(1 - \frac{1}{2N}\right) F_{t-1}, \tag{2.8}$$

because two genes in generation $t-1$ are identical by descent with probability F_{t-1} (Figure 2.6). This linear recurrence equation describes the development of gene identity in the population, and it is iterated in the same way as the recurrence equation (2.4) in the variances (Box 4). However, the calculation here is much simpler, especially if we use the probability $1 - F_t$ that two random genes are not identical to describe the state of the population. Rewriting equation (2.8) in these terms produces the recurrence equation,

$$1 - F_t = \left(1 - \frac{1}{2N}\right)(1 - F_{t-1}),$$

which immediately iterates to the solution

$$F_t = 1 - \left(1 - \frac{1}{2N}\right)^t, \quad t = 0, 1, 2, \dots, \tag{2.9}$$

---

[5]Wright (1931) used a very similar method of partial correlations in line with our evaluations of the variance in gene frequency.

[6]Pick any gene; the second gene is a copy of the same parental gene with probability $\frac{1}{2N}$.

Box 5: Description of consanguinity

Malécot (1948) developed a description of gene identity by descent that applied the concept of gene copying materialized by Watson and Crick's (1953) model of DNA structure and replication. He described two genes as identical when they could be recognized as copies of the same ancestral gene. For instance, the genes you carry are identical to genes in your mother or father. Genes carried in different individuals may be identical if they are blood related, and the concept of gene identity may be used to quantify familial relationships. The probability f_{IJ} of randomly drawing identical copies of a given Mendelian gene from individuals I and J is called the *coefficient of consanguinity* of the two individuals. The coefficient of consanguinity between a child and one of its parents is $f_{PO} = \frac{1}{4}$ (the gene drawn from the parent is the one transmitted to the child with probability $\frac{1}{2}$, and that drawn from the child is the one received from the parent with probability $\frac{1}{2}$). The coefficient of consanguinity between full sibs is also $\frac{1}{4}$, and between half-sibs it is $\frac{1}{8}$. The coefficient of consanguinity of an individual with itself is $\frac{1}{2}$.

The two genes carried by a person may be identical if the parents are blood related. Such an individual is called *inbred*, and the degree of inbreeding is measured by the coefficient of consanguinity of the parents. The *inbreeding coefficient* f_K of an individual K is thus given by $f_K = f_{IJ}$, where I and J are the parents of K. When applied to plants, the inbreeding coefficient of an individual K formed by selfing of the individual I is

$$f_K = \tfrac{1}{2} + \tfrac{1}{2} f_I$$

(with probability $\frac{1}{2}$ the two genes in K are copies of the same gene in I, and with probability $\frac{1}{2}$ they are copies of different genes; the two different genes in I are identical by descent with probability f_I).

Inbred individuals are often disadvantaged, and we will return to this aspect of consanguinity in Chapter 9.

---

if we assume $F_0 = 0$, that is, no pair of genes in generation 0 is identical by descent. The population identity coefficient therefore increases monotonically, and $F_t \to 1$ as $t \to \infty$. *All allelic genes in the population will therefore eventually become identical by descent*, and thus copies of one gene in the original population. Needless to say, all genes will be of the same allelic type. The buildup of gene identity by random genetic drift may be illustrated using the Wright–Fisher animator of Mikkelsen et al. (2006).

Equations (2.4) and (2.5) describing the increase of the variance in gene frequency through time are very similar to equations (2.8) and (2.9), which describe the buildup of identity. In both cases the convergence is described by the factor

$$\lambda = 1 - \frac{1}{2N}.$$

The reason is that the two equations describe not only the same process, but

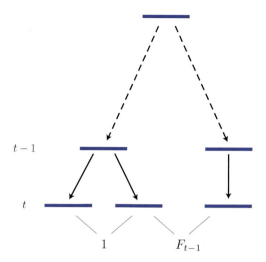

$t-1$

t

1 F_{t-1}

Figure 2.6: Inbreeding in the gamete pool model.

the same property of that process. We will return to this later, but first we shall consider an alternative description of the consequences of the buildup of population gene identity, namely the genealogy of the genes in the population.

2.4 Coalescence of Genes

Random genetic drift produces identity by descent among the genes in the population due to the possibility of proliferating copies of single genes. When this process is allowed to proceed for a long time, all genes in the population will eventually become identical by descent. Then, by definition, all genes in the population will be descendants of a single ancestral gene, and we can, at least in principle, trace the genealogy of the genes all the way back to this common ancestor using Malécot's (1948) description of identity. The result is a structure like the one sketched in Figure 2.6. The genealogical process originates in the present, and time is counted as generations back in time. This view was developed further through the following decades (Kingman 2000, Tavaré 1984), and the current description uses a framework originally developed by J.F.C. Kingman (1982a, 1982c).

The gene genealogy may as well refer to a random sample of genes, our source of knowledge of the population. We can thus discuss the genealogy of the genes we observe. The assemblage of genes forming the entire population is only observable in principle; in practice the analysis of all individuals is usually impossible. In the gamete pool model we therefore consider a random sample of n genes from a population with $2N$ genes—still assuming the Wright–Fisher model. At some point in time these n genes had a common ancestor, and

Figure 2.7 shows a possible pedigree of the genes in the sample leading back to their most recent common ancestor at the top of the diagram.

Two particular genes in this sample of n genes are copies of the same gene in the parental generation with probability $\frac{1}{2N}$. If the population size N is large, this probability is small for any given pair, but we have to mind all the

$$\binom{n}{2} = \frac{n(n-1)}{2}$$

different pairs that we can choose from the hypothetical sample of n genes. Each pair of genes drawn from the population is equally likely to be identical copies of one and the same gene in the parental population. For N sufficiently large ($N \gg n$), however, we can assume that the probability that the sample contains two or more pairs of genes that are copies of the same parental gene is negligibly small, on the order of $(\frac{1}{2N})^2$. Thus, the probability that the sample contains a pair of genes that are copies of one gene in the parental generation is approximately given by

$$\binom{n}{2} \frac{1}{2N} \tag{2.10}$$

when $N \gg n$, and the probability that the sample does not contain such a pair is approximately

$$1 - \binom{n}{2} \frac{1}{2N}. \tag{2.11}$$

The probability (2.11) that the sample does not contain multiple copies of the same parental gene may also be calculated by a simple exact argument (Felsenstein 1971). Number the genes in the sample $1, 2, \ldots, n$. The probability that gene 2 is not a copy of the parental gene of gene 1 is $1 - \frac{1}{2N}$. Given that genes 1 and 2 are copies of different parental genes, the probability that gene 3 is a copy of a different parental gene is $1 - \frac{2}{2N}$. Continuing this argument we get the probability that the n genes are copies of different parental genes as

$$\left(1 - \frac{1}{2N}\right)\left(1 - \frac{2}{2N}\right)\left(1 - \frac{3}{2N}\right) \cdots \left(1 - \frac{n-1}{2N}\right). \tag{2.12}$$

Multiplying out this expression produces equation (2.11) if we neglect terms of the order of $(\frac{1}{2N})^2$ or smaller. Two sources of terms of the order of $(\frac{1}{2N})^2$ appear. The most straightforward contribution of such higher order terms comes from picking copies of the same gene three times, while the second source is to pick two different genes twice.

If two genes are identical because they are copies of the same gene in the parental generation, then the n sampled genes are represented by $n-1$ genes one generation back in time. Viewed in this way, a pair of genes in the sample has *coalesced* into one ancestral gene. The probability (2.10) is thus the probability of a *coalescent event* in a sample of n genes when going one generation back in time. If no coalescent event happens (probability (2.11)), then the sample is still represented by n ancestral genes in the parental population. These events and

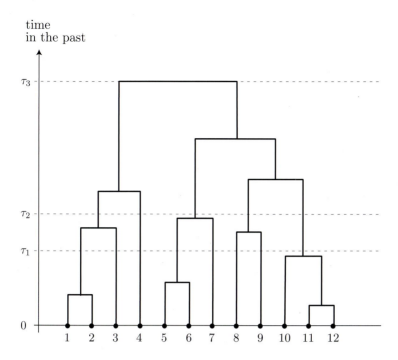

Figure 2.7: The genealogical tree of a sample of $n = 12$ genes.

their probabilities (2.10) and (2.11) define the *coalescence process* of a sample of n genes in a contemporary population.[7]

The pedigree in Figure 2.7 is described in terms of coalescent events where two genes fuse in a single ancestor as we move back in time. The first coalescent event in the figure fuses genes number 11 and 12, and after these two genes have coalesced we need only consider 11 genes in our further analysis of the history of the sample. Figure 2.7 shows the ancestors of the sampled genes at various points in time in the past. At these times the ancestral material giving rise to the 12 genes in the sample is present in a decreasing number of genes. For instance, at time τ_1 before present, genes 1 and 2 have a common ancestor, as do genes 5 and 6, and also genes 10, 11, and 12 have found a common origin. Thus, at time τ_1, the 12 sampled genes have coalesced into eight genes, at time τ_2 into five genes; at time τ_3 all the genes in the sample have coalesced. The time variable τ is chosen to emphasize that we consider time in the past measured from the present, where the sample of the population is taken. From τ_3 and further back the sample is represented by one common ancestor. Time τ_3 is the most recent when this is true, and it is thus called the *coalescence time* of the sample.

---

[7] ∗ The coalescent corresponding to the Wright–Fisher process is a Markov chain (Appendix A.2 on page 357).

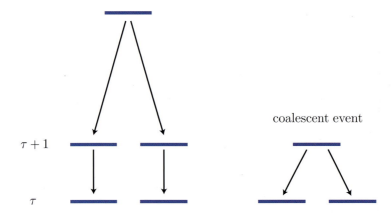

Figure 2.8: Coalescence in the gamete pool model.

The relationship between the n sampled genes and the genes in the parental population is similar to the relationship between the ancestral genes in any generation back in time and their parental genes one generation further back. Suppose in a given generation, say, generation τ before present, that the sample is represented by k ancestral genes, that is, $n-k$ coalescent events have happened between the present when the n genes were sampled, and the generation that we are considering. In generation τ the sample may thus be viewed as one of k genes, and we can trace the origin of these k genes among the parents in generation $\tau+1$ in the same way as we did before (Figure 2.8). The probability of a coalescent event is therefore

$$p_k = \text{Prob(coalescent event in sample of } k \text{ genes)} = \binom{k}{2}\frac{1}{2N}. \qquad (2.13)$$

The probability that the sample is still represented by k ancestral genes in generation $\tau+1$ is $1-p_k$. The probability that the k genes in generation τ retained their integrity in generation $\tau+1$ and have become $k-1$ genes in generation $\tau+2$ by a coalescent event is therefore $p_k(1-p_k)$. Continuing this argument, a coalescent event occurs after one generation with probability p_k, after two generations with probability $p_k(1-p_k)$, and after i generations with probability $p_k(1-p_k)^{i-1}$. Thus, the waiting time T_k until k genes coalesce to $k-1$ has the distribution

$$\text{Prob}(T_k = i) = p_k(1-p_k)^{i-1}, \quad i = 1, 2, 3, \ldots. \qquad (2.14)$$

This waiting-time distribution is the *geometric distribution* (Figure 2.9 and Appendix A.1.4 on page 352). The mean waiting time to the next coalescent event is

$$\text{E}\,T_k = \frac{1}{p_k} = 2N\,\frac{2}{k(k-1)}, \qquad (2.15)$$

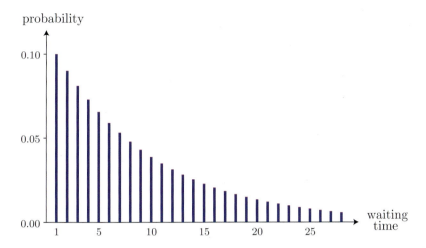

Figure 2.9: The distribution of the waiting time to coalescence when the mean waiting time is 10. About 95 percent of the probability mass is shown.

which is the expected number of generations that the ancestral material of the sample is carried in k genes. For two genes in the sample the probability of a coalescent event is $\frac{1}{2N}$ per generation, and the waiting time is therefore $2N$ generations (Malécot 1948).

The entity in equation (2.15) is interpreted as the average wait between two coalescent events. However, nothing in our argument that led to that equation assumes that a coalescent event occurred when going from generation τ to $\tau+1$. The time we expect to wait until a coalescent event occurs is $\mathrm{E}\,T_k$ whenever the ancestral genes in the current generation number k. This property of no memory of the past, the so-called Markov property, is a characteristic of waiting times that follow the geometric distribution (as shown in Box 6).

The time $T_{n\to1}$ until the sample of n genes coalesce into a single gene is the stochastic variable designating the coalescence time. It is obtained by summing the waiting times to the $n-1$ coalescent events, and the expectation is

$$\mathrm{E}\,T_{n\to1} = 4N \sum_{k=2}^{n} \frac{1}{k(k-1)} = 4N \sum_{k=2}^{n} \left(\frac{1}{k-1} - \frac{1}{k} \right) = 4N \left(\sum_{k=1}^{n-1} \frac{1}{k} - \sum_{k=2}^{n} \frac{1}{k} \right).$$

The expected coalescence time is therefore

$$\mathrm{E}\,T_{n\to1} = 4N \left(1 - \frac{1}{n} \right). \tag{2.16}$$

An impression of the shape of the genealogical tree may be obtained by comparing the expected coalescence time to the time it takes for the last two genes to coalesce. The expected coalescence time is close to $4N$ when the sample of genes is not too small, and the final waiting time is expected to be $\mathrm{E}\,T_2 = 2N$, which is about half the expected coalescence time. The variance of T_2 is the variance in

---

Box 6: Waiting without memory

The lack of memory in the geometric distribution, and hence in the coalescent process, is demonstrated by calculating the probability of a certain waiting time given that the coalescent event had not occurred at a particular time. The event that coalescence did not occur after waiting j generations is $\{T_k > j\}$. We therefore have to evaluate the conditional probability

$$\mathrm{Prob}(T_k = i \mid T_k > j) = \frac{\mathrm{Prob}(T_k = i \text{ and } T_k > j)}{\mathrm{Prob}(T_k > j)} = \frac{\mathrm{Prob}(T_k = i)}{\mathrm{Prob}(T_k > j)}$$

for $i > j$, because for $i \leq j$ the probability is evidently zero. Now,

$$\mathrm{Prob}(T_k > j) = \sum_{\ell=j+1}^{\infty} p_k (1 - p_k)^{\ell-1} = (1 - p_k)^j \sum_{\ell=1}^{\infty} p_k (1 - p_k)^{\ell-1},$$

where the last sum equals one (it is the sum of the probabilities in the geometric distribution). We thus get

$$\mathrm{Prob}(T_k = i \mid T_k > j) = \frac{p_k (1 - p_k)^{i-1}}{(1 - p_k)^j} = \mathrm{Prob}(T_k = i - j),$$

and the probability that coalescence occurs after i generations, given that it had not occurred after j generations, equals the probability that coalescence occurs after $i - j$ generations.

---

a geometric distribution, and it is about $2N^2$. The variance of the waiting time $T_{n \to 2}$ until the sample has coalesced into two ancestral genes is the variance of a sum of independent variables with geometric distributions, and it is therefore the sum of the variances of these variables. Numerical evaluation of this sum shows that unless n is very small ($n < 5$), the approximation $\mathrm{Var}\, T_{n \to 2} \approx \frac{1}{3} N^2$ is very good, and hence $\mathrm{Var}\, T_{n \to 2} \approx \frac{1}{6} \mathrm{Var}\, T_2$. We thus expect that the time until the n genes in a sample are represented by two ancestral genes is about the same as the time it takes for those two genes to coalesce, but the variance in the wait to the final coalescence is about six times the variance in the time it takes to coalesce to two ancestral genes.

The larger the sample, the faster the initial fraction of collapse of the sample. The mean waiting time until the n initial genes have coalesced into $\frac{1}{2} n$ genes is

$$2N \sum_{k=n/2+1}^{n} \frac{2}{k(k-1)} = 4N \left(\sum_{k=n/2}^{n-1} \frac{1}{k} - \sum_{k=n/2+1}^{n} \frac{1}{k} \right) = \frac{4N}{n} \tag{2.17}$$

(for n even, of course). Thus, as n increases this waiting time rapidly decreases, as seen in Figure 2.10.

Our analysis clearly shows that Figure 2.7 does not provide a typical picture of the waiting times between coalescent events. The size of T_2 in particular

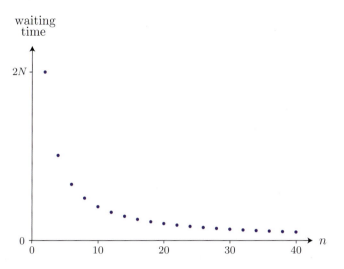

Figure 2.10: The expected waiting time for n genes to coalesce to $n/2$ genes.

seems very short. Nevertheless, such a single deviation is not an argument for an aberrant structure in the pedigree because the distribution of the waiting time to coalescence is very skewed (Figure 2.9). The problem in Figure 2.7 is that not only is the wait to the last coalescence short; the early coalescences actually occur much later than expected. To illustrate this variation through time, I simulated two sets of waiting times $T_{12}, T_{11}, \ldots, T_2$ and inserted them into the genealogy in Figure 2.7, that is, I assumed the same sequence of coalescent events as in that figure. The result is shown in Figure 2.11. The two drawings are given on the same timescale, so the coalescence time in the left simulation is indeed about three times that in the right simulation. Most of the difference is due to the different values of T_2, a difference of about a factor 10, but the probability of getting the left value or longer is about 0.20 and the probability of getting the right value or shorter is about 0.14.

The two drawings in Figure 2.11 show the same sequence of events to ease the comparison—they are said to exhibit the same *topology* of the coalescence tree. A full coalescent simulation produces random waiting times and the topology of the genealogy. A random sequence of events is of course easily performed: given that a coalescent event occurs, any pair of the ancestral genes in the sample is equally likely to coalesce. The simulation is thus finished by coalescing a random pair of ancestral genes every time a coalescent event occurs. This way of constructing the genealogy of the sample is Kingman's (1982a, 1982c) coalescent process (this stochastic process is further discussed in Section 2.6.2 on page 46). Further examples of genealogies produced by the coalescence are obtained from the Wright–Fisher animator (Mikkelsen et al. 2006).

Omitting one gene from the samples in Figure 2.11 does not change the coalescence time. Is that a common occurrence? We may pose this question in

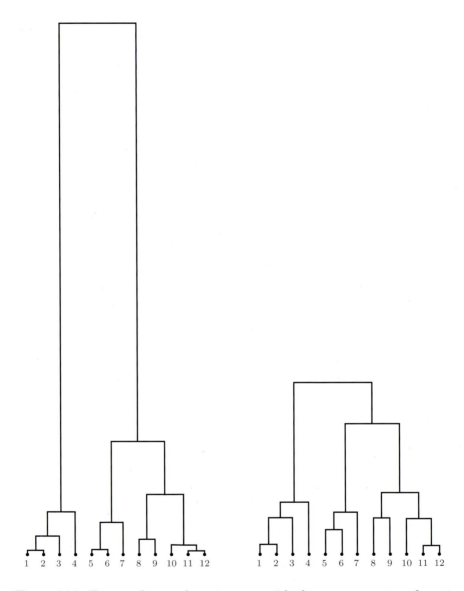

Figure 2.11: Two random coalescence trees with the same sequence of events as that chosen in Figure 2.7. The waiting times are drawn from the geometric distribution (Figure 2.9) with the appropriate probability of coalescence per generation (equation (2.13)).

another way: what is the probability that the inclusion of an extra gene in the sample will change the coalescence time? The number of ways in which the first coalescence involves one of the original n genes is $n(n-1)/2$ out of $(n+1)n/2$ ways among the $n+1$ genes in the augmented sample. The probability that the new gene does not participate in the first coalescent event is therefore

$$\frac{\binom{n}{2}}{\binom{n+1}{2}} = \frac{n-1}{n+1}.$$

Continuing this argument produces

$$\text{Prob}(1,\ldots,n \text{ coalesces before } n+1) = \frac{n-1}{n+1}\frac{n-2}{n}\cdots\frac{2}{4}\frac{1}{3} = \frac{2}{(n+1)n},$$

and this probability decreases rapidly as n increases. Already for $n=4$ it is $\frac{1}{10}$. Thus, in this respect Figure 2.7 is typical.

2.5 Characteristic Time Unit

The waiting times in the coalescent process all have a factor $2N$, that is, the ratio between the various waiting times depends only on the number of genes waiting to coalesce, not on the size of the population. In a sample from a large population the coalescence process proceeds more slowly than in a smaller population, but the structure of the genealogy is the same regardless of population size. This is why units on the vertical time axis in Figure 2.11 are unnecessary. This property is a general property of the Wright–Fisher model in a population of large size. In a large population random genetic drift proceeds more slowly than in a smaller population, but other than that, the general characteristics of the drift process are the same and virtually independent of population size. Indeed, as suggested by the waiting times, the effect of a change in population size parallels a change in the unit used in the measurement of time.

To further try this argument, let us turn to the description of the evolution of the variance in gene frequencies given in equation (2.5):

$$\text{Var}(p_t) = p(1-p)\left(1 - \left(1 - \frac{1}{2N}\right)^t\right), \quad t = 0, 1, 2, \ldots.$$

We seek an approximation to this equation as N becomes large, and it can be obtained by using the mathematical formula

$$\lim_{n\to\infty}\left(1 + \frac{a}{n}\right)^n = e^a \tag{2.18}$$

(see any textbook on analysis—the number e is the base of the natural logarithm; a is any number). The exponential term in the variance may be rewritten as

$$\left(1 - \frac{1}{2N}\right)^t = \left(\left(1 - \frac{1}{2N}\right)^{2N}\right)^{\frac{t}{2N}},$$

and then equation (2.18) provides the approximation

$$\left(1 - \frac{1}{2N}\right)^{2N} \approx e^{-1},$$

which is excellent for large population sizes because the convergence in equation (2.18) is quite rapid. The result of our efforts is thus

$$\mathrm{Var}(p_t) \approx p(1-p)\left(1 - \exp\left(-\frac{t}{2N}\right)\right), \quad t = 0, 1, 2, \ldots. \tag{2.19}$$

The change in the variance of the gene frequency thus becomes independent of the population size if we measure time in units of $2N$ generations, such that one generation is $\frac{1}{2N}$ time units. This is the same property as the one revealed in the analysis of coalescence times, and we can, of course, make the same analysis of the development of population gene identity (equation (2.9) on page 31). We may view this time unit of $2N$ generations as the *characteristic time unit* of random genetic drift, and if we examine our results using this time unit, then we get a simple and robust intuition for the process of random genetic drift: fundamentally we only have *one process of random genetic drift* in the Wright–Fisher model. To describe a particular version of the model we only have to adjust the unit that measures time.

2.6   Diffusion Approximation

The development through time of the distribution in gene frequencies is fairly independent of the population size if time is measured with reference to the characteristic time unit of random genetic drift. The example in Figure 2.3 on page 27 using $2N = 10$ shows the characteristics of the distribution of gene frequencies  in any population with an initial gene frequency of $\frac{1}{2}$. After $2N$ generations, for instance, the distribution of gene frequencies in any population resembles the distribution shown in generation 10 of the example.

The general characterization of the distribution of gene frequencies is done in a large population monitored in characteristic time. Formally we approximate the drift process for $N \to \infty$ and assume a time unit such that the generation time is $\Delta t = \frac{1}{2N}$, and we approximate the genetic drift process by a *diffusion process*. This approximation provides the probability density $\phi(p, y, t)$ of the gene frequency y of allele A after t time units in a population where p is the initial gene frequency. The probability density is given as a solution to a partial differential equation. This equation, the so-called diffusion equation, is given in Box 7 in the two commonly applied versions, the *forward equation* and the *backward equation*. The analysis of genetic drift in the Wright–Fisher model shows a forward-looking point of view, whereas Kingman's coalescent provides a backward description. The two equations have a characteristic and easily recognizable form, and the assumptions of the particular model under scrutiny are expressed in two coefficients.

We will not, however, discuss the analysis of the diffusion equations, but only the results obtained using diffusion approximations.

The diffusion approximation and similar approximations are widely used in population genetics, and their systematic application was championed by Motoo Kimura. The forward diffusion equation was introduced into population genetics in 1935 by Andrey Nikolayevich Kolmogorov working on a model proposed by Sewall Wright. Kolmogorov is one of the founding fathers of the theory of stochastic processes, and he provided the two versions of the equation.

2.6.1 The Wright–Fisher model

Kimura (1955a) formulated and solved the forward diffusion equation for the Wright–Fisher model, thus providing the diffusion approximation for the distribution of gene frequencies produced by the binomial process described in Section 2.1. Figure 2.12 (left) shows the solutions for $p = \frac{1}{2}$ for various characteristic times in the drift process. The figure omits the probabilities of fixation, that is, Prob(A monomorphism) and Prob(a monomorphism)—the "sticks" growing at 0 and 10 in Figure 2.3. Figure 2.12 thus shows only the probability density around polymorphic gene frequencies. The probability of the gene frequency $\frac{i}{2N}$, $i = 1, 2, \ldots, 2N - 1$, is approximated by the area beneath the curve in the interval between $\frac{1}{2N}(i - \frac{1}{2})$ and $\frac{1}{2N}(i + \frac{1}{2})$—found by integrating the diffusion density in that interval. Comparisons with simulations have shown that this approximation is excellent even for quite small populations, even $N < 10$ for not too extreme initial gene frequencies. Integrating the curves in Figure 2.12 from $\frac{1}{4N}$ to $1 - \frac{1}{4N}$ provides the probability that the population is still polymorphic.

After $2N$ generations (characteristic time 1) the distribution is flat and all polymorphic gene frequencies have roughly the same probability, just as we saw in the example with $2N = 10$. After $\frac{1}{10}N$ and $\frac{1}{5}N$ generations the probability of finding very low or very high gene frequencies in the population is very small, and therefore the sticks at 0 and 1 are very small. At the time of $\frac{2}{5}N$ generations (characteristic time $\frac{1}{5}$) low or very high gene frequencies occur, and the sticks therefore must have started to build up.

At N generations the distribution of gene frequencies still shows a clear mode at the initial frequency of $p = \frac{1}{2}$, but at $2N$ generations the distribution for polymorphic populations has "forgotten" where the population came from. This phenomenon occurs for any initial gene frequency, but for a more skewed frequency the flat distribution is reached a little later. For $p = \frac{1}{5}$ (Figure 2.12, right) the distribution is clearly not flat after $2N$ generations, but after $4N$ generations the flat stage is apparently reached. The stick at 0 begins to build up a lot earlier for $p = \frac{1}{5}$, already after $\frac{1}{10}N$ generations, and the probability of being polymorphic is lower throughout the process. This is particularly so when the distributions after $4N$ generations are compared.

Box 7: *Diffusion equations

Let $Y(t)$ be a stochastic variable that describes the gene frequency at the characteristic time t in the population. The change in mean and the variance of the change of Y per time unit are

$$\mu(y) = \frac{1}{\Delta t}\mathrm{E}\big(Y(t+\Delta t) - Y(t)\big|Y(t) = y\big)$$

$$\sigma^2(y) = \frac{1}{\Delta t}\mathrm{Var}\big(Y(t+\Delta t) - Y(t)\big|Y(t) = y\big).$$

In our applications

$$Y(t) = p_{2Nt} = \frac{X_{2Nt}}{2N}.$$

In the Wright–Fisher model the expected change in gene frequency is zero, $\mu(y) = 0$, and the variance per generation is the binomial variance, $\sigma^2(y) = y(1 - y)$.

The diffusion approximation says that $Y(t)$ in the limit as $N \to \infty$ has the probability density $\phi(p, y, t)$, which is a solution to the equation

$$\frac{\partial \phi(p, y, t)}{\partial t} = \frac{1}{2}\frac{\partial^2 \big(\sigma^2(y)\phi(p, y, t)\big)}{\partial y^2} - \frac{\partial \big(\mu(y)\phi(p, y, t)\big)}{\partial y}. \qquad (2.20)$$

This approximation is valid when $\mu(y)$ and $\sigma^2(y)$ are bounded, $\sigma^2(y)$ is strictly positive as $N \to \infty$, and the third central moment of $Y(t)$ tends to zero as $N \to \infty$. The last condition is satisfied for the binomial distribution (and a lot of other reasonable distributions), and it is easy to check its validity. The density of the gene frequency is therefore determined entirely by the infinitesimal mean $\mu(y)$ and the infinitesimal variance $\sigma^2(y)$.

Equation (2.20) is the *forward equation* because it is constructed by considering what happens next, after the population reaches the gene frequency y. The *backward equation* is

$$\frac{\partial \phi(p, y, t)}{\partial t} = \frac{\sigma^2(p)}{2}\frac{\partial^2 \phi(p, y, t)}{\partial p^2} - \mu(p)\frac{\partial \phi(p, y, t)}{\partial p} \qquad (2.21)$$

and is constructed by analyzing the effect of a change in the initial condition specified by the gene frequency p. Kimura was the first to use this equation in population genetics.

The forward diffusion equation for the Wright–Fisher model is thus given by

$$\frac{\partial \phi(p, y, t)}{\partial t} = \frac{1}{2}\frac{\partial^2 \big(y(1 - y)\phi(p, y, t)\big)}{\partial y^2}$$

in the characteristic time for random genetic drift. If we return to time measured in generations we get an additional factor of $\frac{1}{2N}$ on the right side of the equation.

Diffusion equations are partial differential equations. To attain interesting solutions to such equations the model has to be further specified in terms of so-called boundary conditions.

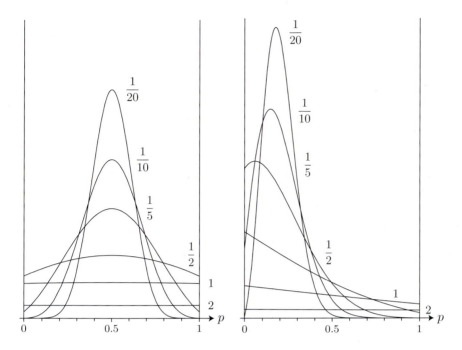

Figure 2.12: Distribution of gene frequencies in the Wright–Fisher model for various characteristic times with the initial gene frequency $p = \frac{1}{2}$ (left) and $p = \frac{1}{5}$ (right).

The flat stage in the distribution of gene frequencies is approximately

$$\phi(p, y, t) \approx K \, \exp\!\left(-\frac{t}{2N}\right) \quad \text{as } t \to \infty, \tag{2.22}$$

where K is a constant and time is measured in generations. The probability in each polymorphic interval is therefore lowered by a fraction of about $\frac{1}{2N}$ per generation.

The diffusion approximation method may be used to find many properties of the random genetic drift process. The diffusion equation can be modified to provide a description of these properties. For instance, a diffusion equation that yields the distribution of the waiting time to fixation or loss of an allele may be constructed. The expected waiting time to loss of A or a in the characteristic time for the drift process is

$$-2\big(p \log p + q \log q\big)$$

(Watterson 1962), in which p and q are the initial frequencies of the alleles (Figure 2.13). The maximum waiting time is $2N \times 1.39$ generations attained for $p = \frac{1}{2}$. The waiting time is about $2N$ generations for $p = 0.2$ and $p = 0.8$, and the closer the gene frequencies are to the fixation states, the steeper the

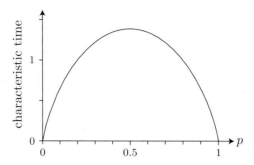

Figure 2.13: Mean waiting time until loss of one of the alleles.

fall in the waiting time. And as p becomes smaller and smaller, the steepness increases without bounds.

2.6.2   The coalescent

In Section 2.4 the effects of random genetic drift were described with reference to a contemporary population by Kingman's (1982a, 1982c) coalescent process. The coalescent process is really the process of random genetic drift going back in time from the present state of the population, and the approximations we used in constructing this process are obtained by assuming a large population size, an assumption similar to that used in the diffusion approximation. Thus, we may seek similar simplifications in the analysis of the genealogical structure of a sample of n genes from a population with $2N$ genes.

The example of a sample of 12 genes (Figure 2.7 on page 35) may be used to develop Kingman's formal description of the coalescent process. The first coalescent event in the figure fuses genes number 11 and 12, written as (11,12), and using this notation, the ancestral material of the genes in the sample at various points of time in the past is described by

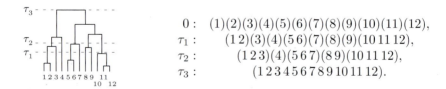

$$
\begin{aligned}
0 :&\quad (1)(2)(3)(4)(5)(6)(7)(8)(9)(10)(11)(12),\\
\tau_1 :&\quad (1\,2)(3)(4)(5\,6)(7)(8)(9)(10\,11\,12),\\
\tau_2 :&\quad (1\,2\,3)(4)(5\,6\,7)(8\,9)(10\,11\,12),\\
\tau_3 :&\quad (1\,2\,3\,4\,5\,6\,7\,8\,9\,10\,11\,12).
\end{aligned}
$$

For instance, at time τ_1, genes 1 and 2 have a common ancestor, as do genes 5 and 6, and also genes 10, 11, and 12. At time τ_3 all genes in the sample have coalesced.

This description of the composition of the sample uses equivalence relations among the genes. The genes within a pair of parentheses are viewed as equivalent, and so the coalesced genes form equivalence classes. The building of the coalescent tree may therefore be viewed as a stochastic process on the set of such equivalence relations among a set of n objects. This is a fairly large and complicated set, so we will not try to describe it in detail. We only need to specify the rules for how the description of our sample behaves as we go back in time. At time τ_2 in Figure 2.7 the sample is at the state

$$\xi = (1\,2\,3)(4)(5\,6\,7)(8\,9)(10\,11\,12).$$

A state η that can be reached from ξ by one coalescence event is called a *follower* of ξ, and this is written $\xi \to \eta$. The state ξ has 10 possible followers, given by

$$
\begin{aligned}
\eta_1 &= (1\,2\,3\,4)(5\,6\,7)(8\,9)(10\,11\,12), \\
\eta_2 &= (1\,2\,3\,5\,6\,7)(4)(8\,9)(10\,11\,12), \\
\eta_3 &= (1\,2\,3\,8\,9)(4)(5\,6\,7)(10\,11\,12), \\
\eta_4 &= (1\,2\,3\,10\,11\,12)(4)(5\,6\,7)(8\,9), \\
\eta_5 &= (1\,2\,3)(4\,5\,6\,7)(8\,9)(10\,11\,12), \\
\eta_6 &= (1\,2\,3)(4\,8\,9)(5\,6\,7)(10\,11\,12), \\
\eta_7 &= (1\,2\,3)(4\,10\,11\,12)(5\,6\,7)(8\,9), \\
\eta_8 &= (1\,2\,3)(4)(5\,6\,7\,8\,9)(10\,11\,12), \\
\eta_9 &= (1\,2\,3)(4)(5\,6\,7\,10\,11\,12)(8\,9), \\
\eta_{10} &= (1\,2\,3)(4)(5\,6\,7)(8\,9\,10\,11\,12),
\end{aligned}
$$

and each of these is obtained by fusing two of the classes of genes in ξ. The number of followers is determined by the number of classes in ξ, written as $|\xi|$, and it is equal to the number of pairs of classes that we may fuse (see page 34). We may thus write the number as

$$\binom{|\xi|}{2} = \frac{|\xi|\,(|\xi| - 1)}{2}.$$

The state reached at time τ_1 in Figure 2.7 has 28 followers, the initial state has 66 followers, and the final state reached at time τ_3 has 0 followers and our interest in the process stops.

The key to the description of the coalescent process is the realization that the $|\xi|$ genes in the ancestral material of the sample are equal in any property that influences their transmission between generations. A given pair of genes coalesce in the parental generation with probability $\frac{1}{2N}$, so measuring time in generations, the rate of coalescence of the pair per time unit is $\frac{1}{2N}$. In building the coalescent process, Kingman assumed that time is measured as the characteristic time for genetic drift, where the time unit is $2N$ generations (Section 2.5). Thus, in characteristic time the rate of coalescence of a given pair of genes per time unit is 1, and therefore the probability that a coalescence of a pair occurs in a small time interval is proportional to the length of the interval if we neglect the possibility that two or more coalescent events may occur. In addition, any of the followers has the same probability of occurring within the

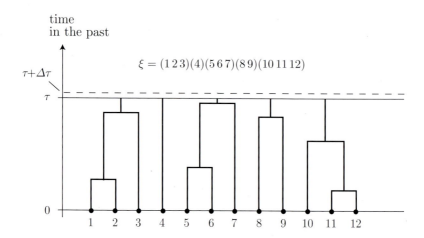

Figure 2.14: The coalescent of a sample of 12 genes at time $\tau = \tau_2$ of Figure 2.7.

next short interval of time $\Delta\tau$ (Figure 2.14). Thus, in a process at state ξ we have for each follower η of ξ that

$$\text{Prob}(\eta \text{ at } \tau + \Delta\tau \,|\, \xi \text{ at } \tau) = \Delta\tau + O\big((\Delta\tau)^2\big). \qquad (2.23)$$

The symbol

$$O\big((\Delta\tau)^2\big)$$

means "terms of the order of $(\Delta\tau)^2$." We assume that at most one event occurs in the short time interval, and the probability that something happens is therefore given by the sum of the probabilities of all the possible changes, that is, by

$$\text{Prob}(\text{follower of } \xi \text{ at } \tau + \Delta\tau \,|\, \xi \text{ at } \tau) = \binom{|\xi|}{2} \Delta\tau + O\big((\Delta\tau)^2\big). \qquad (2.24)$$

The probability that nothing happens is then

$$\text{Prob}(\xi \text{ at } \tau + \Delta\tau \,|\, \xi \text{ at } \tau) = 1 - \binom{|\xi|}{2} \Delta\tau + O\big((\Delta\tau)^2\big), \qquad (2.25)$$

and the probability that two or more events happen in the time interval from τ to $\tau + \Delta\tau$ is very small and of the order of $(\Delta\tau)^2$. Thus, the probability of a coalescent event in a small interval of time increases rapidly with the number of genes in the sample, because the probability is roughly proportional to the square of the number of genes when that number is large.

Kingman's coalescent process is given by equations (2.23) and (2.25). It is a simple stochastic process (called a death process). In such a process the waiting time until an event occurs is exponentially distributed (see Appendix A.1.4) with a mean equal to the reciprocal of the rate at which the event occurs. The exponential distribution shares the property of no memory with the geometric distribution, and it may be viewed as an approximation to the geometric distribution for small event probabilities. The waiting time until a coalescence appears at state ξ is therefore exponentially distributed with mean

$$\frac{2}{|\xi|(|\xi| - 1)},$$

as seen from equation (2.24). Given that a coalescent event occurs at state ξ, any of the $\frac{1}{2}|\xi|(|\xi| - 1)$ followers are equally likely to become the next state in the process.

2.6.3 The ∞-coalescent

The population size disappears from the coalescent process when characteristic time is used. Kingman's coalescent is therefore defined for any number of genes n in the sample, because the approximation used corresponds to assuming an effectively infinite population. Kingman named this process the *n-coalescent*, and reserved the "coalescent" to the precess that obtains as the limit of the *n*-coalescent as $n \to \infty$, the ∞-coalescent, say. In Section 2.4 we showed that the larger the sample, the faster the initial fraction of collapse in the sample. In characteristic time, the waiting time until the n initial genes (for n even) have coalesced to $\frac{1}{2}n$ is $\frac{2}{n}$ (Equation (2.17) on page 38), and as n increases this waiting time rapidly decreases to zero. Thus, in the first split second the infinitely large sample experiences a big crunch to become finite. After that the coalescent behaves like the normal coalescent process in a finite sample. The ∞-coalescent thus describes a very large finite sample. Any coalescent process on a sample of n genes may be viewed as a coalescent on a sample of n genes in the ∞-coalescent. With this simple description of the coalescent process we see that Kingman's coalescent provides a useful approximative description of the genealogy of a sample of $n = 2N$ genes, that is, of the total population. The description obtained for the entire population can never be of a high quality in the Wright–Fisher model, because we cannot neglect the possibility that more than one coalescent event occurs when tracing back all the genes in a given generation to their parental genes.

This description of the coalescent process in a population is very theoretical and very far from the reality of a biological sample, because the time unit in the process is the characteristic time in random genetic drift, i.e., $2N$ generations. Thus, as the sample size n becomes large, so must the population size $2N$, and the process therefore slows as n increases. However, the mathematical description serves to show that the processes are inherently the same in large and small populations if we account for the difference in speed. A large sample reaches the same size as a small sample comparatively fast (in the characteristic time), and after that the two samples go through the same process of genealogical description.

2.7   Moran's Model

The Wright–Fisher model is a discrete generation model. An equally simple model of a population with overlapping generations is the Moran model (Moran 1958). This model assumes a haploid population. To ease comparisons between the models, however, we assume a population size of $2N$ genes in both.

At discrete times $t = 1, 2, \ldots$ a random individual in the population breeds and produces one offspring, and immediately after the breeding event a random adult individual dies (the newly produced offspring cannot die). For instance, in a population with ten individuals a breeding event might look like this:

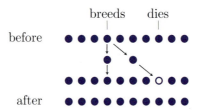

These events can change the number of A genes by either adding one or subtracting one, or they can leave the number of As unchanged. With i A genes, the probabilities of these changes are

$$
\begin{aligned}
P_{i \to i+1} &= \frac{i}{2N}\left(1 - \frac{i}{2N}\right), \\
P_{i \to i-1} &= \left(1 - \frac{i}{2N}\right)\frac{i}{2N}, \\
P_{i \to i} &= \left(\frac{i}{2N}\right)^2 + \left(1 - \frac{i}{2N}\right)^2.
\end{aligned}
\tag{2.26}
$$

The first equation gives the probability that an A breeds and an a dies, the second is vice versa, and the last one specifies the probability that the dying individual is of the same type as the breeding individual.

As a stochastic process the Moran model is much simpler than the Wright–Fisher model. As a biological model, however, the Wright–Fisher model is more attractive because of its potential to include the description of the mating structure in the population—a question we return to in Chapter 3. The two models are nevertheless sufficiently similar to supplement each other in the description of random genetic drift.

Assume that the state of the population at time t is described by X_t, the number of individuals that carry allele A. The change through time in the variance of the gene frequency in Moran's model is given by

$$
\mathrm{Var}(X_t) = 4N^2 p(1-p)\left(1 - \left(1 - \frac{1}{2N^2}\right)^t\right),
\tag{2.27}
$$

where p is the initial gene frequency in the population. The change in variance thus corresponds to that of a Wright–Fisher model with a population size of N^2. The change in variance per unit time is therefore very slow, and progressively slower the larger the population size. The Moran model, however, assumes that breeding events only involve one individual in the population, and that most of the individuals in the population survive to breed again. The difference between the Wright–Fisher model and the Moran model is therefore the way in which we measure time rather than the probability that a given gene is passed on to the next generation.

To focus on the unit of time we compare equations (2.5) and (2.27) by using the approximation (2.18) on page 41. For the Moran model we get

$$\mathrm{Var}(p_t) \approx p(1-p)\left(1 - \exp\left(-\frac{t}{2N^2}\right)\right), \quad t = 0, 1, 2, \ldots. \tag{2.28}$$

The characteristic time unit is thus $2N^2$ breeding events. The population sizes in the Moran model and the Wright–Fisher model are the same, however, and therefore both models are described by equation (2.19) if we view the Moran model as having a generation time of N time units . This is reasonable because in N time units, N individuals breed, and this equals the number that breed in one breeding event in the discrete generation model.[8] The development of gene identity in a Moran population is therefore very similar to that in a Wright–Fisher population with this change in the timescale.

Description of the coalescence structure is a lot simpler in Moran's model. The difficulty in the Wright–Fisher model was that the coalescence description is an approximation that works only in large populations, allowing us to neglect multiple coalescent events in the same generation. In the Moran model only one coalescent event can occur in each breeding event. The probability that two genes chosen from the population are copies of the same parental gene prior to the last breeding event is

$$\left(1 - \frac{1}{2N}\right)\frac{2}{2N(2N-1)} = \frac{1}{2N^2},$$

because coalescence can only happen if the breeding gene did not die. Only one pair of genes may coalesce at that event, namely the breeder and the offspring. Thus, we once again get the rate of one coalescence per characteristic time unit.

The probability of a coalescent event in a sample of n in the breeding event immediately prior to the time of sample is

$$\binom{n}{2}\frac{1}{2N^2}$$

(Kingman 1982b), and even if we sample the entire population this formula is still valid and the probability becomes

$$1 - \frac{1}{2N}.$$

---

[8]This works. Note, however, that in the Wright–Fisher model N diploid individuals breed, whereas in the Moran model $2N$ haploid individuals breed.

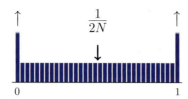

Figure 2.15: Distribution of gene frequencies as t becomes large.

Thus, our arguments in Section 2.4 become exact for the Moran model for any size of the population and any sample size. *The Moran model n-coalescent is an exact description of a sample of n genes from a population of 2N haploid individuals.*

Exercises

Exercise 2.2.1 Find the variance $\mathrm{Var}(p_{A_1 t} + p_{A_2 t})$ using the formulae (2.6) and (2.7).

Exercise 2.7.1 Argue that the transition probabilities (2.26) do indeed describe the Moran model, and show that they sum to one.

Exercise 2.7.2 Show that $\mathrm{E}(X_t | X_{t-1} = i) = i$.

Exercise 2.7.3 Show that $\mathrm{Var}(X_t | X_{t-1} = i) = \dfrac{i}{N}\left(1 - \dfrac{i}{2N}\right)$.

Exercise 2.7.4 * Show that equation (2.27) is correct.

Exercise 2.7.5 Compare the expression (2.27) for $\mathrm{Var}(X_t)$ in the Moran model with $\mathrm{Var}(p_t)$ in the Wright–Fisher model (equation (2.5) on page 28). What is the reason for the difference?

Exercise 2.7.6 Show that the form of the distribution in Figure 2.15 (given by equation (2.22) on page 45) is conserved from generation to generation in the Moran model. Demonstrate this by showing that if all the polymorphic gene frequencies are equally probable at time t, then they are also equally probable at time $t + 1$.

Chapter 3

Diploid Populations

The gamete pool or Wright–Fisher model is simplified to allow description of the genetic dynamics entirely in terms of transmission of genes between generations. This corresponds to assuming a population of haploid organisms. Higher organisms with sexual reproduction have a life cycle that oscillates between haploid and diploid life stages, and in most animals and plants the diploid stage is the most conspicuous, while the haploid, or gametic, stage leads a more inconspicuous life. The gametes—in humans, eggs and sperm—unite to form the zygote that develops into the individual, which at maturity breeds and transmits its genetic material through the production of gametes. To make a diploid population genetic model where we can specify the genotypic frequencies, we must specify the rules of mating, which in turn establish the way gametes are united. We presuppose that the considered allelic variation in genes and genotypes does not affect survival and reproduction, implying gene frequency conservation. A natural extension of this is to assume that the variation does not affect mating in the population, that is, we assume mating is independent of genotypes. The simplest such model is *random mating*.

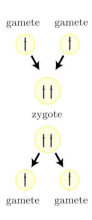

In a very large population the genetic outcome of random mating is particularly simple. Choose an individual to function as a female and consider one of her eggs. To fertilize this egg random mating stipulates the choice of a random male, who delivers the required sperm. The two homologous genes of the male have an equal probability of making it to the chosen sperm, and the sperm thus carries a randomly chosen gene in the population of individuals that can act as males. In conclusion, zygotes are made by *random union of gametes*.

The genotypic frequencies in a two-allele polymorphism in a large randomly mating population of hermaphrodites are

$$\mathsf{AA}:\ p^2, \quad \mathsf{Aa}:\ 2pq, \quad \mathsf{aa}:\ q^2, \tag{3.1}$$

referred to as being in *Hardy–Weinberg proportions*. These frequencies stipulate independence of the two genes in the genotype, and a sample of n individuals from the population is therefore equivalent to a random sample of $2n$ genes. The number of A genes in the sample is hence binomially distributed $b(2n, p)$. Without Hardy–Weinberg proportions the distribution of the observed gene frequency is more complicated.

The assumption of large population size ensures that the difference in gene frequencies among fertilizing sperm and fertilized eggs is negligibly small. Also, in a large population we may ignore the differences between the genotypic frequencies and the probabilities of choosing a sperm carrying, for instance, allele A independently of the allele in the egg. With distinct sexes the *random-union-of-gametes proportions* of a two-allele polymorphism therefore are

$$\text{AA}: \; p_\female p_\male, \quad \text{Aa}: \; p_\female q_\male + q_\female p_\male, \quad \text{aa}: \; q_\female q_\male, \tag{3.2}$$

which corresponds to the Hardy–Weinberg proportions, when the gene frequencies in the sexes are equal, $p_\female = p_\male$. For alleles that segregate according to Mendel's law the random-union-of-gametes proportions are equal in the two sexes, and the gene frequency is $p = \frac{1}{2}(p_\female + p_\male)$. In the following generations the genotypic proportions are thus expected to be in Hardy–Weinberg proportions. With gene frequency conservation, random mating produces genotypic frequencies in the same Hardy–Weinberg proportions in each generation. A population in this perpetuated state is said to be in *Hardy–Weinberg equilibrium*.

We may thus assume random mating and move on to the discussion of the transmission of genes. Random mating is, however, a complicated process to model in a finite population; we have to specify random mating of what and with whom. So before moving on we must scrutinize the intricacies of random mating in a finite population.

3.1   Random Mating

The simplest model of mating interprets the gamete pool model very literally. The parents shed their gametes into a gamete pool, and offspring are formed by the fusion of two randomly chosen gametes from the pool. We may call this model of random mating the *gamete mating model*, and we shall assume this mating rule for the gamete pool model. The gamete mating model reflects the mechanics of mating in, for instance, oysters, corals, pine, and rye.

The description of the development of gene identity in the gamete pool model (Section 2.3 on page 31) can be further interpreted in a diploid population. In this model the population coefficient of gene identity F_t has several interpretations. It describes the average level of consanguinity among the individuals in the population, and it is thus the population coefficient of consanguinity, where the consanguinity coefficient of two individuals is the probability that two allelic genes, one from each, are identical by descent (see Box 5 on page 32). Consanguinity between mates allows an offspring to be an identical homozygote, that is, the two genes carried in a particular locus are identical by descent. An individual with consanguineous parents is called inbred, and its degree of inbreeding is described as the consanguinity coefficient of the parents (see Box 5). Accordingly, when referring to the offspring, this coefficient is called the inbreeding coefficient of the individual. In the gamete pool model the population coefficient of gene identity is also the probability that an individual is identical homozygote, and F_t is also referred to as the population inbreeding coefficient.

The other obvious models of random mating are the *individual mating models*, where individuals are picked at random and their gametes fused. This is a very complicated class of models that can only be imbedded in the gamete pool model under rather special circumstances. The simplest individual mating model is monogamous mating, to which we return in Section 3.1.1. We will not pursue these models further, but instead look at some simple mating restrictions in the gamete pool model. Strictly speaking, the gamete pool model treats all gametes as equal, which for higher organisms corresponds to assuming a population of hermaphrodites. The simplest individual mating model in this context excludes self-fertilization. Then the two genes in an individual cannot be copies of the same gene in the parental population, whereas two genes chosen from different individuals can. Thus, the population coefficient of consanguinity and the population inbreeding coefficient must be different.

Let F_t designate the probability that two random genes chosen from different individuals in generation t are identical by descent, and let G_t be the population inbreeding coefficient in generation t. The probability that two genes chosen from different individuals in generation t are copies of the genes in gametes from the same individual in generation $t-1$ is $\frac{1}{N}$. Two genes taken from the same individual are copies of the same gene in that individual with probability $\frac{1}{2}$ (Mendel's law). They are copies of the two different genes carried by the individual with probability $\frac{1}{2}$, and therefore identical with probability $\frac{1}{2}G_{t-1}$. We thus have the recurrence equations

$$F_t = \frac{1}{N}\left(\tfrac{1}{2} + \tfrac{1}{2}G_{t-1}\right) + \left(1 - \frac{1}{N}\right)F_{t-1}, \tag{3.3}$$

$$G_t = F_{t-1}, \tag{3.4}$$

where the second is the definition of the population inbreeding coefficient. Like equation (2.8) on page 31, these equations describe the process of random genetic drift. If selfing is allowed, then $G_t = F_t$ and we recover the gamete pool equation. At any rate, random genetic drift proceeds in building up gene identity, and the difference between allowing selfing or not is manifested in the rate at which the level of gene identity increases.

The two-dimensional linear recurrence equations (3.3) and (3.4) may be analyzed in much the same way as before. The only complication is dimensionality, as seen by the arguments given in Box 8. The rate of convergence with selfing excluded is a bit slower than that in the model with random mating of gametes, and the convergence is described by the factor

$$\lambda \approx 1 - \frac{1}{2N}\left(1 - \frac{1}{2N}\right).$$

Box 8: Consanguinity with no selfing

By rewriting equations (3.3) and (3.4) in a manner corresponding to the one used in the gamete pool model (equation (2.3) on page 31) we get

$$
\left\{ \begin{array}{c} 1 - F_t \\ 1 - G_t \end{array} \right\} = \left\{ \begin{array}{cc} 1 - \frac{1}{N} & \frac{1}{2N} \\ 1 & 0 \end{array} \right\} \left\{ \begin{array}{c} 1 - F_{t-1} \\ 1 - G_{t-1} \end{array} \right\}, \tag{3.5}
$$

and this iterates to

$$
\left\{ \begin{array}{c} 1 - F_t \\ 1 - G_t \end{array} \right\} = \left\{ \begin{array}{cc} 1 - \frac{1}{N} & \frac{1}{2N} \\ 1 & 0 \end{array} \right\}^t \left\{ \begin{array}{c} 1 \\ 1 \end{array} \right\}. \tag{3.6}
$$

Such a matrix power may be evaluated by using the spectral resolution of the matrix (found in any textbook on linear algebra), and as before we get convergence of F_t and G_t to 1 as $t \to \infty$. An equivalent and easier way to construct the solution in this case is to turn the equations into a second-order recurrence equation by inserting $G_{t-1} = F_{t-2}$, that is,

$$
1 - F_t = \left(1 - \frac{1}{N} \right)(1 - F_{t-1}) + \frac{1}{2N}(1 - F_{t-2}), \tag{3.7}
$$

and look for solutions of the form $1 - F_t = a\lambda^t$, where λ describes the rate of change in the coefficient of consanguinity between individuals in the population (a is a constant describing the initial condition). This parameter is the numerically largest solution to the equation

$$
\lambda^2 - \left(1 - \frac{1}{N} \right)\lambda - \frac{1}{2N} = 0, \tag{3.8}
$$

which is known as the *characteristic equation* corresponding to equation (3.7), or to the matrix in equation (3.5). Any root λ of the characteristic equation is an *eigenvalue* of the corresponding matrix. We then seek the largest solution, or the leading eigenvalue, which is

$$
\lambda = \tfrac{1}{2}\left(1 - \frac{1}{N} + \sqrt{1 + \frac{1}{N^2}} \right) \approx 1 - \frac{1}{2N}\left(1 - \frac{1}{2N} \right),
$$

where the approximation assumes N to be large.

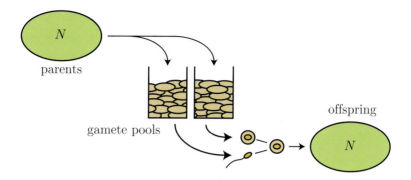

Figure 3.1: The diploid gamete pool model.

Comparison of this measure of the rate of convergence to the factor $1 - \frac{1}{2N}$ in the gamete pool model facilitates the biological interpretation of the assumed deviation from the Wright–Fisher model. The gamete pool model with random mating of gametes is thus chosen as a reference model, and we represent the non-selfing model by applying the Wright–Fisher model with the same rate of convergence to complete identity. The rate of convergence in the reference model is determined entirely by the population size, and we may therefore assign this population size to the non-selfing model. This descriptive population size is named *the effective population size N_e*, and it is in general defined as

$$1 - \frac{1}{2N_e} = \lambda, \tag{3.9}$$

where λ describes the rate of convergence towards genetic homogeneity of the model in question (Wright 1931). In the model that excludes selfing this becomes

$$1 - \frac{1}{2N_e} \approx 1 - \frac{1}{2N}\left(1 - \frac{1}{2N}\right),$$

giving the effective population size $N_e \approx N + \frac{1}{2}$. Thus, the convergence of the population coefficient of consanguinity is approximately like that in the gamete pool model if we add an extra gene to the population. Unless N is very small, $N_e \approx N$ is a good approximation.

A further biologically reasonable complication is to allow for two sexes with different population sizes $N_\female$ and $N_\male$. This requires two gamete pools, one for female gametes and one for male gametes (Figure 3.1), and these pools contribute equally to the offspring population because each offspring is produced by fusing one gamete from each pool. Thus, two allelic genes picked at random from different individuals in the offspring population have the probability $\frac{1}{4}$ of originating from females among the parents, $\frac{1}{4}$ of being of male origin, and the probability $\frac{1}{2}$ of one originating from a female and one from a male. The probability that two random genes originate from the same parent individual is therefore

$$\frac{1}{4N_{\female}} + \frac{1}{4N_{\male}} = \frac{1}{N_{tot}},$$

say, and in this respect the population looks like a population with N_{tot} individuals in the gamete pool model without selfing. Males and females are formed in the same way, so the population inbreeding coefficient G_t is the same for both sexes. G_t is the probability of drawing identical male and female genes from the two pools from which generation t was formed. The population coefficient of consanguinity F_t, that is, the probability of drawing identical genes from different individuals, is also independent of sex. The recurrence equations for these variables are

$$F_t = \frac{1}{N_{tot}} \left(\tfrac{1}{2} + \tfrac{1}{2}G_{t-1} \right) + \left(1 - \frac{1}{N_{tot}} \right) F_{t-1}, \qquad (3.10)$$

$$G_t = F_{t-1}. \qquad (3.11)$$

These equations are indistinguishable from equations (3.3) and (3.4), and the effective population size is therefore $N_e \approx N_{tot} + \tfrac{1}{2}$. If the population sizes are not too small, we therefore get the effective population size given by

$$\frac{1}{N_e} \approx \frac{1}{4N_{\female}} + \frac{1}{4N_{\male}} \qquad (3.12)$$

(Wright 1931). The rate of convergence to homozygosity is thus influenced the most by the sex with the lowest population size (see Exercise 3.1.1).

3.1.1   Mating of individuals

Individually based models of random mating have an effect on two levels: mating affects the gene frequencies among offspring, and given the gene frequencies it likewise affects the genotype frequencies. The simplest individual mating is monogamy, where individual females and males form exclusive pairs for the breeding period. We restrict attention to discrete nonoverlapping generation models where individuals breed only once. Monogamous mating is therefore equivalent to allowing at most one mating for each individual. The numbers of breeding males and females are necessarily equal, that is, $N_{\female} = N_{\male} = \tfrac{1}{2}N$ to retain the size N of the breeding population.

In a population with the genotypes AA, Aa, and aa mating commences by drawing at random i AA, j aa homozygotes, and $\tfrac{1}{2}N - i - j$ heterozygotes as breeders among females, and k, $\tfrac{1}{2}N - k - \ell$, and ℓ as breeding males. The genotypic frequencies in the population are z_{AA}, z_{Aa}, and z_{aa} ($z_{AA} + z_{Aa} + z_{aa} = 1$), and the genotypic numbers of breeding females (X_{AA}, X_{Aa}, X_{aa}) and males (Y_{AA}, Y_{Aa}, Y_{aa}) each have a multinomial distribution (Appendix A.1.3 on page 351). For instance, the genotypic distribution among females is

$$\mathrm{Prob}\big((X_{\mathsf{AA}}, X_{\mathsf{Aa}}, X_{\mathsf{aa}}) = (i, \tfrac{1}{2}N - i - j, j)\big)$$

$$= \begin{pmatrix} \tfrac{1}{2}N \\ i \ \tfrac{1}{2}N - i - j \ j \end{pmatrix} z_{\mathsf{AA}}^{i} z_{\mathsf{Aa}}^{\frac{1}{2}N - i - j} z_{\mathsf{aa}}^{j}. \tag{3.13}$$

The expected values of the frequencies of A among breeding females and males are evidently the same and equal to $p = z_{\mathsf{AA}} + \tfrac{1}{2} z_{\mathsf{Aa}}$. The variances are also equal, and that for females is

$$\mathrm{Var}\, p_{\female} = \mathrm{Var}\left(\frac{X_{\mathsf{AA}} + \tfrac{1}{2} X_{\mathsf{Aa}}}{\tfrac{1}{2}N}\right) = \frac{pq(1 + \phi)}{N}, \quad \text{where} \quad \phi = -\frac{z_{\mathsf{Aa}} - 2pq}{pq}$$

is the deviation from Hardy–Weinberg proportions:

$$z_{\mathsf{AA}} = p^2 + \phi pq, \quad z_{\mathsf{Aa}} = 2pq - 2\phi pq, \quad \text{and} \quad z_{\mathsf{aa}} = q^2 + \phi pq. \tag{3.14}$$

For positive ϕ we therefore have an excess of homozygotes compared to the Hardy–Weinberg proportions, and for ϕ negative we have a deficit.

The expected values of the frequencies of A in the two sexes are equal, but we nevertheless expect the actual frequencies to be different:

$$\mathrm{Var}(p_{\female} - p_{\male}) = \frac{2pq(1 + \phi)}{N}.$$

This alone will produce a deviation ϕ' from the Hardy–Weinberg proportions among offspring, and most likely the effect is an excess of heterozygotes (see Exercise 3.0.1). A finite number of breeding individuals will also be an additional source of variation in the genotypic proportions among offspring (Box 9).

3.1.2 Random mating and selfing

Many higher plants have hermaphroditic flowers, that is, flowers that possess both female (carpels) and male (anthers) organs. Pollen are produced on the anthers, and the female gamete, the oozyte, resides in the carpel. Pollen are deposited on the stigma, a scar on the carpel usually placed on an outgrowth, the style. The pollen form a tube from the stigma down through the style to deliver the male gamete to the oozytes.

With random mating pollen are deposited on the stigma from a randomly chosen plant, a process often referred to as outbreeding. Many plants in addition allow pollen from the same plant, or even the same flower, to fertilize their oozyte, a process called selfing (see Brown 1990). Even wind-pollinated species may show a probability of selfing higher than that expected from the gamete pool model (Muona 1990).

If the probability of self-fertilization of an oozyte, say, α, is the same for all genotypes, then this model assumes mating to be independent of genotype. In

---

Box 9: Random mating with monogamy

The pairs formed by $\frac{1}{2}N$ females and $\frac{1}{2}N$ males in the genotypic numbers i, $\frac{1}{2}N - i - j$, and j, and k, $\frac{1}{2}N - k - \ell$, and ℓ have the distribution

$$\binom{N}{P} z_{\mathsf{AA}}^{i+k} z_{\mathsf{Aa}}^{N-i-j-k-\ell} z_{\mathsf{aa}}^{j+\ell},$$

where $\boldsymbol{P}$ is the matrix given in the table:

♀\♂	AA	Aa	aa
AA	b_{AA}	$i - b_{\mathsf{AA}} - b_{\mathsf{Aa}}$	b_{Aa}
Aa	$k - b_{\mathsf{AA}} - b_{\mathsf{aA}}$	$N - i - j - k - \ell$ $+ b_{\mathsf{AA}} + b_{\mathsf{Aa}} + b_{\mathsf{aA}} + b_{\mathsf{aa}}$	$\ell - b_{\mathsf{Aa}} - b_{\mathsf{aa}}$
aa	b_{aA}	$j - b_{\mathsf{aA}} - b_{\mathsf{aa}}$	b_{aa}

Given the genotypic frequencies in males and females, the effect of individual mating is that it adds an extra source of variation, namely the pair formation process described by the four numbers b_{AA}, b_{Aa}, b_{aA}, and b_{aa}. The probability of the mating pattern $\boldsymbol{P}$ given the genotypic frequencies in males and females is thus

$$\frac{\binom{N}{P}}{\binom{N}{i\ \ N-i-j\ \ j} \binom{N}{k\ \ N-k-\ell\ \ \ell}},$$

which is a hypergeometric distribution.

---

a population where the genotypic frequencies are z_{AA}, z_{Aa}, and z_{aa}, selfing then produces offspring in the frequencies

$$\mathsf{AA}:\ z_{\mathsf{AA}} + \tfrac{1}{4}z_{\mathsf{Aa}}, \quad \mathsf{Aa}:\ \tfrac{1}{2}z_{\mathsf{Aa}}, \quad \mathsf{aa}:\ z_{\mathsf{aa}} + \tfrac{1}{4}z_{\mathsf{Aa}}.$$

The fraction $1-\alpha$ of offspring produced by random mating is in Hardy–Weinberg proportions. The selfed offspring are inbred, those formed from foreign pollen are outbred. In a large population the genotypic frequencies among offspring are then

$$(1 - \alpha)p^2 + \alpha(z_{\mathsf{AA}} + \tfrac{1}{4}z_{\mathsf{Aa}}), \quad 2(1 - \alpha)pq + \tfrac{1}{2}\alpha z_{\mathsf{Aa}}, \quad (1 - \alpha)q^2 + \alpha(z_{\mathsf{aa}} + \tfrac{1}{4}z_{\mathsf{Aa}}),$$

where $p = z_{\mathsf{AA}} + \tfrac{1}{2}z_{\mathsf{Aa}}$. Assuming no selection, and thus gene frequency conservation, we get that the change in frequency of heterozygotes between generations is given by the recurrence equation:

$$z_{\mathsf{Aa}}' = 2(1 - \alpha)pq + \tfrac{1}{2}\alpha z_{\mathsf{Aa}}$$

The genotypic frequencies thus change unless the frequency of heterozygotes

---

Box 10: Random genetic drift with selfing

In the system of equations (3.16) and (3.17) the eigenvalues of convergence to homozygosity are for $\alpha \gg \frac{1}{2N}$ approximately

$$1 - \frac{1}{2N} + \frac{\alpha}{2N(2-\alpha)} \quad \text{and} \quad \frac{\alpha}{2} - \frac{1}{N(2-\alpha)},$$

where we neglected terms of the order of magnitude $(\frac{1}{N})^2$. The first describes the leading eigenvalue, and the second is recognized as the eigenvalue of convergence to the characteristic genotypic frequencies for given gene frequencies. For small α the first is close to the leading eigenvalue in the Wright–Fisher model, and the eigenvalue increases as the frequency of selfing increases. When α approaches one, the leading eigenvalue approaches one, and our approximation breaks down (we cannot conclude that it is exactly one because we neglected small terms). The case $\alpha = 1$ is special, however, because the gamete pool becomes empty.

---

equals $\hat{z}_{\mathsf{Aa}} = 2pq(1 - \hat{F})$ with the equilibrium level of gene identity given by

$$\hat{F} = \frac{\alpha}{2 - \alpha}. \tag{3.15}$$

The recurrence equation in the frequency of heterozygotes may then be transformed into the homogeneous equation:

$$z'_{\mathsf{Aa}} - \hat{z}_{\mathsf{Aa}} = \tfrac{1}{2}\alpha(z_{\mathsf{Aa}} - \hat{z}_{\mathsf{Aa}}).$$

The genotypic frequencies therefore converge very rapidly to the equilibrium proportions, corresponding to the gene frequency p and the population inbreeding coefficient $\hat{F}$.

Random genetic drift in such a population with random mating and partial selfing may be analyzed in the same way as random mating with no selfing:

$$F_t = \frac{1}{N}\left(\tfrac{1}{2} + \tfrac{1}{2}G_{t-1}\right) + \left(1 - \frac{1}{N}\right)F_{t-1}, \tag{3.16}$$

$$G_t = \alpha\left(\tfrac{1}{2} + \tfrac{1}{2}G_{t-1}\right) + (1 - \alpha)F_{t-1}. \tag{3.17}$$

The equation for the consanguinity coefficient is exactly the same equation (3.3) with no selfing, whereas the equation for the inbreeding coefficient has an extra term compared to equation (3.4), namely the term with the factor α. For $\alpha = 0$ we thus recover equation (3.4), as we should. The convergence to genetic homogeneity is for $1 > \alpha \gg \frac{1}{2N}$ approximately described by the effective population size

$$N_e = N\,\frac{2 - \alpha}{2(1 - \alpha)}$$

(Box 10), and the rate of convergence to homogeneity is diminished by a factor of $1 - \hat{F}$.

Complete selfing ($\alpha = 1$) is special, because the plants exchange no gametes, and fairly quickly every plant in the population becomes a homozygote, but many different homozygotes may nevertheless be present. The effective population size therefore approaches infinity, corresponding to the state of a population with permanent genetic variation. The development of the consanguinity coefficient, on the other hand, is poorly modeled because of the lack of coherence of the population. In real life, however, the population is expected to converge to homogeneity due to so-called demographic stochasticity caused entirely by the variation in number of offspring produced by each individual.

3.2   Effective Population Size

When considering the effective size of a population, we always have to remember that effective size is *not* a property of the population. It is a number that helps us describe some of the effects of random genetic drift in the population, in particular the convergence to genetic homogeneity, in that the factor λ describing this convergence is specified as

$$\lambda = 1 - \frac{1}{2N_e}.$$

The effective population size is a reference to the gamete pool model that will give us an intuitive feel for various properties of the population.

The rules of mating have an effect on the buildup of consanguinity in the population (Section 3.1), and we measured this effect by using the effective population size. The concept of effective population size may be used to describe the pace of random genetic drift in other situations where the life of the population deviates from the simplistic gamete pool model. We will consider the effective population size in two examples, varying population size and varying reproductive success among individuals.

If the population size varies such that the population size in generation t is N_t, then the effective population size over a period of T generations is approximately

$$N_e \approx \left(\frac{1}{T} \sum_{t=1}^{T} \frac{1}{N_t} \right)^{-1} \tag{3.18}$$

(Box 11), when reproduction is described by the gamete pool model and population sizes are not too small (Wright 1938). The expression for N_e is a type of average known as the *harmonic average* (the familiar average is called the *arithmetic average*).

Variation in the number of offspring produced per parent gives an expression for $\mathrm{E}\big(\mathrm{Var}(p_t|p_{t-1})\big)$, which is different from that obtained assuming the gamete pool model. This effect can also be expressed in terms of an effective population size. With constant population size in a diploid population the mean number of offspring per parent is 1, and the mean number of genes transmitted per parent

---

Box 11: Variation in population size

The change per generation in the population inbreeding coefficient in a Wright–Fisher population with varying size is again given by equation (2.8), and after T generations we therefore have

$$1 - F_{T+1} = (1 - F_1) \prod_{t=1}^{T} \left(1 - \frac{1}{2N_t}\right), \quad \text{where}$$

$$\prod_{t=1}^{T} \left(1 - \frac{1}{2N_t}\right) = \left(1 - \frac{1}{2N_1}\right)\left(1 - \frac{1}{2N_2}\right) \cdots \left(1 - \frac{1}{2N_{T-1}}\right).$$

The effective population size is then given by

$$1 - \frac{1}{2N_e} = \left(\prod_{t=1}^{T} \left(1 - \frac{1}{2N_t}\right)\right)^{\frac{1}{T}},$$

and the quantity on the right side is known as the *geometric average* or the *logarithmic average*:

$$\log\left(1 - \frac{1}{2N_e}\right) = \frac{1}{T} \sum_{t=1}^{T} \log\left(1 - \frac{1}{2N_t}\right).$$

For small x we have the approximation $\log(1 - x) \approx -x$, and using this produces equation (3.18).

---

is therefore 2. A useful approximation of the effective population size is in this case

$$N_e \approx \frac{2N - 1}{1 + \frac{1}{2}\sigma^2}, \tag{3.19}$$

where σ^2 is the variance in number of genes transmitted by a parental individual to the offspring population (Wright 1938).

In the gamete pool model the probability that a chosen gamete is produced by a given individual is $\frac{1}{N}$. The number of gametes contributed by an individual is therefore binomially distributed $b(2N, \frac{1}{N})$. The variance in number of genes transmitted by an individual is then $2(1 - \frac{1}{N})$, and formula (3.19) produces

$$N_e \approx \frac{2N - 1}{1 + (1 - \frac{1}{N})} = N.$$

Thus, the approximation is accurate for the Wright–Fisher model.

An extreme assumption is that all parents produce exactly one offspring each, corresponding to a contribution of two gametes to the offspring population. This gives a minimal variance, $\sigma^2 = 0$, and therefore $N_e \approx 2N - 1$. Thus, if the variation in number of offspring is very small, then the effective population

size may be up to twice the actual number of individuals in the population.
In nature such a low level of variation is unexpected. Variation in litter size
is usually large. Even in shorebirds, where a clutch of four eggs is almost
universal, variation emerges during the rearing and maturation of the offspring.
In a discrete generation model *the number of offspring produced by a parent is
the number of its offspring that breed in the following generation.*

The opposite extreme is that only the offspring of one individual become
parents in the next generation. In this case $\sigma^2 = \frac{1}{N}(2N)^2 - 2^2 = 4N - 4$ (formula A.3 on page 347), and therefore $N_e \approx 1$. In real panmictic populations
the rate of convergence of a population to monomorphism will fall somewhere
between these two extremes. Although the variation in effective population size
among species and populations is expected to be as large as the variation in any
other variable related to demography, a rule of thumb is that the effective population size is usually lower than the actual population size, often considerably
lower.

3.2.1   Lose variation or gain identity

In the description of random genetic drift, the correspondence between the development of variance in gene frequencies and consanguinity relationships holds for
a wide spectrum of population models, although not for all reasonable models.
For systematic changes in the population size over a period the two descriptions
differ, however. The population size N_t in the *offspring* generation appears in
the recurrence equation (2.4) for the variance,

$$\mathrm{Var}(p_t) = \frac{p(1-p)}{2N_t} + \left(1 - \frac{1}{2N_t}\right)\mathrm{Var}(p_{t-1}), \qquad (3.20)$$

whereas the recurrence equation (2.8) depends on the population size in the
parental population N_{t-1},

$$F_t = \frac{1}{2N_{t-1}} + \left(1 - \frac{1}{2N_{t-1}}\right)F_{t-1}. \qquad (3.21)$$

This has prompted the definition of two effective population sizes: the *variance
effective population size* $N_{e(V)}$ and the *inbreeding effective population size* $N_{e(F)}$
(Crow 1954). Each of these is defined by a comparison of the recurrence equation
with the corresponding equation in the gamete pool model. Whether one or
the other effective population size is relevant depends on the application for
which it is intended. In the genealogical description in Section 2.4, for instance,
the inbreeding effective population size is a good indicator of the coalescence
probability.

3.2.2   An exponentially growing population

The most straightforward example of a difference between the variance effective
population size $N_{e(V)}$ and the inbreeding effective population size $N_{e(F)}$ is a

population undergoing exponential growth. Thus, we assume

$$N_O = \Psi N_P,$$

where N_P and N_O are the sizes of the parent and offspring populations, respectively, and where Ψ is the factor by which the population is multiplied each generation. In such a population the population size changes as

$$N_t = \Psi^t N,$$

where $N = N_0$, the initial size of the population. Then the recurrence equations (3.20) and (3.21), with a minor rewrite, become

$$\frac{\text{Var}(p_t)}{p(1-p)} = \frac{1}{2N\Psi^t} + \left(1 - \frac{1}{2N\Psi^t}\right)\frac{\text{Var}(p_{t-1})}{p(1-p)}, \quad \text{and}$$

$$F_t = \frac{1}{2N\Psi^{t-1}} + \left(1 - \frac{1}{2N\Psi^{t-1}}\right)F_{t-1}.$$

We get the Wright–Fisher model when $\Psi = 1$, $N_t \to \infty$ when $\Psi > 1$, and N_t decreases when $\Psi < 1$. Of course, the case of $\Psi > 1$ is only of biological relevance for a limited number of generations, so let us consider the recurrence equations for $t < T$.

The consanguinity equation provides from equation (3.18) on page (62) the inbreeding effective population size

$$N_{e(F)} \approx N \left(\frac{1}{T} \sum_{t=1}^{T} \frac{1}{\Psi^{t-1}}\right)^{-1} \tag{3.22}$$

for sufficiently large population sizes, and $N_{e(V)}$ is obtained by replacing Ψ^{t-1} with Ψ^t. Thus, we get

$$N_{e(V)} = \Psi N_{e(F)}. \tag{3.23}$$

The variance effective population size is therefore a factor Ψ larger than the inbreeding effective population size, quite in accordance with our immediate intuition that $N_{e(F)}$ is related to the population size among parents and $N_{e(V)}$ to that among offspring.

In an exponentially growing population the effective population size from equation (3.22) is, however, suitable for approximation of neither the buildup of consanguinity, nor the loss of variation in the population. The value obtained will depend strongly on the time interval in which it is calculated. Equation (3.18) is useful only to evaluate the influence of stationary year to year variation in the size of the breeding population. The gene genealogy in an exponentially growing population thus cannot be approximated by the coalescent using an effective population size. The main problem is systematic changes in the population size, and hence, the rate of coalescence is different early and late in the genealogy. The effect of minor oscillations about a characteristic value may often be captured by the effective population size, but very different population sizes at different time periods may produce aberrant genealogies. The genealogy of a sample from a population undergoing exponential growth may, however, be described by a modified coalescent process.

In a population of constant size ($\Psi = 1$) the process is described by equation (2.13):

$$p_k = \text{Prob(coalescent event in sample of } k \text{ genes)} = \binom{k}{2}\frac{1}{2N} \cdot$$

With exponential growth this probability changes with time, and if the sampled population is of size N, then the size at generation τ back in time is $\Psi^{-\tau}N$. The probability that a pair of genes in a sample of k genes in generation τ find a common ancestor among their parents is thus given by

$$p_{k\tau} = \binom{k}{2}\frac{1}{2N\Psi^{-\tau-1}},$$

because the population size among the parents is $2N\Psi^{-\tau-1}$. The population size of the parents of the sampled population is $N_P = N\Psi^{-1}$. The denominator is therefore equal to $2N_P\Psi^{-\tau}$, which is the size of the population of parents of the population τ generations back in time.

Without passing to characteristic time, the coalescent equation (2.24) in a constant population is

$$\text{Prob(follower of } \xi \text{ at } \tau + \Delta\tau \,|\, \xi \text{ at } \tau) \approx \binom{|\xi|}{2}\frac{1}{2N}\Delta\tau.$$

This defines a process in continuous time, and so exponential growth should be expressed in continuous time:

$$N_t = e^{\psi t}N \quad \text{and} \quad N_\tau = e^{-\psi\tau}N,$$

where $\psi = \log\Psi$ is the growth rate of the population. The equation describing coalescence in a population undergoing exponential growth then becomes

$$\text{Prob(follower of } \xi \text{ at } \tau + \Delta\tau \,|\, \xi \text{ at } \tau) \approx \binom{|\xi|}{2}\frac{1}{2N_P e^{-\psi\tau}}\Delta\tau.$$

Passing to characteristic time in a constant population corresponds to using the time τ' where $\Delta\tau' = \frac{1}{2N}\Delta\tau$. A similar time change in the exponentially growing population uses the time $\tilde{\tau}$, where

$$\Delta\tilde{\tau} = \frac{1}{2N_P e^{-\psi\tau}}\Delta\tau$$

(Slatkin and Hudson 1991). This corresponds to the transformation of time given by

$$\tilde{\tau}(\tau) = \frac{1}{2N_P\psi}\left(e^{\psi\tau} - 1\right), \tag{3.24}$$

and in this time frame we get a coalescent process, in that

$$\text{Prob(follower of } \xi \text{ at } \tilde{\tau} + \Delta\tilde{\tau} \,|\, \xi \text{ at } \tilde{\tau}) \approx \binom{|\xi|}{2}\Delta\tilde{\tau}$$

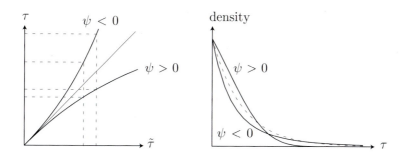

Figure 3.2: Waiting times in an exponentially growing population. Left: The transformation (3.25). The straight solid line is the identity line, corresponding to Kingman's coalescent. Right: A sketch of the waiting time distributions. The dashed curve is the exponential distribution ($\psi = 0$).

is indistinguishable from equation (2.24), the coalescent equation in a constant population. The waiting time to the next coalescent event is therefore exponentially distributed in $\tilde{\tau}$ time.

A given waiting time can be transformed into real time by the reverse of the transformation (3.24):

$$\tau(\tilde{\tau}) = \frac{1}{\psi} \log(2N_P \psi \tilde{\tau} + 1) \quad \left(\text{for } \tilde{\tau} < \frac{-1}{2N_P \psi} \text{ when } \psi < 0\right). \qquad (3.25)$$

In a growing population ($\psi > 0$) the waiting times are always shortened compared to those in the coalescent in a constant population with the current population size, because historic populations were smaller than the present ones. In addition, the waiting time to the first coalescent event is shortened much less than later waiting times (see Figure 3.2), which are expected to be longer. In a declining population ($\psi < 0$) the waiting times are always longer than those in a constant population with the current population size, because historic populations were larger. The waiting time to the first coalescent event is lengthened much less than later (and expected longer) waiting times (see Figure 3.2). Examples of genealogies produced by the coalescence in an exponentially growing population are obtained from the Hudson animator (Christensen et al. 2006).

Exponential growth in a real population cannot be sustained for an extended period of time, and exponential decline cannot extend beyond the extinction of the population. The coalescent analysis of a sample cannot assume arbitrary growth rates in the past for the sampled population. A positive growth rate can only be assumed for a time much less than $\psi^{-1} \log N$ (the number of generations ago that the population consisted of just one individual).

3.2.3   Overlapping generations

For organisms whose life histories differ qualitatively from that of the Wright–Fisher model, the concept of effective population size should be used with caution. Random genetic drift in organisms with overlapping generations is basically the same process as in the Wright–Fisher population, but a comparison based on the population size alone is not productive (see Section 2.7 on the Moran model). The reason is that the time unit used in the comparison with the Wright–Fisher model is questionable.

Exponential growth in a population with discrete nonoverlapping generations is very similar to that in a population with overlapping generations and continuous breeding. Demographic descriptions of human populations routinely use models assuming discrete overlapping generations even though continuous breeding is the practice. Accordingly, exponential growth in the Wright–Fisher model is described in the diffusion approximation of that model. The Wright–Fisher model, however, is a discrete nonoverlapping generations model, and it is not obvious that real populations with overlapping generations and possibly continuous breeding are well described by this model. The Moran model presented in Section 2.7 is equally simple and has overlapping generations, thus illustrating the complications of such models.

The evolution of the variance of the gene frequency in Moran's model may be expressed in terms of an effective population size $N_e = N^2$ in the Wright–Fisher model, but this disregards the important differences in the time units applied in the two models. In Section 2.7 we succeeded in defining a generation time that rendered the two models equivalent. We chose N generations as the generation time and arrived at complete correspondence between the two models for the same number of genes, that is, $N_e = N$.

The demographic generation time is not a well-defined quantity in a population with overlapping generations, but the average age of parents when offspring are produced is often used. The age distribution in a Moran population is geometric, the probability of breeding is the same for all age classes, and the average age of a breeding parent is therefore $2N$. The result is $2N$ breeding events and $2N$ offspring per generation, and $N_e = \frac{1}{2}N$. If this works, then we should get agreement with formula (3.19) with σ^2 as the variance in number of genes transmitted by an individual to the population during $2N$ breeding events. This variance is

$$\sigma^2 = 2\left(1 - \frac{1}{2N}\right),$$

and we therefore get

$$N_e = \frac{2N(2N-1)}{4N-1} \approx N$$

when N is not too small.

The correspondence between the models is therefore satisfactory also when a more biological comparison is made. Such a comparison always has to be made in the two tempi applied here, namely an adjustment of timescales and an evaluation of the effective population size corresponding to the population with overlapping generations in the new timescale. In an age-structured population of fixed size Sagitov and Jagers (2005) found the effective population size in terms of the age-specific mortality and fecundity in the population.

3.3 Genotypic Frequencies

The genotypic frequencies in a two-allele polymorphism in a large randomly mating population are in Hardy–Weinberg proportions. The genotype frequencies among individuals formed by the union of pairs of randomly chosen gametes from the gamete pool in the Wright–Fisher model at generation t is therefore expected to be in the Hardy–Weinberg proportions:

$$\text{AA}: \ p_t^2, \quad \text{Aa}: \ 2p_t q_t, \quad \text{aa}: \ q_t^2. \tag{3.26}$$

The actual numbers of the three genotypes follow a multinomial distribution of a sample of size N drawn from a large population with the genotypes in these frequencies (see equation (3.13) on page 59 or Appendix A.1.3 on page 351).

In a large randomly mating population with inbreeding, described by the population inbreeding coefficient F, the two chosen gametes carry identical genes with probability F, and among these the genotypic frequencies are

$$\text{AA}: \ p, \quad \text{Aa}: \ 0, \quad \text{aa}: \ q.$$

Among the gamete pairs that do not carry identical genes the genotypic frequencies are in Hardy–Weinberg proportions. Thus, the genotypic frequencies in a two-allele polymorphism are

$$\text{AA}: \ p^2(1-F) + pF, \quad \text{Aa}: \ 2pq(1-F), \quad \text{aa}: \ q^2(1-F) + qF.$$

Simple rearrangements change these to

$$\text{AA}: \ p^2 + Fpq, \quad \text{Aa}: \ 2pq - 2Fpq, \quad \text{aa}: \ q^2 + Fpq.$$

Thus, in generation t in the gamete pool model we expect to find the genotype frequencies

$$\text{AA}: \ p^2 + F_t pq, \quad \text{Aa}: \ 2pq - 2F_t pq, \quad \text{aa}: \ q^2 + F_t pq. \tag{3.27}$$

The individuals in the population were formed by random mating, however, and we therefore expect Hardy–Weinberg proportions (3.26). An apparent contradiction! Both expectations are nevertheless correct in that equation (3.27) gives the overall expectation and equation (3.26) provides the expected genotype frequencies given the gene frequency in generation t. The average of the expectation should reconcile the two, and because

$$\mathrm{E}\, p_t^2 = (\mathrm{E}\, p_t)^2 + \mathrm{Var}(p_t) = p^2 + \mathrm{Var}(p_t)$$

we expect the genotype frequencies

$$\text{AA}: \ p^2 + \mathrm{Var}(p_t), \quad \text{Aa}: \ 2pq - 2\mathrm{Var}(p_t), \quad \text{aa}: \ q^2 + \mathrm{Var}(p_t) \tag{3.28}$$

in generation t. This expectation deviates from Hardy–Weinberg proportions just as it did in the frequencies (3.27) argued from a description of consanguinity, and we therefore achieve correspondence between our two approaches to describing genotype frequencies with random genetic drift, in that

$$\text{Var}(p_t) = F_t pq, \quad t = 0, 1, 2, \ldots, \tag{3.29}$$

where $\text{Var}(p_t)$ is given by equation (2.5) and F_t is given by equation (2.9) on pages 28 and 31, respectively.

A simple result like this must have a simple biological origin. This will indeed become clear if we take a closer look at genotype frequencies in Wright's island

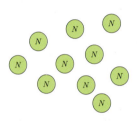

model. Mixing this population in generation t produces a so-called Wahlund effect (to which we will return in Section 5.3). This effect causes an excess of homozygotes compared to the Hardy–Weinberg proportions, and it is due to the variation in gene frequencies among the islands. Its magnitude is $\text{Var}(p_t)$ because the genotypic frequencies in this mixed population are given by equation (3.28). In the mixed population F_t reflects a genuine consanguineous relationship among individuals in the population at time t, and the genotypic frequencies are therefore given by equation (3.27).

Exercises

Exercise 3.0.1 Show that the frequency of the homozygote AA in the random-union-of-gametes model with distinct sexes is lower than or equal to the corresponding Hardy–Weinberg proportion with gene frequency $p = \frac{1}{2}(p_{\female} + p_{\male})$.

Exercise 3.1.1 Let $N_{\female} = 1000$ and draw a figure that shows the variation in N_e when $N_{\male}$ varies between 10 and 1000.

Exercise 3.1.2 Show that the genotypic frequencies z_{AA}, z_{Aa}, and z_{aa} may be parametrized as given in equation (3.14) in terms of the gene frequencies p and q, and ϕ as the deviation from the corresponding Hardy–Weinberg proportions.

Exercise 3.2.1 Consider a species of butterfly with a spring and a summer generation. Assume that the gamete pool model provides a good description of the breeding in the population, and assume that the two generations each year have the population sizes N_1 and N_2, respectively. (1) Describe the development of consanguinity in the population, and (2) find N_e and an approximation for N_1 and N_2 large.

Exercise 3.2.2 In Exercise 3.2.1, let $N_1 = 1000$ and draw a figure that shows the variation in N_e when N_2 varies between 10 and 1000.

Chapter 4

Mutation and Variation

The replication mechanism of the DNA molecule ensures conservative transmission of genetic information from the mother to the two daughter cells resulting after cell division. Errors do occur, however. An error in the transmission of genetic information is called a *mutation*. Mutation is random. In a functional genetic element mutation occurs independently of its function and efficacy—an implementation of the central dogma of molecular biology.

One of the assumptions in the law of gene frequency conservation is that the genes do not change, that is, mutation is absent. Mutation is in fact never absent, but the assumption is that effects of mutation may be neglected. For deterministic arguments assuming an infinitely large population this assumption is quite all right, but our conclusion of ultimate monomorphism for the evolution of a real and finite population cannot be correct with mutation. A population cannot remain monomorphic when mutation occurs, and in this sense mutation generates variation in the population.

Mutations are in general rare. They are seen as sporadic deviations from the usual results in crosses, but only special crosses allow for observations of mutations. Offspring of a Mendelian cross, for instance between *Pisum sativum* plants from a true-breeding line with yellow peas and a line of green peas, should always be yellow peas (genotype Aa), but if a mutation occurred that changed the gene of allele type A to one of allele type a, then a green pea would appear in the pod. If this happened on a plant from the green-pea line pollinated by a yellow-pea plant, the odd pea might be due to accidental contamination with pollen from the green-pea parent. The observation of the mutation would therefore be much more convincing if the deviant pea was observed in a pod on a yellow-pea parent. Reliable observations of rare events are difficult even for highly skilled observers. Even in bacteria, where a large number of progeny may be screened, care is essential in the design of experiment and interpretation of results (Luria and Delbrück 1943).

Mutation gives rise to variation. The variations on the characters that Mendel studied are mutations that humans have bred into varieties of the domesticated *Pisum sativum*. Domesticated animals also show variation in characters that reveal Mendelian inheritance. Such variation is the raw material of evolutionary change, and Darwin used variation in domesticated organisms to illustrate the breadth of potential phenotypic variation that can be maintained by differential breeding.

4.1   Mutation

The simplest kinds of mutation are errors in DNA replication (see Figure 1.1 on page 13). A wrong base may be inserted, resulting in a bad but acceptable base pairing of the DNA, and after repair of the mismatch, or later by the next replication, pairing is corrected. This might result in the correct pairing of the wrong base, producing a stabilized change of the DNA sequence. The base pairing during replication is actually quite accurate because it is proofread immediately, but the four bases can change into rare forms, where a pyrimidine mimics the alternative pyrimidine or a purine mimics the alternative purine, and this may cause errors where T and C or A and G are interchanged. Such a mutation is called a *transition*. A mutation where a pyrimidine is substituted for a purine, or vice versa, is called a *transversion*. Such errors do occur in DNA replication, but they are considerably rarer than transitions. The per base mutation rate in higher organism is typically on the order of 10^{-9} or lower per generation.

Another class of replication errors occurs when the correct sequence of base pairings fails. This error is particularly common in stretches of DNA with repetitive structures. If the template strand has the sequence

$$T-G-A-G-T-T-T-T-T-T-T-G-C-T$$

and the DNA synthesis runs from left to right, then after a while the copy reads

$$A-C-T-C-A-A-A-A-A$$
$$T-G-A-G-T-T-T-T-T-T-T-G-C-T .$$

Now suppose that the end of the copy briefly loosens its attachment to the template and then reattaches. The As and Ts may pair erroneously in this situation, leaving an orphan A or T within the paired stretch, for instance:

$$A-C-T-C-A-A-A-A-A-A$$
$$T-G-A-G-T-T-T-T-T-T-T-G-C-T ,$$

and continuing the synthesis yields

$$A-C-T-C-A-A-A-A-A-A-A-C-G-A$$
$$T-G-A-G-T-T-T-T-T-T-T-G-C-T .$$

The poly-A stretch has thus been elongated by one base—the sequence has been changed by an *insertion*. The slip of the pairing may have worked the other way round leaving an orphan T, and hence the copy sequence changes by a *deletion*. A deletion or insertion of one base in a coding region will disturb the *reading frame*, that is, the definition of codons in the sequence. In the example given above, the DNA-codon sequence (... T, GAG, TTT, TTT, TGC, T...), say, will change to (... T, GAG, TTT, TTT, TTG, CT...). The codons after the insertion will thus have shifted one base, a *frameshift* mutation, and this will often have profound impact on the function of the resulting protein.

This mechanism for producing insertions and deletions also works for dinucleotide repeats, for instance the sequence CAGTATATATATATTC. These are called *microsatellites*. A microsatellite may thus show variation in the number of repeats, which is counted on a DNA-sequence gel (see page 18). They are known to show high frequencies of insertions and deletions of a single motif, leading to mutation rates in the number of repeats of up to 10^{-3}. The insertion and deletion rates naturally increase with the length of the repeat, and some inherited diseases are caused by long repeats of a motif consisting of a few nucleotides. Such repeats may elongate during the development of the individual and cause failure of a gene or its regulation. One example of this is the fragile-X syndrome (Oostra and Chiurazzi 2001), which is the most common form of inherited mental retardation in humans (the frequency is about 10^{-4} in males and 10^{-7} in females—males have only one X chromosome, so the condition is more exposed in males). The repeat occurs within a gene on the X chromosome. The unit of the repeat is the trinucleotide CGG, and individual genes have between 6 and 54 repeats with a mode at 29. Female carriers who transmit the disease to their male offspring typically have a chromosome with several hundred repeats, and as the rate of repeat expansion increases with length, male offspring soon suffer a severely reduced expression of the gene. Another example of this kind of disease is Huntington chorea (Rosenblatt et al. 2001).

A third type of mutation is caused by spontaneous changes of the bases caused by the environment in which the cell lives. Some of these parallel the induced mutations discussed below, and an ambient cause of damage is oxidation, in particular by superoxide radicals, hydroxyl radicals, and hydrogen peroxide—all produced in aerobic cells.

A related source of mutation is a consequence of methylation of DNA, a mechanism for controlling expression of genes. In mammalian genomes the cytosine is methylated[1] in about three fourths of 5′–CG–3′ motifs, referred to as CpG motifs. The methylated cytosine mutates to thymine (CpG → TpG) comparatively often, causing a depauperate frequency of the dinucleotide in mammalian genomes.

Table 4.1 shows 15 sequences about 500b downstream[2] of the human β-hemoglobin gene on chromosome 11. The variation among the sequences in the first

---

[1] A hydrogen (–H) on one of the carbon atoms in cytosine is replaced by a methyl group (–CH$_3$) to produce methylcytosine.

[2] Distance on a DNA sequence is expressed in number of DNA base pairs (unit b), or the derived units, kilobases (1kb = 1000b) and megabases (1Mb = 1000kb).

Table 4.1: Fifteen noncoding sequences on human chromosome 11

```
T A T G T G T G T A C A T A T A C A C A T A T A T A T A T A T A T T T T T - - - - T T T C T T
T A T G T G T G T A T A T A T A C A C A T A T A T A T A T A T A T A T T T - - - - T T T T T
T A T G T G T G T A C A T A T A T A C A T A T A T A T A T A T A T T T - - T T T C T T
T A T G T G T G T A T A T A T A C A C A T A T A T A T A T A T A T T T T T - - T T T C T T
T A T G T G T G T A C A T A T A C A C A T A T A T A T A T A T A T A T T T - - T T T C T T
T A T G T G T G T A T A T A T A C A T A T A T A T A T A T A T A T T T - - - - T T T C T T
T G T A T G T G T G T A T A T A T A C A T A T A T A T A T A T A T T T - - - - T T T C T T
T A T G T G T G T A T A T A T A C A T A T A T A T A T A T A T T T T T - - - - T T T C T T
T A T G T G T G T A T A T A T A C A T A T A T A T A T A T A T A T T T - - - - T T T C T T
T A T G T G T G T A T A T A T A C A C A T A T A T A T A T A T A T T T - - - - T T T C T T
T A T G T G T G T A T A T A T A C A T A T A T A T A T A T A T A T A T A T T T T C T T
T A T G T G T G T A T A T A T A C A C A T A T A T A T A T A T A T T T T T - - T T T C T T
T A T G T G T G T A T A T A T A C A C A T A T A T A T A T A T A T T T T T - - - - T T T C T T
T A T G T G T G T A C A T A T A C A C A T A T A T A T A T A T T T T T T T T - - - - T T T C T T
T A T G T G T G T A C A T A T A C A C A T A T A T A T A T A T A T T T - - T T T C T T
T · T · T G T G T · · A T A T A · A · A T A T A T A T A T · T · T · T · · · · T T · T T
```

From GenBank (2006). Accession numbers AF186606–AF186620 made available by
Fullerton et al. (2000). Alignment with default settings in MultAlin (Corpet 1988).

25 of the sites is in terms of so-called single nucleotide polymorphisms (SNPs,
pronounced "snips"), where the variable site only segregates two bases (marked
by a · in the consensus sequence at the bottom of the table). The alleles in these
SNPs all differ by a transition, maintaining the pattern of alternation between
purines and pyrimidines. Close to the end of the displayed sequences they
differ by insertions or deletions—called *indels* in sequence comparisons because
of inability to decide whether an observed difference is due to an insertion or
a deletion. Indels are marked relative to the longest sequence (the eleventh
sequence in the Table) by dashes at positions in the shorter sequences. The
alignment of the sequences, however, becomes uncertain in regions with indels.
For instance, the three transversions before the marked indels disappear if they
are viewed as indels on the border between the dinucleotide repeat of TA and a
repeat of the nucleotide T (Table 4.2).

4.1.1   Induced mutation

In 1930 Herman Muller described X rays as mutagenic. This was the first
discovery of an exogenic mutation agent. Later on, it was realized that any kind
of ionizing radiation[3] produces genetic changes. Muller also devised a way to
quantify the effect of a mutagen. He showed that the effect of ionizing radiation
is fairly simple in that, for a given kind of applied radiation, doubling the dose
produced twice as many mutations. The mutagenic response is therefore in
general a linear function of the dose applied. Deviations from linearity occur
for high doses given at high rates (i.e., over a short time interval), because
of the saturation of the DNA repair system in the cells, and because of the
lethal effect of extensive radiation damage. For small to moderate doses given
at low rates the linearity of the dose-response curve is firmly established. For

---

[3]Any type of energetic electromagnetic radiation, namely ultraviolet rays, X rays, and
gamma rays. In addition, energetic particle rays, for instance, β, α, and neutron radiation
from radioactive decay.

Table 4.2: Fifteen noncoding sequences on human chromosome 11

```
TATGTGTGTACATATACACATATATATATATAT----TTTTTTTCTT
TATGTGTGTATATATACACATATATATATATATAT----TTTTTTTT
TATGTGTGTACATATACATATATATATATATATAT--TTTTTCTT
TATGTGTGTATATATACACATATATATATATATAT--TTTTTTTCTT
TATGTGTGTACATATACACATATATATATATATAT--TTTTTCTT
TATGTGTGTATATATACATATATATATATATATAT-----TTTTCTT
TGTATGTGTGTATATATACATATATATATATATAT---TTTTTTCTT
TATGTGTGTATATATACATATATATATATATATAT-----TTTTTTCTT
TATGTGTGTATATATACATATATATATATATATAT----TTTTTCTT
TATGTGTGTATATATACACATATATATATATATAT----TTTTTCTT
TATGTGTGTATATATACACATATATATATATATATATATATTTTTCTT
TATGTGTGTATATATACACATATATATATATATAT--TTTTTTTCTT
TATGTGTGTATATATACACATATATATATATAT----TTTTTTTCTT
TATGTGTGTACATATACACATATATATATATAT----TTTTTTTTTCTT
TATGTGTGTACATATACACATATATATATATATAT--TTTTTTCTT
T·T·TGTGT··ATATA·A·ATATATATATAT········TTTT·TT
```

From GenBank (2006). Accession numbers AF186606–AF186620 made available by
Fullerton et al. (2000).

very small doses, inference on the relation between dose and response becomes difficult due to the noisy background of spontaneous mutation and the natural background radiation.

The simple dose-response relation allows straightforward comparisons of the effects of different types of ionizing radiation. The biological effect is quantified by the unit Sievert (Sv, J/kg), which is a measure of the amount of energy deposited in tissue, organ, or person. This is a rough measure of the amount of damage to living tissue inflicted by the radiation (in acute irradiation of humans about 50 percent die if untreated when they receive a dose on the order of magnitude of 1Sv).

Charlotte Auerbach and James Robson discovered in 1947 that chemicals could be highly mutagenic. They studied the effect of mustard gas by spraying it at 10-second intervals into a stream of air passing male *Drosophila melanogaster*. This exposure produced a number of mutations equivalent to an X–ray dose of about 30Sv (adult flies can survive quite high radiation doses, as long as the dose rate is low enough to avoid cooking the flies).

Many chemicals are even more potent as mutagens than mustard gas. Chemical analogs of the nucleotide bases that are less stable can induce changes similar to those caused by the instabilities described for the four nucleotide bases. Some of these mutagens are highly specific in the kind of damage they cause.

4.1.2 Somatic mutation

The only heritable mutations are those that occur in germline cells, that is, cells that may give rise to gametes. In humans the germline commences with the fertilized egg (the zygote) and continues in the cells that eventually form the gamete-producing tissue of the gonads—the ovaries and the testes. This is the germline in all higher animals, whereas most tissue in many lower animals, multicellular plants, and fungi are reproductively competent. Trees, for instance, can commonly produce reproductive organs on all their branches, and at least

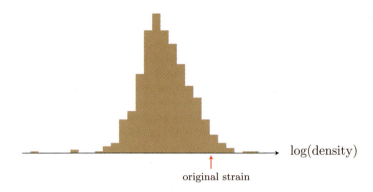

log(density)

Figure 4.1: Variation in growth rate among 567 new mutants in *Pseudomonas fluorescens*. The growth rates are inferred by measuring optical density in cultures of the mutant strains. For explanation see Box 12. From data made available by Rees Kassen and Thomas Bataillon.

some of the cells in vegetative buds therefore belong to the germline.

Mutations may of course occur in cells outside the germline in what is called the soma. Such mutations cannot be transmitted to offspring and are therefore not heritable. Their effect is limited to the individual in which they occur, but these somatic mutations may influence the function of the organism. They give rise to sectors in the body with a genotype that differs from that specified by the genotype of the zygote. Some such heterogeneities have noticeable phenotypic effects—some are benign, and some may produce detrimental effects, for instance cancer. The development of a cancer in the body involves genetic changes that liberate a cell and its descendants from some of the usual constraints of the tissue they reside in, leading to uncontrolled proliferation of specific, more or less differentiated cell types. Malignant tumor cells have a disease in their genes, often several genes—so-called oncogenes (for further details see Cooper 1995). Heritable cancers often look like ordinary somatic cancers, but an individual predisposed for a cancer will either have certain unstable genes or will carry a mutation in one of the oncogenes.

4.1.3   Effects of mutations

Random mutational changes in the sequence of a gene may cause a change in gene function that alters the phenotype, and ultimately, the survival and reproduction of the individual carrying the mutated gene (Figure 4.1). These effects can be drastic, causing disease and death, or they can be minor, for instance by influencing the ability to meet environmental challenges (Kassen and Bataillon 2006). The mutants with the growth rates shown in Figure 4.1 are discovered as resistant to an antibiotic, that is, they have an *a priori* phenotypic effect. We will return (in Section 8.4) to the discussion of these kinds of genetic

---

Box 12: Growth rate of mutations

Kassen and Bataillon (2006) investigated the distribution of the growth rate of mutants in *Pseudomonas fluorescens*. They isolated mutants resistant to an antibiotic from a culture of a susceptible strain dubbed wild type. Samples from cultures of the wild type were spread on agar plates containing the antibiotic in a density that allowed bacteria carrying resistance mutations to found discrete colonies. 567 strains were founded from these colonies, and special care was taken to guarantee that the corresponding mutations emerged independently. The growth rate of a strain was estimated from the change in optical density of a culture grown from a small inoculum for 24 hours. Figure 4.1 shows the distribution of the resulting growth rates in a medium without the antibiotic, and the position of the wild type strain is indicated. Kassen and Bataillon (2006) also measured the growth rates of the mutant strains in the same medium with the antibiotic and found a fairly similar distribution of growth rates.

---

variation and here restrict our attention to mutations that do not affect the fitness of the carrier. We will consider so-called *neutral mutations*. These form alleles that obey the law of gene frequency conservation, of course as long as they are undisturbed by further mutation. Thus, we will analyze how mutation affects the basic rules of population genetics developed in Chapter 2.

Two different kinds of mutations that change a DNA base exist in protein coding genes. *Synonymous* mutations change the codon without causing a change in the corresponding amino acid. The alternative *nonsynonymous* mutations change the amino acid. The sequences shown in Table 4.3 vary in the second codon after the translation initiation codon ATG, in that CAC and CAT are found. Both code for histidine, and we expect the original mutation that created the variation to be neutral. The sixth codon after ATG is present as GAG and GTG coding for different amino acids because a change in the second position always changes the amino acids. Nonsynonymous mutations may significantly alter gene function and cause disease, but on the other hand, variation in the amino acid sequence need not have discernable effects on the individual carrier of the variants (Kimura 1983*a*).

4.2 The Two-Allele Model

Gene frequency conservation and random genetic drift lead to loss of genetic variability, as most prominently expressed in the limiting distribution of the gene frequency in the two-allele Wright–Fisher model (Figure 2.5 on page 30). The assumption of conservative transmission of alleles neglecting rare mutations is all right in a polymorphic population, but the deduced result of eventual monomorphism is evidently not robust to that assumption. Mu-

Table 4.3: Fifteen sequences around the start codon in the β-hemoglobin gene

```
CAGACACCATGGTGCACCTGACTCCTGAGGAGAAGTCTGCCGTTACT
CAGACACCATGGTGCACCTGACTCCTGTGGAGAAGTCTGCCGTTACT
CAGACACCATGGTGCACCTGACTCCTGAGGAGAAGTCTGCCGTTACT
CAGACACCATGGTGCACCTGACTCCTGAGGAGAAGTCTGCCGTTACT
CAGACACCATGGTGCACCTGACTCCTGAGGAGAAGTCTGCCGTTACT
CAGACACCATGGTGCACCTGACTCCTGAGGAGAAGTCTGCCGTTACT
CAGACACCATGGTGCACCTGACTCCTGAGGAGAAGTCTGCCGTTACT
CAGACACCATGGTGCACCTGACTCCTGAGGAGAAGTCTGCCGTTACT
CAGACACCATGGTGCACCTGACTCCTGTGGAGAAGTCTGCCGTTACT
CAGACACCATGGTGCACCTGACTCCTGAGGAGAAGTCTGCCGTTACT
CAGACACCATGGTGCACCTGACTCCTGAGGAGAAGTCTGCCGTTACT
CAGACACCATGGTGCATCTGACTCCTGAGGAGAAGTCTGCCGTTACT
CAGACACCATGGTGCATCTGACTCCTGAGGAGAAGTCTGCCGTTACT
CAGACACCATGGTGCATCTGACTCCTGAGGAGAAGTCTGCCGTTACT
CAGACACCATGGTGCACCTGACTCCTGAGGAGAAGTCTGCCGTTACT
CAGACACCATGGTGCA·CTGACTCCTG·GGAGAAGTCTGCCGTTACT
                ATG
```

From GenBank (2006). Accession numbers AF186606–AF186620 made available by
Fullerton et al. (2000). Alignment by MultAlin (Corpet 1988).

tations generate variation, and neutral mutation will therefore counteract the loss of genetic variability due to random genetic drift.

The effect of mutation may be studied by amending the Wright–Fisher process described in Section 2.2, and although the classical two-allele model is biologically unrealistic because of the ubiquitous presence of multiallelic variation, it is a simple and transparent model for illustrating the effects of mutation. In addition, the results are robust in describing the frequency of neutral alleles subject to mutation. The two-allele Wright–Fisher model with mutation should therefore be viewed as a learning model rather than a descriptive one.

The simplified assumption is that mutation from A to a occurs at a frequency of μ_A per generation, and mutation from a to A with the frequency μ_a. No other mutations are allowed. The change in gene frequency between parents and offspring is in this model expected to be given by the recurrence equation

$$p' = (1 - \mu_A)p + \mu_a q. \tag{4.1}$$

Mutation will thus in general change the gene frequency except when $p = \hat{p}$, where

$$\hat{p} = \frac{\mu_a}{\mu_A + \mu_a}. \tag{4.2}$$

This is a gene frequency equilibrium of the recurrence equation (4.1) in a very large population, often referred to as a mutation–balance equilibrium, because the two directions of mutation are expected to balance out exactly when the gene frequency is $\hat{p}$. Equation (4.1) is a linear recurrence equation, which may be iterated in the same way as equation (2.8) on page 31 by considering the deviation $p_t - \hat{p}$ of the gene frequency in generation t from the equilibrium. The gene frequency after t generations is thus given by

$$E p_t = \hat{p} + (1 - \mu_A - \mu_a)^t (p - \hat{p}), \tag{4.3}$$

where p is the initial gene frequency of allele A. The factor $1 - \mu_A - \mu_a$ is a tiny bit smaller than one, and we therefore expect the gene frequency to slowly converge to the mutation–balance equilibrium $\hat{p}$ given by equation (4.2). Such slow convergence is conveniently approximated by a continuous time process, namely $\mathrm{E}\, p_t \approx \hat{p} + \exp(-\mu t)\,(p - \hat{p})$, where the rate of convergence to equilibrium is $\mu = \log(1 - \mu_A - \mu_a) \approx \mu_A + \mu_a$. In a very large population this provides the deterministic description that the gene frequency of allele A converges to $\hat{p}$ ($p_t \to \hat{p}$ for $t \to \infty$) at the rate μ.

The slow convergence to the mutation–balance equilibrium poses the question of whether it can prevail under the influence of random genetic drift, where the long-term push is towards extreme gene frequencies. Mutation is introduced into the Wright–Fisher model by assuming that the changes occur in the gamete pool. The size of the pool is assumed to be very large, effectively infinite, so the change in gene frequency due to mutation is given by the deterministic equation (4.1). The Wright–Fisher model with mutation is therefore the gamete pool model where the probability parameters in the binomial distribution are modified by mutation, that is,

$$P_{i \to j} = \mathrm{Prob}(X_O = j \mid X_P = 2Np) = \binom{2N}{j} (p')^j (q')^{2N-j}; \qquad (4.4)$$

here p' is given by equation (4.1) and $q' = 1 - p'$. Thus, the conservation of gene frequencies expressed in equation (2.3) on page 25 is replaced by equation (4.1), the usual deterministic model.

The only result that we will consider here is the size of the variance after sufficiently long time, where we may assume that the population approaches equilibrium between mutation and random genetic drift. The result is not a fixed value of the gene frequency as in a deterministic model, but rather a *steady state*, where the distribution of the gene frequency does not change through time.[4] The approximate result for the steady state variance is

$$\mathrm{Var}(p_t) \to \frac{\hat{p}\hat{q}}{1 + 4N\mu} \quad \text{for } t \to \infty, \qquad (4.5)$$

when the population size is large and mutation rates are small. The rate of convergence of the variance to its limiting value is the same as the rate of convergence $\mu = \mu_A + \mu_a$ of the gene frequency in equation (4.3).

When the expected number of mutations per generation is low, the variance approaches the variance obtained with random genetic drift in a population starting with gene frequency $\hat{p}$. In this case we therefore expect the population to be nearly fixed for one of the alleles: for A with probability $\hat{p} = \mu_A/\mu$ and for a with probability $\hat{q} = \mu_a/\mu$. Conversely, when $2N(\mu_A + \mu_a)$ is large the variance becomes small, and we expect to find the population close to its equilibrium gene frequency $\hat{p}$. Thus, the expected appearance of the population is highly dependent on the number of mutations per generation.

---

[4] $*$ The Wright–Fisher process with mutation is a recurrent Markov chain that possesses a unique stationary distribution (Appendix A.2 on page 357).

4.2.1   Distribution of the gene frequency

The distribution of the gene frequency at the steady state, the so-called *stationary distribution*, may be obtained by using the diffusion approximation introduced in Section 2.6. This limiting distribution describes a balance between the loss of variation due to random genetic drift and creation of variation due to mutation at the steady state. Balance is reached when the joint action of drift and mutation does not change the distribution of the gene frequencies. This distribution does not depend on t, and in terms of the density $\phi(p, y, t)$ it may be formally expressed as

$$\frac{\partial \phi(p, y, t)}{\partial t} = 0.$$

Thus, the stationary distribution $\hat{\phi}(p, y)$ is the solution to a simpler diffusion equation as sketched in Box 13. The result is the Beta distribution[5] of the gene frequencies, namely $B(4N\mu_A, 4N\mu_a)$ with the density

$$\hat{\phi}(p, y) = \frac{\Gamma(4N\mu)}{\Gamma(4N\mu_A)\Gamma(4N\mu_a)} \, y^{4N\mu_A - 1}(1 - y)^{4N\mu_a - 1}, \qquad (4.6)$$

where Γ is the Gamma function.[6] This is often referred to as Wright's (1931) formula.

Examples of the shape of this distribution are shown in Figure 4.2. The interpretation of the distribution approximation is like that on page 43. An approximation of the probability of the gene frequency $\frac{i}{2N}$ in a population of size $2N$ is found by integrating the diffusion density in the interval between $\frac{1}{2N}(i - \frac{1}{2})$ and $\frac{1}{2N}(i + \frac{1}{2})$. The monomorphic "sticks" of the distribution are included, however, and the probabilities of the monomorphic states are found as the masses in the intervals between 0 and $\frac{1}{2}\frac{1}{2N}$ and between $1 - \frac{1}{2}\frac{1}{2N}$ and 1. The left-hand graphic assumes $\mu_A = \mu_a$, so the deterministic equilibrium gene frequency is $\hat{p} = \frac{1}{2}$. For $2N\mu = 20$, 5, and 2, the distribution has a mode at the equilibrium frequency, and for $2N\mu = 20$ in particular the population is expected to show minor deviations from the deterministic equilibrium frequency. For low values of $2N\mu$, $\frac{1}{2}$ and $\frac{1}{10}$, the distribution is qualitatively different. Here the majority of the probability mass in the distribution piles up close to and in the fixation states (the intervals close to 0 and 1). Thus, the population has a very high probability of being fixed or of being close to fixation. The turning point is at $2N\mu = 1$, where the distribution is uniform and the population has the same probability of being at any of the possible gene frequencies.

The same contrast between $2N\mu > 1$ and $2N\mu < 1$ appears for more extreme gene frequencies. The right-hand graphic in Figure 4.2 assumes $\mu_A = 4\mu_a$, which gives a deterministic equilibrium gene frequency of $\hat{p} = \frac{1}{5}$. This drawing is a bit more messy because more probability mass is concentrated at low frequencies

---

[5]The Beta distribution is defined and discussed in elementary textbooks in statistics, e.g., Blæsild and Granfeldt (2003).

[6]The Gamma function is a standard mathematical function. The constant coefficient in the Beta distribution is sometimes referred to as the Beta function value $B(4N\mu_A, 4N\mu_a)$, not to be confused with the Beta distribution.

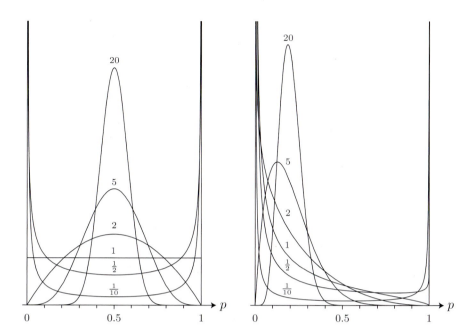

Figure 4.2: Distribution of gene frequencies in the Wright–Fisher model with mutation for various values of $2N\mu$. The left-hand graphic assumes $\hat{p} = \frac{1}{2}$ and the right-hand graphic assumes $\hat{p} = \frac{1}{5}$. The values of $2N\mu$ are 20, 5, 2, 1, $\frac{1}{2}$, $\frac{1}{10}$. For $\hat{p} = \frac{1}{2}$ the distribution has a mode at the equilibrium frequency for $2N\mu = 20$, 5, and 2; the distribution is uniform for $2N\mu = 1$; for $2N\mu = \frac{1}{2}$ and $\frac{1}{10}$ the distribution is U-shaped. For $\hat{p} = \frac{1}{5}$ the distribution is also U-shaped for $2N\mu = \frac{1}{2}$ and $\frac{1}{10}$.

of allele A, but for $2N\mu = \frac{1}{2}$ and $2N\mu = \frac{1}{10}$ the distribution is U-shaped, and for $2N\mu = 20$ and $2N\mu = 5$ the distribution has a clear mode. The mean of each of the distributions is $\frac{1}{5}$.

In terms of Wright's island model, the gene frequencies in the populations cluster around the deterministic equilibrium for $2N\mu \gg 1$, and for $2N\mu \ll 1$ the gene frequencies in the populations cluster around the fixation states.

What does the stationary distribution tell us about the evolution of a single population? If at a certain point in time it has the gene frequency p, then its development is influenced by random genetic drift and mutation, and this may be described by the time-dependent distributions, as we did in Section 2.6.1 (see, for instance, Crow and Kimura 1970). However, the stationary distributions (4.6) can give us an impression of the changes in gene frequency through time in a population where we may assume stationarity, that is, the interplay between the mutation and drift processes has reached the steady state. If we focus on a randomly chosen island in Wright's island model, then at any point in the future, we will predict its gene frequency to follow the stationary dis-

---

Box 13: * Diffusion equations

The change in gene frequency per generation from equation (4.1) is given by

$$p' - p = -\mu_A p + \mu_a q,$$

and we therefore get the change per time unit in characteristic time as

$$\mu(y) = \lim_{N \to \infty} 2N\left(-\mu_A y + \mu_a(1 - y)\right).$$

A prerequisite for using the diffusion approximation was that this limit is finite, and we therefore make the approximation on the assumption that $2N\mu_A$ and $2N\mu_a$ are constants as $N \to \infty$. This implies the assumption that $\mu_A \to 0$ and $\mu_a \to 0$ as $N \to \infty$. This assumption is easier to express if we make the rewrite

$$\mu(y) = 2N\mu(\hat{p} - y)$$

and consider $2N\mu$ a constant as $N \to \infty$. The expected change in the gene frequency per generation is therefore very small and the variance in one generation again becomes the binomial variance, $\sigma^2(y) = y(1 - y)$.

The diffusion equation used to obtain an approximation of the probability density $\phi(p, y, t)$ as $N \to \infty$ is

$$\frac{\partial \phi(p, y, t)}{\partial t} = \frac{1}{2}\frac{\partial^2 \left(y(1 - y)\phi(p, y, t)\right)}{\partial y^2} - \frac{\partial \left(2N\mu(\hat{p} - y)\phi(p, y, t)\right)}{\partial y}.$$

The equation has been solved and it is possible to draw figures like those in Figure 2.12 to show the development in the population. We will focus on another aspect of the process, however, namely the distribution reached after a long time as $t \to \infty$. This stationary distribution is reached when the distribution does not depend on t, that is,

$$\frac{\partial \phi(p, y, t)}{\partial t} = 0.$$

Thus, the stationary distribution $\hat{\phi}(p, y)$ is the solution of

$$\frac{1}{2}\frac{\partial^2 \left(y(1 - y)\hat{\phi}(p, y)\right)}{\partial y^2} = \frac{\partial \left(2N\mu(\hat{p} - y)\hat{\phi}(p, y, t)\right)}{\partial y}. \qquad (4.7)$$

The solution of this equation is the Beta distribution (4.6).

tribution (4.6)—on the condition, of course, that we do not look at the gene frequency in the population when we choose it. Thus, given enough time the gene frequencies that the population passes through will follow the stationary distribution, while the change from generation to generation will be quite small.

Using this observation the stationary distribution can tell us a great deal about the long-term dynamics of a population. Random genetic drift will tend to cause the gene frequencies of the population to wander around. For $2N\mu \gg 1$ the tendency will effectively be counteracted by mutation, so that excursions away from the deterministic equilibrium will be rather small. The effect of mutation is an input of several mutants per generation. For $2N\mu \ll 1$ the effect of mutation is too small to significantly impact the genetic drift. Mutants enter the population several generations apart. Thus, a polymorphic population will go through random genetic drift, taking it to extreme gene frequencies (Figures 4.2). If the number of mutations is very small, the population will be fixed most of the time, but a mutant 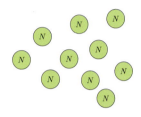 may invade occasionally. Due to the conservation of gene frequencies, this mutant has the high probability $1 - \frac{1}{2N}$ of being lost, but even then it has a positive probability of becoming fixed. Thus, the population will usually be fixed, but once in a long while it will go through a period of polymorphism, which in some instances may eventually lead to the fixation of the alternative allele. Such a population state is called *transient polymorphism*.

From this description the U-shaped distributions in Figure 4.2 give an intuitive feel for how often the population is in a polymorphic transition period and how often it is fixed or almost fixed for one allele. For $\mu_A = \mu_a$ and $2N\mu = 1$, the two processes are exactly balanced. Mutation keeps the population from spending more time in a fixation state than in any other possible state, but at the same time, mutation cannot keep the gene frequency close to the deterministic equilibrium.

The extension of the two-allele results to multiple alleles is straightforward (see Crow and Kimura 1970). For instance, Wright's formula for multiple alleles just involves the generalization of the Beta distribution to more types, the so-called Dirichlet distribution (Wright 1949). The results are rather unamenable, mainly because of the need to describe mutation among the alleles (for k alleles $k(k-1)$ mutation rates have to be specified). A closer investigation of the nature of multiallelic variation requires models capable of reflecting in a simple way the mutational characteristics of natural variation. We will commence the discussion of such models in Section 4.5. However, the two-allele model provides an adequate basis for an intuitive description of the dynamics of the frequency of a given allele as a balance between the effects of random genetic drift and mutation to and from the allele. The present results are therefore good indicators of the conditions for polymorphism or monomorphism in a population.

4.3   Genealogy and Variation

The coalescent is the cornerstone of contemporary developments in the genetic analysis of populations. It serves two main functions, the description of genealogy and the description of variation. The description of genealogy was the theme in Chapter 2. We will return to the description of variation in a population, but let us briefly consider an example of how genetic variation produced by mutation can be used as a tool for inference on genealogy—a description that resembles the classical description of evolutionary relationships among species.

The material is the sample (introduced on page 19) of twelve genes whose DNA sequence variation exhibits five alleles. Many phylogenies can describe the relationship among these alleles, but using Occam's razor,[7] we would prefer the relationship that postulated the smallest number of mutational events, as such events are presumed to be rare. A reasonable phylogenetic relationship of the five alleles is therefore given by

	$\mid\ \mid$ $\mid$
1:	G A C C T G A T T C T $\cdots$
2:	G A C A T G A T T C T $\cdots$
3:	G A C C T G A T C C T $\cdots$
4:	G A C C T G A T C C T $\cdots$
5:	G A C G T G A T T C T $\cdots$
6:	G A C A C G A T T C T $\cdots$
7:	G A C C T G A T T C T $\cdots$
8:	G A C C T G A T T C T $\cdots$
9:	G A C A T G A T T C T $\cdots$
10:	G A C C T G A T T C T $\cdots$
11:	G A C C T G A T T C T $\cdots$
12:	G A C C T G A T C C T $\cdots$
	$\mid\ \mid$ $\mid$

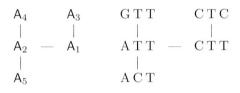

if we list only the variable sites (positions 3, 4, and 8). This presumes four mutational events, which is clearly the minimal number of possible changes that would make five entities different. The argument for preferring the above relationship among the alleles is in evolutionary analysis referred to as the principle of *parsimony*.

A possible genealogy of the sample using the suggested allele phylogeny is shown in Figure 4.3. This genealogy designates allele A_1 as the ancestral allele in the sample, with alleles A_2, A_3, A_4, and A_5 arising at times τ_2, τ_3, τ_4, and τ_5 before present. The phylogeny of the alleles agrees with the gene genealogy as it should, but many other genealogies among the genes in the sample are also consistent with the relationship above among the alleles (Figure 4.4). Both these suggestions of a coalescent tree for the sample are based on the assumption that one of the observed alleles is the ancestral allele. The five alleles may of course be descended from a sixth allele, say, A_0, which may not be present in the population. Except for the trivial case that it is an ancestor of the most recent common ancestor of the observed sequences, the inclusion of allele A_0 in the genealogy calls for an extra mutation event. The most recent common ancestor of the sampled sequences thus defines the ancestral allele, which in turn defines the *root* of the allele phylogeny. The coalescent process may be used to find the

---

[7]The principle that the simplest of competing explanations should be preferred to the more complex. This philosophical rule is credited to William of Occam (approximately 1287–1317).

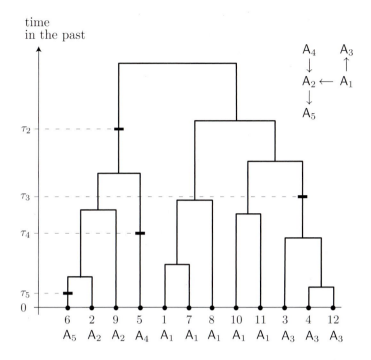

Figure 4.3: The coalescent of a sample of 12 sequences with mutations indicated by heavy bars. The numbers are those of the sampled genes.

likelihood of these genealogies based on a model that describes the process of mutation. We will examine such models in later sections where a population genetic description of mutation is pursued.

The simplest model specifies that at any site in the sequence mutation occurs independently with probability μ per gene per generation. If we know the coalescent of the 12 sequences, then we have observed the total length of the coalescent tree given by $12T_{12} + 11T_{11} + \cdots + 3T_3 + 2T_2$ gene generations. During this time we observed four mutations, and we can therefore estimate the mutation rate as 4 divided by the total length of the coalescent tree.

This simple model assumes that the mutation rate is independent of the composition of the collection of genes we consider. If the mutation rate varies among alleles, inference based on population samples becomes complicated. For instance, if the observed alleles mutate with the probabilities μ_1, μ_2, μ_3, μ_4, and μ_5, then the probability of detecting a mutation by comparing them to the parental genes of the sample is $\frac{1}{12}(5\mu_1 + 2\mu_2 + 3\mu_3 + \mu_4 + \mu_5)$ in the phylogeny in Figure 4.3, whereas before τ_2 it was μ_1. We will therefore assume in the following that the mutation rate is the same for all alleles.

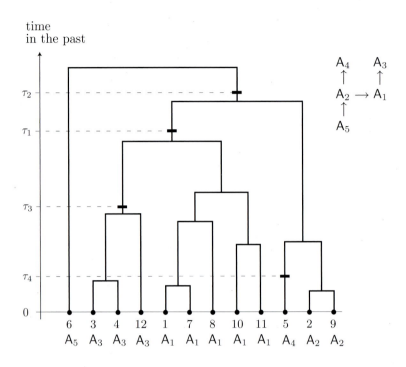

Figure 4.4: The coalescent with mutations indicated by heavy bars. An alternative coalescent of the sample of 12 sequences with A_5 as the ancestral allele.

4.3.1   Alcohol dehydrogenase in *Drosophila melanogaster*

Most natural populations of *Drosophila melanogaster* are polymorphic for alcohol dehydrogenase when individuals are analyzed using protein electrophoresis. Two alleles, Adh^F and Adh^S, are observed.[8] Kreitman (1983) investigated sequences from five populations that represent the range of the species (Washington, Florida, Burundi, France, and Japan). Two sequences of the Adh^S allele were determined in flies from Florida and one from each of the other populations. The variation of the translated sequence of that allele is shown in Table 4.4. The coding region of 765 base pairs is split in four exons. No variation was found in exon 1. The variable sites in the remaining exons are shown in the order they occur in the sequence. The variable sites are SNPs. None of the site variations changes the amino acid of the corresponding protein, and they are therefore called *silent* polymorphisms. The sequences define four alleles within the electrophoretic allele Adh^S, and these are silent alleles. The allele observed in Washington is also seen in Florida, and the Burundi allele is sampled in France too. The unique Florida allele is rather close to the Burundi allele (2 SNP distance), and these are equidistant from the Wash-

---

[8]F and S designate proteins showing fast and slow migration in electrophoresis.

Table 4.4: Variation among Adh^S alleles in *Drosophila melanogaster*

Washington	T	T	T	A	A	C	C	C	A	G
Florida-1	T	T	T	A	A	C	C	C	A	G
Florida-2	A	C	C	C	C	T	C	C	A	G
Burundi	T	C	C	C	C	T	C	C	A	A
France	T	C	C	C	C	T	C	C	A	A
Japan	T	C	C	C	C	T	T	T	C	A

Data from Kreitman (1983).

ington allele (6 SNP distance). Finally, the sequence from Japan is closest to the Burundi allele (3 SNP distance). This produces the simple relationship expressed by the distance tree shown in the figure to the right. This is an additive tree, in that the distance between any two alleles can be read from the tree by adding the length of the branches that connect them.

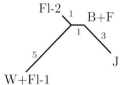

The distance tree suggests an allele phylogeny that may be embedded in coalescent trees with different topologies. The root in particular may be placed on any edge of the tree, and the coalescent process with added mutation events can provide the probabilities of these alternative descriptions of the sequence evolution of the Adh^S allele. Alternatively, the root of the allele tree may be placed by comparisons with homologous alleles in closely related species (see, e.g., Felsenstein 2004).

4.4   Mutation and Coalescence

We can for any particular gene ask how far back that gene mutated, thus assuming its present sequence. In Figure 4.5, for instance, gene number 4 mutated at time τ_2 in the past and gene 5 at τ_5. If the probability that the gene differs from its parental gene is μ, then the waiting time S_1 to mutation has a geometric distribution, that is,

$$\text{Prob}(S_1 = i) = \mu(1 - \mu)^{i-1}, \quad i = 1, 2, 3, \ldots, \qquad (4.8)$$

and the average number of generations that we have to go back before finding the mutation is μ^{-1}. The probability that a gene in a sample of n differs from its parental gene is about $n\mu$ if we assume that μ is sufficiently small to ignore the possibility of more than one mutation in a given generation. In the gamete pool model $n\mu$ will be the probability of detecting a mutation in any generation back in time, unless a coalescent happens before we detect a mutation event. If a coalescent event does occur, we thus only follow $n - 1$ genes and the probability of a mutation event is then $(n - 1)\mu$ per generation. The waiting time to a mutation is the same for any gene at any time (the waiting time distribution has no memory), but when we observe a collection of genes, the

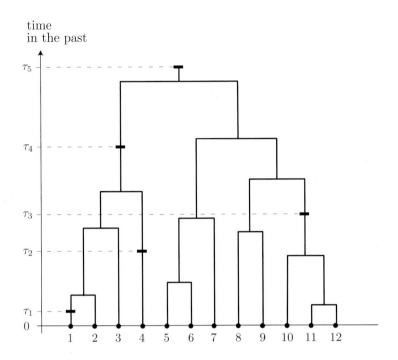

Figure 4.5: The coalescent with mutation.

coalescence process among those genes changes the per generation probability of detecting a mutation event. Thus, we have to contemplate the two processes simultaneously.

Suppose that at a given time in the past the genes in our sample have coalesced into k ancestral genes. The probability that a coalescent event or a mutation event happens when going back to the parents of these k genes is

$$p_k + k\mu = \binom{k}{2}\frac{1}{2N} + k\mu, \tag{4.9}$$

when both the coalescent probability p_k (equation (2.13) on page 36) and the mutation probability are so small that we can ignore the possibility of concurrent events. The distribution of the waiting time to an event is thus the geometric waiting time distribution with event probability $p_k + k\mu$. This waiting time to the first event is the stochastic variable $\min(T_k, S_k)$, where T_k is the waiting time to a coalescent event and S_k is the waiting time to a mutation event among k genes. After the wait a coalescent event occurs if $T_k < S_k$, and a mutation event occurs if $T_k > S_k$. In Appendix A.1.4 the probabilities of these events are shown to be the relative event probabilities, that is,

$$\text{Prob(coalescent event occurs first)} = \text{Prob}(T_k < S_k) \approx \frac{p_k}{p_k + k\mu},$$

$$\text{Prob(mutation event occurs first)} \; = \; \text{Prob}(T_k > S_k) \; \approx \; \frac{k\mu}{p_k + k\mu} ,$$

when the event probabilities are small. Inserting p_k gives

$$\frac{p_k}{p_k + k\mu} = \frac{\frac{1}{2}k(k-1)\frac{1}{2N}}{\frac{1}{2}k(k-1)\frac{1}{2N} + k\mu} = \frac{k-1}{k-1+4N\mu} = \frac{k-1}{k-1+\theta} ,$$

where we have introduced the parameter $\theta = 4N\mu$. The parameter θ is the traditional population genetic parameter that expresses the effect of mutation. We can recognize it as twice the mutation rate measured in the characteristic time of random genetic drift. Thus, we get

$$\text{Prob(coalescent event occurs first)} \; \approx \; \frac{k-1}{k-1+\theta} \quad \text{and} \qquad (4.10)$$

$$\text{Prob(mutation event occurs first)} \; \approx \; \frac{\theta}{k-1+\theta} . \qquad (4.11)$$

In case of a coalescent event the ancestral sample size changes from k to $k-1$ with the ensuing change in the event probability (4.9). In case of a mutation event the probability (4.9) stays unchanged, showing that mutations are just placed onto the coalescent. Thus, given the coalescent of the sample, mutations may simply be added by the mutation process given by the waiting time distribution (4.8). Of course, similar results are obtained by adding mutation to the continuous time coalescent (developed in Box 14).

The probability of sampling two identical genes from the population is the coefficient of gene identity F, and this equals the probability that a coalescent event precedes mutation in a sample of two genes:

$$F = \frac{1}{1+\theta} \qquad (4.12)$$

(Watterson 1975). When $\theta \gg 1$, F is close to zero and the probability of finding identical pairs of genes in the population is small. The probability that a mutation event occurs before a coalescent event is large, and so a sample will be dominated by single mutations like those to genes 1 and 4 in Figure 4.6. When $\theta \ll 1$, F is close to one and the probability of a mutation event before a coalescent is small. Thus, there is a high probability that the entire sample coalesces before a mutation event occurs.

The coalescent approach allows a more accurate description of the composition of a sample. The probability of sampling n genes that are identical by descent from the population equals the probability that $n-1$ coalescent events happen before the first mutation event, namely

$$F^{(n)} = \frac{n-1}{n-1+\theta} \times \frac{n-2}{n-2+\theta} \times \cdots \times \frac{2}{2+\theta} \times \frac{1}{1+\theta}$$

(Watterson 1975). This probability is an extension of the coefficient of gene identity F, in that $F = F^{(2)}$, and it is traditionally written as

$$F^{(n)} = \frac{(n-1)!\,\theta}{S_n(\theta)} , \qquad (4.13)$$

Box 14: Kingman's coalescent with mutation

The coalescent process was described in Section 2.6.2 as one in which the state ξ in a short interval of time changes to the follower η. Using characteristic time and taking all followers into account gives

$$\text{Prob}\big(\text{coalescent in } (\tau, \tau + \Delta\tau)\,\big|\,\xi \text{ at } \tau\big) \approx \binom{|\xi|}{2}\Delta\tau.$$

Now any gene is assumed to mutate with probability μ per generation, which in characteristic time corresponds to the mutation rate of $2N\mu$ per time unit. Any gene present at time τ thus mutates in the time interval from τ to $\tau + \Delta\tau$ with the probability $\frac{1}{2}\theta\Delta\tau$, where we—as above—have ignored terms of the order $(\Delta\tau)^2$. Therefore we get

$$\text{Prob}\big(\text{mutation in } (\tau, \tau + \Delta\tau)\,\big|\,\xi \text{ at } \tau\big) \approx \tfrac{1}{2}|\xi|\theta\Delta\tau.$$

The probability that something happens in the interval from τ to $\tau + \Delta\tau$ is therefore

$$\text{Prob}\big(\text{a change in } (\tau, \tau + \Delta\tau)\,\big|\,\xi \text{ at } \tau\big) \approx \binom{|\xi|}{2}\Delta\tau + \tfrac{1}{2}|\xi|\theta\Delta\tau.$$

Simplifying this expression yields

$$\text{Prob}\big(\text{a change in } (\tau, \tau + \Delta\tau)\,\big|\,\xi \text{ at } \tau\big) \approx \tfrac{1}{2}|\xi|\big(|\xi| - 1 + \theta\big)\Delta\tau.$$

The properties of the coalescent process with mutation may now be described as before. When in the state ξ we wait an average of

$$\frac{2}{|\xi|\big(|\xi| - 1 + \theta\big)}$$

time units before something happens. When something does happen, a coalescent and a mutation occur with the probabilities

$$\frac{|\xi|\big(|\xi| - 1\big)}{|\xi|\big(|\xi| - 1 + \theta\big)} \quad \text{and} \quad \frac{|\xi|\theta}{|\xi|\big(|\xi| - 1 + \theta\big)},$$

respectively. These probabilities are obtained as the relative rates at which the two kinds of events occur (see Appendix A.3.1). Taking away a factor of $|\xi|$ we get the two probabilities as in equations (4.10) and (4.11):

$$\frac{|\xi| - 1}{|\xi| - 1 + \theta} \quad \text{and} \quad \frac{\theta}{|\xi| - 1 + \theta}.$$

where

$$S_n(\theta) = \theta(1 + \theta)(2 + \theta) \cdots (n - 1 + \theta)$$

(a factor θ has been included in both the numerator and the denominator).

The probabilities (4.10) and (4.11) may be interpreted as providing an immediate description of the structure of the sample (Ewens 1972). Suppose that the first $i - 1$ genes in our sample are identical by descent. Now pick an extra gene and ask, what is the probability that these i genes are identical by descent? This conditional probability may be calculated as

Prob(genes $1, 2, \ldots, i$ identical | genes $1, 2, \ldots, i - 1$ identical)

$$= \frac{\text{Prob(genes } 1, 2, \ldots, i \text{ identical)}}{\text{Prob(genes } 1, 2, \ldots, i - 1 \text{ identical)}} = \frac{F^{(i)}}{F^{(i-1)}} = \frac{i - 1}{i - 1 + \theta}, (4.14)$$

which is equal to the probability (4.10). Thus, the probability

Prob(genes $1, 2, \ldots, i$ not identical | genes $1, 2, \ldots, i - 1$ identical)

$$= 1 - \frac{i - 1}{i - 1 + \theta} = \frac{\theta}{i - 1 + \theta} \tag{4.15}$$

is the probability that the extra gene picked is of a new mutant type when the first $i-1$ were identical, and it equals (4.11). These arguments may be extended to a very useful description of the structure of a sample, but before developing them further, it is convenient to introduce simple models for genetic variation.

4.4.1 The appearance of a sample

The typical state of the sampled population depends on the value of θ ($4N\mu$ in Section 4.2.1), in that polymorphism is expected when $\theta \gg 1$ and monomorphism when $\theta \ll 1$. This result is recalled in the event probabilities (4.10) and (4.11), and in the probability $F^{(n)}$ that a sample of n genes does not show variation, that is, $F^{(n)} \to 1$ for $\theta \to 0$. Even for a low θ the population will occasionally go through an episode of transient polymorphism where the probability of observing variation in a sample is high. The scaled mutation rate θ determines the probability of observing a population with variation. All else being equal, the coalescent tree of the sequences of the sampled genes has the same properties whether variation is observed or not, so the estimation of θ will at best be highly biased if it is based only on samples with variation. Single nucleotide polymorphisms provide examples of this problem. The human genome is littered with these, and an explanation of their sporadic occurrence is that they reflect transient polymorphisms due to mutation–mutation balance and random genetic drift with a low θ. Embedded in this hypothesis is that the SNP sites are not fundamentally different from many monomorphic sites.

To circumvent this bias one suggestion is to take more samples, but that option may not exist in an evolutionary analysis, because if one population is in a rare episode of variation at a given site in the genome, chances are that other populations of the species will also show variation. Repeated sampling through the history of the population is only possible for microbes at best. The estimate of the mutation rate based on the total length of the coalescent tree discussed on page 85 is therefore of questionable quality when the per site mutation rate is of interest. The only way to clear this hurdle is to find a way to obtain independent samples at present that provide information on the mutation process.

In parallel to the SNP description above we could base an evolutionary analysis on assumptions of homogeneity in the evolutionary processes through the positions in the sequence. Jukes and Cantor (1969) used this idea in their analysis of amino acid sequences. Translated to DNA sequences their model is given by the mutation rate matrix in Table 4.5. If reasonable, Jukes and Cantor's

Table 4.5: Jukes–Cantor model

from	A	T	C	G
A		μ	μ	μ
T	μ		μ	μ
C	μ	μ		μ
G	μ	μ	μ	

model allows us to estimate the per site mutation rate from data on sequence variation. The homogeneity among the rates of the various kinds of changes ignores, however, the molecular difference between transitions and transversions (see page 72). Kimura (1980) included this distinction in the model in Table 4.6, where μ describes the transition rate and ν the transversion rate. These

Table 4.6: Kimura's model

from	A	T	C	G
A		ν	ν	μ
T	ν		μ	ν
C	ν	μ		ν
G	μ	ν	ν	

two models are highly symmetric, and the mutation rate is independent of the base. Thus, at their evolutionary equilibrium the DNA bases are expected to be equally frequent. This deviates from observations, and Kimura's model is often used with an ad hoc modification that includes the base frequencies in the sequence as parameters (Hasegawa et al. 1985, Tamura and Nei 1993).

The Jukes–Cantor family of models maintains a homogeneity among sites which seems unreasonable for protein coding sequences. Here the neutral mutation rate is likely to depend on whether the mutation changes an amino acid or not—thus the distinction between synonymous and nonsynonymous mutations. Goldman and Yang (1994) introduced this distinction among mutations into the modified Kimura model, resulting in codon-based analysis of the sequence. But even richer structures in the sequence may be introduced into the analysis by applying hidden Markov models. Such models may incorporate the molecular structure of a sequence, namely coding sequences, introns, UTRs, etc. (Pedersen and Hein 2003).

The analysis of genetic sequence data is not our concern here. Applications in molecular evolution are covered by Felsenstein (2004) and those in population genetics by Hein et al. (2005). In terms of the models in this section the parameters in Kimura's model are mutation rates. In relation to the analysis of observed sequence variation, however, they are rates of substitution. Given that the variation we are observing is neutral, this distinction is immaterial. A new mutant exists in the frequency of $\frac{1}{2N}$ in the population, and so the probability that it becomes fixed is $\frac{1}{2N}$. The number of mutants produced per generation is $2N\mu$, so the average number of mutants fixed per generation equals $2N\mu \times \frac{1}{2N} = \mu$. The rate of fixation of neutral mutations per generation is therefore equal to μ. In other words, *the neutral substitution rate equals the neutral mutation rate* (Kimura 1968, King and Jukes 1969). This produces a *molecular clock* that ticks at the rate μ through evolution. When the mutation rate varies and depends on the base composition of sequences, then no time-homogeneous molecular clock exists (see page 85).

The molecular clock provides the observable time in the coalescent of a sample because the passing of time is revealed by emergence of mutations (see Section 4.3). The coalescence is based on the known sample size n, but the population size N is unknown, wherefore inference on the sample is served in characteristic time.

4.4.2   Gene identity and mutation

The probability F_t that two randomly chosen genes are identical must be lower with than without mutation because a mutation event interrupts the line of identity by descent. Suppose that the gene we are studying mutates at a frequency of μ per generation in the gamete pool model (Figure 2.1 on page 24). For example, two genes are copies of the same gene in generation $t-1$ with probability $\frac{1}{2N}$ if neither has mutated. Thus, we get the recurrence equation

$$F_t = (1 - \mu)^2 \left(\frac{1}{2N} + \left(1 - \frac{1}{2N} \right) F_{t-1} \right), \qquad (4.16)$$

which may be iterated to

$$F_t = \hat{F} - \left((1 - \mu)^2 \left(1 - \frac{1}{2N} \right) \right)^t (F_0 - \hat{F}), \quad t = 0, 1, 2, \ldots \qquad (4.17)$$

(Malécot 1948, Kimura and Crow 1964), where $\hat{F}$ is approximately given by equation (4.12) when the population size is large and the mutation rate is small (this is shown in Box 15). No matter the initial value F_0, the change in the population coefficient of gene identity is monotonic, and $F_t \to \hat{F}$ as $t \to \infty$. The population will eventually reach a level of identity that depends only on

---

Box 15: Development of gene identity

The linear recurrence equation (4.16) is iterated in the same way as the recurrence equation (2.4) in the variances on page 28 (see Box 4), but the equilibrium is no longer total homozygosity. Equilibrium is reached when $F_t = F_{t-1}$ and the equilibrium value is therefore given by

$$\hat{F} = \frac{(1-\mu)^2 \frac{1}{2N}}{1 - (1-\mu)^2 \left(1 - \frac{1}{2N}\right)} \approx \frac{\frac{1}{2N}}{\frac{1}{2N} + 2\mu} = \frac{1}{1 + 4N\mu},$$

where the approximation assumes that μ and $\frac{1}{2N}$ are small. The rewrite

$$F_t - \hat{F} = (1-\mu)^2 \left(1 - \frac{1}{2N}\right)(F_{t-1} - \hat{F})$$

iterates to the solution (4.17).

---

the number $2N\mu$ of mutations produced in the population each generation. When $2N\mu$ is large, $\hat{F}$ is small and $\hat{F} \to 0$ as $N \to \infty$. Thus, in a very large population a sample rarely contains identical genes. When $2N\mu$ is small $\hat{F}$ is large, and even though $\hat{F} \to 1$ as $\mu \to 0$, any real population is not expected to remain monomorphic for an evolutionarily extended period.

In classical genetics two genes, one of which has just mutated, cannot be identical by descent, but when we observe nucleotide sequences we can still trace the origin of simple mutants (Figure 4.5 on page 88). Mutation gives rise to a new identifiable sequence, which we may then consider as a new allele (for simplicity we ignore the possibility of producing equal sequences by different mutations). In this view, genes of the same allelic class are identical by descent, and the genealogy of an allele may be studied by examining the ancestors of the genes of that particular allele type in the sample. This corresponds to arresting the coalescent process when a mutation event occurs. The mutation thereby stops a line of identity by descent going back in time, and the coalescent with mutation therefore reveals a series of coalescent trees that all end at a mutation (Figure 4.6). The expected time to the first coalescent within an allelic class of k genes in a sample of n is

$$\mathrm{E}\,T_k = \frac{2Np}{\binom{k}{2}},$$

where p is the gene frequency of the allele at the time of sampling. The genealogy of the observed sequences of an allele does not, however, adhere to the coalescent because the distribution of the waiting time is not exponential. Patterson (2005) established this result using a diffusion approximation, which also produced approximations of the waiting time distribution (Figure 4.7).

The description of random genetic drift by the N_e reference to the Wright–Fisher model extends to models that include neutral mutation, and in general

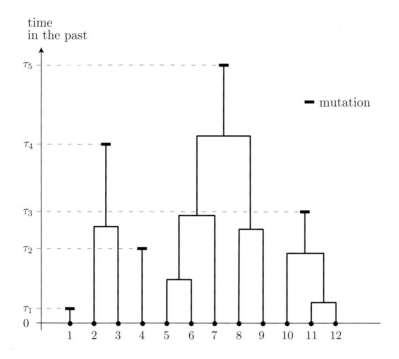

Figure 4.6: Line of descent of the various mutants in Figure 4.5.

we have

$$\hat{F} \approx \frac{1}{1 + 4N_e\mu}.$$

Using Wright's island model again we can conclude that the equilibrium value of the variance in the two-allele mutation balance model corresponds to the behavior of the consanguinity coefficient considered here (see Section 4.2). When studying the effects of mutation, we need only replace the actual population size N with the effective population size N_e, that is, $\theta = 4N_e\mu$.

4.5   The Infinite Alleles Model

Two alleles are generally not enough to describe genetic variation. Many more are needed, but multiallelic models are complicated, and we therefore need simplifications. As a first approximation we can assume the number of possible alleles of the gene of interest to be so high that we can ignore the possibility that mutation produces an allele already present in the population. Any newly mutated allele is different from those that existed before. The resulting model was suggested by Malécot (1948) and Kimura and Crow (1964), and it is known as the *infinite alleles model*. This is a classical way to construct a model. The simplest model of genetic variation is the two-allele model, but other finite alleles models are fairly complicated, and the case of an unlimited number of

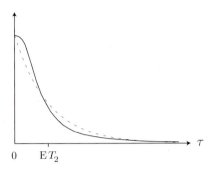

Figure 4.7: The distribution of the waiting time until coalescent of two sequences of the same allele type (solid curve). The average waiting time is $\mathrm{E}\,T_2 = 2Np$, where p is the frequency of the allele in the population. For comparison the dashed curve shows the exponential distribution with the same average. After Patterson (2005).

alleles is easier to simplify so that it yields biologically usable results. We thus count the alleles in the primitive manner as one, two, and many.

The model is built for the purpose of discussing data on variation in proteins revealed by electrophoresis, and the unit is therefore the allele. Its counterpart for sequence data is the infinite sites model discussed in Section 4.6, and the results for the infinite alleles model are useful for that model as well. In the analysis of the data example in Section 4.3 we assumed that each of the five alleles had arisen only once by mutation. This assumption is equivalent to viewing the infinite alleles model as the basis of our data analysis.

The gene identity description in Section 4.4.2 is well suited for describing certain properties of the infinite alleles model of mutation applied to the gamete pool model of reproduction. Genes of the same allelic type may be viewed as identical because they are all copies of the original mutant. The probability of sampling two allelic genes from the gamete pool in a given generation is the population gene identity coefficient F. In terms of the gene frequencies $p_1, p_2, \ldots$ of alleles A_i, $\mathsf{A}_i,\ldots$ in the gamete pool, this probability is

$$F = \sum_{i=1}^{\infty} p_i^2 \qquad \left(\text{and} \quad \sum_{i=1}^{\infty} p_i = 1\right). \tag{4.18}$$

The number of potential alleles is infinite, but the number of alleles in the population cannot exceed $2N$, of course. The above sums therefore only have a finite number of positive terms, the rest being zero.

The behavior of F through time is given by equation (4.17) on page 93. At equilibrium we have

$$\hat{F} \approx \frac{1}{1+\theta}, \quad \theta = 4N\mu \quad \left(\text{or } \theta = 4N_e\mu\right).$$

This suggests that observations of the gene frequencies and calculation of F

using equation (4.18) may provide information on the parameter θ under the hypothesis that variation in the population is maintained as an equilibrium between mutation and drift.

However, this is an example where an immediate solution to an inferential problem turns out to be utterly wrong. This was shown by Warren Ewens in 1972 in a paper that laid the foundation for modern molecular population genetics. He addressed the problem from a statistical point of view, and his work is a prime example of a situation where a proper statistical analysis completely changes attitudes towards data within a scientific field. His arguments reflect a deep insight into the process of evolution by mutation and random genetic drift, and the statistical theory is beautiful—although you may need a statistician to assist you in the aesthetic evaluation. Ewens' analysis is a predecessor of Kingman's coalescence theory, and it is the first step towards the development of the retrospective evolutionary analysis of molecular population genetic data. His arguments were based on a direct calculation of the coefficients $F^{(i)}$ of gene identity, $i = 2, 3, \ldots, n$ (equation (4.13) on page 89), and the result is *Ewens' sampling formula* (Ewens 1972).

Great discoveries are often based on observations that afterwards seem trivial. Investigation of a sample of size n disclosed the alleles A_1, A_2, $\ldots$, A_k in the frequencies p_1, p_2, $\ldots$, p_k $(p_1 + p_2 + \cdots + p_k = 1)$. With reference to a model that allows a lot more variation than we may expect to find in a population and a lot more variation than we are likely to observe in a sample, an important observation is that we observed k alleles in our sample of size n. Ewens showed that all the information about θ in the sample is contained in k, the number of observed alleles. Knowing k and n thus produces an estimate of θ, regardless of what gene frequencies are observed. The value of F calculated from the observed gene frequencies using the frequency of homozygotes (4.18) therefore depends only on θ through k. The total frequency of homozygotes depends on the number of alleles, but that is a very weak dependence (see Exercise 4.5.1).

An immediate question is whether the infinite alleles model provides a good description of the observed pattern of allelic variation. Experience has deemed it very useful in the analysis and discussion of allelic data. The resolution of the question and the usefulness of the model require a goodness-of-fit test, however. The observed gene frequencies in the sample provide an excellent basis for a test because Ewens showed that their distribution for a given k is independent of θ. The actual construction of a test is a statistical problem, and statistics is not our concern here. We can still study the distribution of gene frequencies to get a feel for how it should look in a typical sample.

Ewens (1972) gave three examples of gene frequencies in a sample with $n = 350$ and four alleles, $k = 4$ (Figure 4.8). Sample B has a typical spectrum of gene frequencies from the distribution in the infinite alleles model. Samples A and C are possible but deviate significantly from the typical spectrum of gene frequencies expected in the infinite alleles model. Sample A deviates significantly because the gene frequencies are too even, and sample C deviates because one allele is too dominant in its frequency. An example of a deviation with gene frequencies that are too even is the MHC polymorphism in mammals

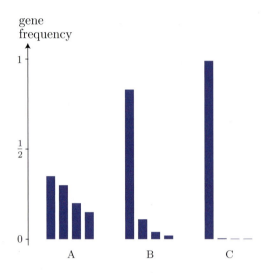

Figure 4.8: Three examples of gene frequencies given by Ewens (1972). The alleles in each example are ordered by decreasing frequency.

(Hedrick et al. 1991, Hughes and Nei 1988, Nei and Hughes 1991).

Kingman (1982c) showed that the distribution of the sizes of the clusters of related genes in the coalescent (Figure 4.6 on page 95) follows Ewens' sampling formula. This parallel between Ewens' sampling formula and the coalescent with mutation provides a good intuitive reason for the appearance of the gene frequency distributions in the infinite alleles model (Figure 4.8). The length of the branches in the coalescent tree (Figure 4.3 on page 85) indicates the total gene × time available where mutations may occur, and a large fraction of this gene-time occurs close to the present unless the sample is very small (the waiting time to the final coalescent event is about half the coalescence time of the sample). A rule of thumb is that the mutations may be viewed as uniform over the coalescent tree, and this uniformity leads to the characteristic form of the gene frequency distribution where a few common and, typically, a lot of more rare alleles are present. The rule is not absolute; mutations in the older part of the tree may be undetectable due to later mutations.

The coalescent description of a sample is retrospective. We place ourselves in the present and look back at the genetic processes that shaped the sample. In statistical terms, the description of the sample is thus conditional upon the event that we observed the sample. Ewens' sampling formula and the ensuing statistical analysis are retrospective, reflecting the situation of any analysis of evolutionary phenomena.

4.5.1 Heterozygosity in the infinite alleles model

With random union of gametes, the consanguinity description provides the genotypic composition of a population that evolves according to the infinite alleles model of mutation and the gamete pool model of reproduction. The total expected frequency of homozygotes in the population equals the probability of sampling two identical genes from the gamete pool—the probability F from equation (4.18). At equilibrium between drift and mutation this frequency of homozygotes becomes $\hat{F} \approx 1/(1 + \theta)$. For historical reasons the amount of variation in a population is commonly expressed by the expected frequency of heterozygotes $H = 1 - F$,

$$H = \sum_{i=1}^{\infty} \sum_{\substack{j=1 \\ j \neq i}}^{\infty} p_i p_j \tag{4.19}$$

(note that the genotype $\mathsf{A}_i\mathsf{A}_j$ occurs twice in this sum, namely as $\mathsf{A}_i\mathsf{A}_j$ and as $\mathsf{A}_j\mathsf{A}_i$; this may seem awkward, but the sum is simpler to handle when written this way—a trick of the trade). The total frequency of heterozygotes of a gene is often referred to as the *heterozygosity* of the gene, and its equilibrium value in the gamete pool model is $\hat{H} \approx \theta/(1 + \theta)$.

It has become customary in the population genetic literature to use heterozygosity as a measure of variation. Based on observations heterozygosity is calculated for each gene, and the overall amount of variation in the population is then judged from the average heterozygosity over genes

$$\bar{H} = \frac{1}{L} \sum_{\ell=1}^{L} H_\ell,$$

where L is the number of genes investigated and H_ℓ is the heterozygosity at gene number ℓ. This average is only a crude indication of variation in a population, even when the infinite alleles model provides a good description of the variation at any one of the considered genes. That is because the mutation rate is expected to vary among genes. For instance, the mere size of a gene very likely influences the mutation rate if we posit that mutations originate as errors in the conservative replication of the bases at single sites. Such an effect was observed in investigations of protein variation (Koehn & Eanes 1978, see Figure 4.9). The mutation rate will not be a simple function of the size of the coding region of the gene because the per site mutation rate may vary. In the infinite alleles model, however, we follow only a subset of the possible mutations, namely those that may be considered neutral, and the fraction of neutral mutations is likely to show substantial variation among genes. The observed average heterozygosity will thus be dictated by the genes investigated. Comparisons among populations or species of the level of variation should therefore only be done using homologous genes, and even then the statistical properties of $\bar{H}$ are very complicated.

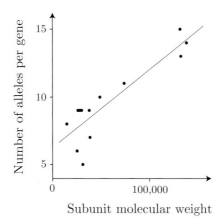

Figure 4.9: The number of electrophoretically defined alleles in species of the *Drosophila willistoni* group as a function of the molecular weight of the subunits of the corresponding protein. Redrawn after Koehn and Eanes (1978); data on variation from Ayala et al. (1974).

4.6   The Infinite Sites Model

A variant of the infinite alleles model is the *infinite sites model* originally formulated by Kimura (1969a). This is the infinite alleles model augmented by a structure that reflects the structure of the gene. The assumption that the number of alleles is so high that any mutant is new is implemented by assuming that we can ignore the possibility that mutation occurs at a site that already varies in the population. Any newly mutated site is different from those that mutated before, and in theory we therefore allow an unlimited number of sites. The two models are very similar as models of the evolutionary effects of mutation, but the theoretical analysis of the infinite alleles model is simpler. The

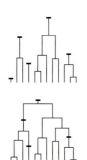

description of the infinite alleles model corresponds to considering the partitioned coalescent in Figure 4.6. DNA sequence data contribute information on the full coalescent (Figure 4.5), and the infinite sites model therefore adds sufficient structure to the infinite alleles model to allow reconnection of the coalescent.

The models differ in their description of mutation, however. The infinite alleles model assumes a probability μ of mutation of any gene in the population in any generation. The infinite sites model describes mutation as changes in one of a large number of sites, and the mutation rate in this model may therefore differ from that in the infinite alleles model. Given that differing sequences are viewed as different alleles, an infinite sites model nevertheless produces an infinite alleles model with the same mutation rate.

The alleles of the constructed data example in Sections 1.1.2 and 4.3 are structured into sequences, but the infinite sites model cannot be applied because the first of the variable sites exhibits three bases, necessitating two mutations at that site. In the infinite sites model any mutation in the genealogy of the sampled genes is apparent in the sample as a site segregating two bases, that is, as a SNP. The variation seen in the Adh^S allele of *Drosophila melanogaster* may thus be described by the infinite sites model (Section 4.3.1). More than a million SNP sites were determined based on very few individuals in the initially published version of the human genome (Sachidanandam et al. 2001).

```
G T T          C T C
  |              |
A T T   —      C T T
  |
A C T
```

The coalescence theory developed in Section 4.4 may therefore be applied immediately to the infinite sites model. Equations 4.10 and 4.11 on page 89 are particularly useful for describing the properties of a sample. The probability of observing no variation in a sample of two sequences is the probability F that the first event is a coalescence (equation (4.12)). The probability of observing no variation in a sample of n genes is $F^{(n)}$, the probability that $n-1$ coalescence events happen before the first mutational event (equation (4.13)). These probabilities are of course the same as those pertaining to the infinite alleles model. The description of variation is different, however, because the alleles in the infinite sites model exhibit their genealogical relationship. The probability of observing i differences between two genes is the probability that i mutational events precede the coalescence of the two genes, that is, for $i = 0, 1, 2, \ldots$ we get

$$\text{Prob}(i \text{ site differences between two sequences}) = \frac{1}{1+\theta}\left(\frac{\theta}{1+\theta}\right)^i \quad (4.20)$$

(Watterson 1975). Sequence differences are often called *mismatches*, and the distribution (4.20) is the mismatch distribution. The mismatch distribution is a geometric distribution with mean θ and variance $\theta(1+\theta)$ (or rather the alternative formulation of the geometric distribution given on page 355).

The coalescent and mutation events in the genealogy of the twelve sequences in Figure 4.5 on page 88 have the probability

$$F^{(12)} \times \frac{\theta}{11+\theta}\frac{\theta}{7+\theta}\frac{\theta}{4+\theta}\frac{\theta}{2+\theta},$$

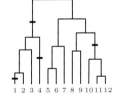

because the first mutation in (1) happens when the sample consists of 12 genes, the next in (4) among 8 genes, that in (10,(11,12)) among 5 genes, and finally (((1,2),3),(4)) among 3 genes. The mutations result in observation of five alleles where allele pairs differ at one, two, or three sites. Other placements of the mutations in the genealogy may result in a different number of alleles. For instance, if the mutation that hits (4) instead hits (1,2) or (10,(11,12)), four alleles would be observed, and among these a pair would differ at all variable sites. The genetic structure of a sample in the infinite sites model is thus richer than that in the infinite alleles model, because it reflects all mutations in the genealogy and is contingent upon the placement of the mutations.

A richer structure is also a more complicated one. A simple property is the number of site differences observed in a sample. With n sequences sampled, the distribution of the number of SNPs, $i = 0, 1, 2, \ldots$, is given by Watterson (1975):

$$\text{Prob}(i \text{ site differences among } n \text{ sequences}) \;=\; F^{(n)}$$

$$\times \sum_{j_2=0}^{i} \sum_{j_3=0}^{i-j_2} \cdots \sum_{j_n=0}^{i-j_2-j_3-\cdots-j_{n-1}} \left(\frac{\theta}{n-1+\theta}\right)^{j_n} \left(\frac{\theta}{n-2+\theta}\right)^{j_{n-1}} \cdots \left(\frac{\theta}{1+\theta}\right)^{j_2},$$

where j_ℓ is the number of mutations, while the ancestral material is represented by ℓ genes.[9] Denote the number of site differences among n sequences by $\mathcal{S}_n$, and the equation then provides the distribution of $\mathcal{S}_n$: $\text{Prob}(\mathcal{S}_n = i)$, $i = 0, 1, 2, \ldots$.

This distribution reflects the simple argument on mutation on page 85 in Section 4.3. If we knew the coalescent of a sample of n sequences, then we would have observed the total length of the coalescent tree as

$$\mathcal{L}_n = \sum_{k=2}^{n} k T_k$$

gene generations, where T_k is the length of time interval in which the sample is represented by k ancestral genes. Mutation occurs at a site in the sequence with probability μ per sequence per generation, and therefore, if $\mathcal{L}_n = \ell$ the number of mutations is Poisson distributed with mean $\ell\mu$, that is,

$$\text{Prob}(\mathcal{S}_n = i \,|\, \mathcal{L}_n = \ell) \;=\; \frac{(\ell\mu)^i e^{-\ell\mu}}{i!}. \tag{4.21}$$

The expected number of mutations in a given coalescent is thus given by the total length ℓ of the tree, that is, by $\text{E}(\mathcal{S}_n \,|\, \mathcal{L}_n = \ell) = \ell\mu$. The expected number of SNPs is therefore $\text{E}\,\mathcal{S}_n = \mu \text{E}\,\mathcal{L}_n$. The distribution of $\mathcal{L}_n$ is somewhat complicated, but the calculation of its mean is straightforward. The average length of the coalescent tree is

$$\text{E}\,\mathcal{L}_n = \sum_{k=2}^{n} k\,\text{E}\,T_k = \sum_{k=2}^{n} k\,\frac{4N}{k(k-1)} = 4N \sum_{k=1}^{n-1} \frac{1}{k}.$$

The expected number of segregating sites in a sample of n genes is thus given by

$$\text{E}\,\mathcal{S}_n \;=\; \theta \sum_{k=1}^{n-1} \frac{1}{k}. \tag{4.22}$$

---

[9]The summation means that the js are summed from zero to n in such a way that their sum equals i. The alternative and shorter way to write this is to write one sum of $j_2, j_3, \ldots, j_n$ from 0 to i with the condition $j_2 + j_3 + \cdots + j_n = i$.

For $n = 2$ this is the expected value θ of the geometric distribution (4.20). Thus, we expect the final waiting time T_2 to provide a large contribution to the number of segregating sites, but even though the first waiting times $T_n, T_{n-1}, \ldots$ are very short in a large sample, they cover many gene generations. The result is that $\mathrm{E} \, \mathcal{L}_n$ increases without bounds as the sample size n increases. The length of the ∞-coalescent is therefore infinite, and unless the sample is very small, a large fraction of the total length is situated early in the process.[10] Early mutations are guaranteed to define alleles of low frequency in the sample, so in a sufficiently large sample we expect to observe many rare alleles.

Watterson (1975) suggested an estimator of θ for the infinite alleles model:

$$\hat{\theta}_W = \frac{\mathcal{S}_n}{\sum_{k=1}^{n-1} \frac{1}{k}} \,,$$

and from equation (4.22) we have $\mathrm{E} \, \hat{\theta}_W = \theta$. An alternative estimator suggested by Tajima (1983) is based on pairwise comparisons, that is on $\mathcal{S}_2(ij)$, $i, j = 1, 2, \ldots, n$, the number of mismatches between sequences i and j. $\binom{n}{2}$ such comparisons can be made, and Tajima suggested the estimator

$$\hat{\theta}_T = \frac{2}{n(n-1)} \sum_{i=1}^{n} \sum_{j=i+1}^{n} \mathcal{S}_2(ij) \,,$$

which again has the mean θ. Both estimators are easy to calculate, neither use all information present in data, and they are both inferior to the maximum-likelihood estimator (see Hein et al. 2005). Tajima (1989a, 1989b) noted, however, that the two estimators react differently if there are deviations from the assumptions in Kingman's coalescent. Watterson's estimator just counts the number of SNPs seen in the n sequences, and the mutation events may be moved around in the coalescent tree without any effect. Tajima's estimator is sensitive to such perturbations. Suppose, for instance, that in a sample of $n = 20$ sequences $\mathcal{S}_n = 10$ SNPs are observed, giving $\hat{\theta}_W = 2.8$. If the 20 sequences are 10 identical pairs, and any two of these 10 pairs differ at one site, then $\hat{\theta}_T = 18/19 = 0.95$. On the other hand, if the sample contains two sequences differing at 10 sites and each is sampled 10 times, then $\hat{\theta}_T = 100/19 = 5.3$. Comparison of the two estimators may therefore be used to test whether or not the assumptions in Kingman's coalescent are compatible with observed variation. Exponential growth of the population size makes recent branches of the coalescent tree longer and older ones shorter, causing a majority of mutations to be recent. The sequences sampled thus resemble those considered in the second example, and indeed, Tajima's estimator is expected to be larger than Watterson's (Merriwether et al. 1991). Exponential decline of the population makes recent branches comparatively short, producing a sample resembling that in the first example, and the expectation is in general $\hat{\theta}_T < \hat{\theta}_W$.

---

[10]Remember that "early" in this connection means in the recent past.

4.7   Age of an Allele

The alleles in a sample emerged at different times in the past, and a classical question in population genetics is: What is the relation between gene frequency and age? Assume in our token coalescent with $n = 12$ in Figure 2.7 (page 35)

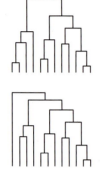

that precisely one mutation occurred after the common ancestor of the genealogy. The sample thus shows two alleles, the ancestral A and the mutation a. A very recent mutation would give $q = \frac{1}{12}$, and an old one $q = \frac{1}{3}$ or $q = \frac{2}{3}$.

A recent allele therefore seems to be associated with a low frequency in this genealogy, but a rare allele observed only once may be the oldest if the mutated allele coalesces with the ancestral alleles of the remaining $n-1$ alleles. We may designate individual 1 in our sample of $n = 12$ to play that role, and an allele observed with gene frequency $\frac{1}{12}$ may indeed be the ancestral allele.

The probability that an allele with frequency p is the ancestral allele in the population is p (Kimura and Ohta 1973, Watterson 1976). Donnelly (1986) found the probability that i genes of the oldest allele in the sample ($i = 1, 2, \ldots, n$) are observed:

$$\frac{\theta}{n}\binom{n}{i}\binom{n + \theta - 1}{i}^{-1},$$

assuming the infinite alleles model. This probability of course equals $F^{(n)}$ for $i = n$ $\big($see equation (4.13) on page 89$\big)$. To return to our example, the probability that the oldest allele occurs once is

$$\frac{\theta}{n - 1 + \theta},$$

which equals the probability that the last sampled gene is of an allele type different from that of the first $n-1$ genes $\big($see equation (4.15)$\big)$. The probability that the oldest allele occurs $n - 1$ times is $\theta F^{(n)}$, which can be viewed as the probability that one among n sampled genes mutates within a characteristic time unit (which incidentally also is the average time to coalescence of the last two genes in the sample genealogy).

The probability of sampling one gene of the oldest allele type is greater than that of the entire sample being ancestral alleles when $\theta > 1$, and the opposite is true when $\theta < 1$. In fact, when $\theta > 1$, the probability of i genes of the oldest allele decreases for $i = 1, 2, \ldots, n$, when $\theta < 1$ they increase, and for $\theta = 1$ the distribution is uniform, with the probability equal to $\frac{1}{12} = 0.08$ for $i = 1, 2, \ldots, n$ (Figure 4.10). This result is expected from our interpretation of the distributions in Figure 4.2 given by Wright's formula. For θ small, monomorphism is the most common state of the population, and if variants do occur they are usually recent mutations. For θ large, polymorphism is the rule, and though it takes

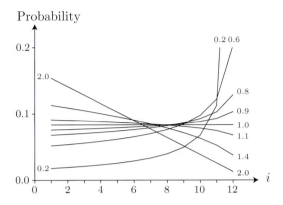

Figure 4.10: The distribution of the number i of the oldest allele in a sample of 12 sequences for various θ between 0.2 and 2. The value of θ is given on the curves.

a while for an allele to drift to appreciable frequencies, it takes a comparable amount of time for it to be lost again. The balance between these processes of introduction and loss of an allele is apparently that the oldest allele has a low frequency. Referring, however, to Ewens' (1972) results, as illustrated in Figure 4.8 on page 98, the conclusion is, rather, that the oldest allele is not always the most common one in the sample. Thus, the typical coalescent for θ small looks like that in the figure. For θ larger than one the coalescent is more like that in Figure 4.3, where genes 5–9 are ancestral.

4.8 Numerical Simulations

Simulation studies of evolution by random genetic drift are quite straightforward in the classical Wright–Fisher model: the offspring population is formed by one random draw from the binomial distribution with the gene frequency of the parent population. However, when mutation is allowed such studies become quite tedious. We need to get an impression of the properties of variation in the population, that is, we want to describe a "typical" population, and we rarely know anything about the history of a population. The initial state of a simulation has long-term influence on the calculated pattern of variation in the population. This is unfortunate unless it is chosen to reflect a biological phenomenon. When we do not want the initial condition of the population to reflect on the results, it should be determined by a draw from the stationary distribution of the process being studied. This stationary distribution has to be determined, and that usually means making numerous simulations for a lot of generations—often an inconvenient prerequisite. An alternative way to avoid the effect of start values on the results is to run simulations for many generations (at least $t > 4N$ and $t > \mu^{-1}$ generations) to reach steady state and find an initial condition for the "real" simulation job. The reference for the simulation studies is current natural populations, and due to our ignorance about their evolutionary genetic history, steady state seems to be the prudent assumption.

Retrospective analysis of samples provides a more limited and well-defined time scope. The coalescent process gives an easy and straightforward way to study the appearance of extant populations, and due to our ignorance about the size of the sampled population, time is conveniently measured in units of $2N$ generations, that is, in characteristic time. A random realization of the n-coalescent is produced by a simulation where the $n-1$ waiting times to coalescent events are chosen by drawing randomly from the appropriate geometric or exponential distributions. The $n-1$ coalescent events are then fixed in time, and the total coalescence time is specified. Finally, we need to fix the coalescent events. At the first coalescence time two genes among the n sampled genes are randomly chosen and coalesced. This is continued for the $n-1$ ancestral genes at the next coalescence time, and so on until the ancestors of the sampled genes coalesce into one gene after the total coalescence time. The end result is a coalescent tree specifying the genealogy of the sample.

Both the infinite sites and the infinite alleles model of mutation are easily implemented on this tree. The number of mutants is determined by a random draw from a Poisson distribution with mean $\theta\ell$, where ℓ is the total length of the tree in characteristic time, and these mutants are then distributed at random over the tree. It may suffice to fix mutants to the branches of the tree, and the number of mutants on a branch is a random draw from a Poisson distribution with mean $\theta\ell_b$, where ℓ_b is the length of the branch.[11] Alternatively, the total number of mutants may be placed on the branches by a random draw from a multinomial distribution with probability parameters ℓ_b/ℓ (see Appendix A.1.5 on page 355). A sample from the infinite alleles model is then obtained by following each gene back through the coalescent to the first mutational event, which in turn determines its allelic type. To produce a sample from the infinite sites model, a new site is mutated at each mutational position on the tree.

To frame a sample from the infinite sites model into a sample of DNA sequences showing SNP variation, the mutations may be placed at randomly chosen sites in the sequence—chosen "without replacement" to make sure that each site mutates only once. This simple way of placing the mutants corresponds to the Jukes–Cantor model (Table 4.5 on page 92). Further discussions of simulations based on the coalescent are given by Hudson (1991), and simulation tools using these principles are found in Christensen et al. (2006).

Finally, simulations based on the Wright–Fisher model may be used to construct the coalescent of a sample. Starting from a steady state population the simulation proceeds for many generations (a multiple of $4N$), keeping track of identity relations among the genes. At the chosen close of the simulation a sample of n genes is taken, and its coalescent is reconstructed from the record of identity relations.

---

[11]The sum of these random draws for each branch is a random draw from a Poisson distribution with mean $\theta\ell$, giving all the mutations in the tree (see Appendix A.1.5 on page 355).

Exercises

Exercise 4.2.1 Show that $\hat{p}$ in equation (4.2) is an equilibrium of the recurrence equation (4.1).

Exercise 4.2.2 Show that equation (4.3) provides the solution to the recurrence equation (4.1).

Exercise 4.5.1 Draw a picture showing the relationship between the average number of mutants per generation $2N\mu$ and the probability that two randomly drawn genes from the gamete pool are of the same allelic type. Draw a similar picture for the heterozygosity.

Exercise 4.5.2 What is the maximal heterozygosity in a population with two alleles? —with three alleles? —with k alleles?

Chapter 5

Migration

The genetic variation we find within a species is not necessarily present in local populations. Individuals in a local population tend to have more similar geno-types, and the remaining variation is manifested as differences in the genetic composition of the various populations of the species. An example of variation at different levels can be found in morphological characters in the fish *Zoarces viviparus*, which was subject to one of the first population genetic investigations by Johannes Schmidt (1917). The number of vertebrae was counted in samples of individuals from locations along the coasts of northern Europe. The observed distributions of individual counts at four locations in Mariager Fjord, Denmark, is shown in Figure 5.1. The distributions have the characteristic bell-shaped form found for many quantifiable morphological characters, so-called quantitative characters (Chapter 7). The locations closest to the head of the fjord show low counts with rather similar distributions, whereas the sample from the mouth of the fjord clearly shows higher average count. In the middle of the fjord Schmidt analyzed a sample that revealed a local population that appeared to be a mixture of fish from the populations at the head and mouth of the fjord. The majority of individuals caught at Hadsund resemble those caught in the innermost populations, but an excess of high counts forming a shoulder

Figure 5.1: Variation in vertebra count of *Zoarces viviparus* in four locations in Mariager Fjord, Denmark. For each sample a histogram of the counts indicates the location in the fjord (the geographical scale is not linear). The dashed line shows a count of 110 vertebrae. Data from Schmidt (1917).

and a tail on the distribution is probably caused by immigrants from the mouth of the fjord. Thus, geographical variation among local populations exists, and individuals seem to move between locations.

The variation is heritable in that counts of offspring vertebrae resemble that of the mother.[1] The character typically shows a correlation within a population of about 0.4 between the traits of mothers and their offspring (Smith 1921). The genetic basis of this is not known, but the character is probably influenced by allelic variation in many genes (we return to the discussion of such variation in Chapter 7). To assess the impact of migration on patterns like those in Figure 5.1, we study the general genetic effects of migration among populations.

5.1 Population Genetics of Migration

The genetic impact of migration is given as the fraction of immigrants among individuals in the local population that by breeding form the next generation. In the long run offspring of emigrants may however return to their population of origin, but with reference to the discussions in the previous three chapters, we will focus on the changes in the local population. Migration is thus viewed as genetic interaction among the subdivisions of a distributed population. Initially we reason about a very simple model of the effects of recurrent immigration, namely the classical island model, where immigrants are assumed to originate from a very large population, the mainland population, which is undisturbed by the presence of the island population—the focus of interest. This population is described by the Wright–Fisher model, and the influence of migration is described as a disturbance that corresponds to 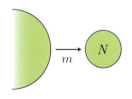 the way in which we described the influence of mutation. To underscore this parallel we make the simplifying assumption that the effects of mutation can be ignored. Breeding is thus described by the gamete pool model, and by adding the assumption of random union of gametes, we may consider it as describing a diploid population.

Like mutation, low levels of immigration may be seen as a disturbance of an isolated population—an *introgression* of foreign genetic material. To describe the development of the island population, the mutation frequency μ is simply replaced by the immigration frequency m, and our analysis of the effects of mutation is a good description of the effects of low frequencies of immigration. For instance, equation (4.16) on page 93 becomes

$$F_t = (1 - m)^2 \left(\frac{1}{2N} + \left(1 - \frac{1}{2N} \right) F_{t-1} \right), \tag{5.1}$$

which iterates to the solution (4.17). The assumption of the same population

---

[1]The fish is viviparous, and the number of vertebrae can be counted in the foeti carried by the mother.

size N generation after generation amounts to a simplification just like that in the mutation model of equation (4.16), and variations around a stationary value may of course be allowed by using the effective population size. The deterministic change in gene frequency of allele A between generations becomes

$$p' = (1 - m)p + m\hat{p}, \tag{5.2}$$

where $\hat{p}$ is the gene frequency on the mainland, and iteration of this produces an equation

$$p^{(t)} = \hat{p} + (1 - m)^t (p - \hat{p})$$

similar to equation (4.3). In a very large population we have, as expected, $p^{(t)} \to \hat{p}$ for $t \to \infty$, but our main interest is in the balance between random genetic drift and migration. The analysis of the variance in gene frequency with mutation thus carries to this simple island model, and the results are

$$F_t \to \frac{1}{1 + 4Nm} \quad \text{and} \quad \text{Var}(p_t) \to \frac{\hat{p}\hat{q}}{1 + 4Nm} \quad \text{for } t \to \infty, \tag{5.3}$$

where we approximated the limits by assuming that the population size N is large and the migration rate m is small.

The Wright–Fisher model with mutation and the classical island model are similar and thus yield similar diffusion equations. The steady state solution is thus again given by the Beta distribution in Wright's formula (4.6):

$$\hat{\phi}(p, y) = \frac{\Gamma(4Nm)}{\Gamma(4Nm\hat{p})\Gamma(4Nm\hat{q})} \, y^{4Nm\hat{p}-1}(1 - y)^{4Nm\hat{q}-1} \tag{5.4}$$

(Wright 1931). Figure 5.2 gives two examples, with the gene frequency in the mainland population $\hat{p} = \frac{1}{2}$ and $\hat{p} = \frac{1}{5}$, respectively.

Figure 5.2 shows the same probability distributions as those for mutation in Figure 4.2 on page 81, but they convey a different message. Wright's island model may again aid our interpretation of the stationary distributions. The most immediate possibility is that immigrants are introduced from the mainland population into each of the gamete pools in the island populations, just like mutants. We may use a more realistic reference model, however. The gene frequency in the total of the populations in the original model, Wright's island model without mutation, is conserved. Hence we can use the assembly of populations as the constant source of immigrants.

Wright's island model can therefore be extended to model a simple distributed population. A full migration model could be that the same number of emigrants leave each population, are collected in a pool of migrants, mixed, and then redistributed evenly among the island populations. The important property is that the immigrants contribute a fraction m to the gamete pool in each population (Figure 5.3). The gene frequency in the gamete pool of a chosen population, say, number i, is thus given by

$$p'_i = (1 - m)p_i + m\hat{p}$$

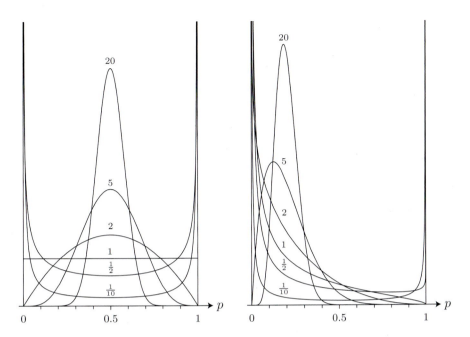

Figure 5.2: Distribution of gene frequencies in the classical island model with immigrants from a mainland population with gene frequency $\hat{p}$. The left-hand graphic assumes $\hat{p} = \frac{1}{2}$ and the right-hand graphic assumes $\hat{p} = \frac{1}{5}$. The values of $2Nm$ are 20, 5, 2, 1, $\frac{1}{2}$, and $\frac{1}{10}$.

in which $\hat{p}$ is the gene frequency in the entire population. The drawing on the left in Figure 5.2 shows that in a Wright's island model with migration and $\hat{p} = \frac{1}{2}$, the gene frequencies in the islands cluster around the mean in the archipelago when $2Nm \gg 1$. For $2Nm \ll 1$ most islands show gene frequencies at or near 0 or 1, and occasional islands with polymorphic gene frequencies exist. With this interpretation $2Nm$ determines the amount of geographical variation in gene frequencies.

The classical island model focuses on genetic variation in the local population, and from this point of view the parallel between mutation and migration becomes reasonable. The effect of random genetic drift in a small population is loss of genetic variation. Mutation and migration both introduce variation into the population, the only difference being the source of variation. Mutation creates genetic variants *de novo*, and the steady state result describes the amount of genetic variation that can be retained in the population. Migration introduces variation already present in surrounding conspecific populations, and the steady state describes the part of the prevailing variation that we can expect to see in the local population. Thus, the steady state describes the appearance of the local population as compared to a description of genetic variation over a broader area.

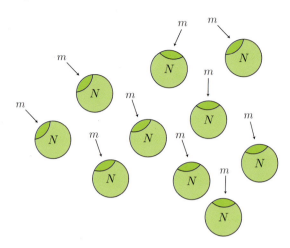

Figure 5.3: Wright's island model with migration.

In investigations of small local populations, a typical situation is that mutation occurs at a considerably lower frequency than migration, and that the number of mutants per generation in the local population is small. The local population is therefore expected to be monomorphic most of the time, if it is isolated from surrounding populations. However, the regional population may be sufficiently large for mutation to maintain polymorphism. In terms of the infinite alleles model, this would predict diminished genetic variation in the local population and genetic differentiation among local populations. We will return to the description of the contrast between local and global variation in Section 5.3.1 after some more detailed discussions of genetic variation in a structured population, but first we shall consider the coalescence structure of a sample from such a population.

5.1.1   Genealogy in a structured population

The difference between mutation and migration becomes even more pronounced when we examine the coalescent structure of a sample, and we need only study the simplest situation: A sample of n genes is taken from a population that exchanges migrants with a neighboring population. This refers to a two-island model, where reproduction in both populations is supposed to follow the gamete pool model. The sampled population consists of N breeding individuals and the neighbor has population size N'. The probability that a gene in the sample finds its parental gene in the other population is m, and the probability of a local parent is thus $1 - m$ (Figure 5.4). The probability that a gene in the sample has a foreign parental gene is about nm, if we assume that m is sufficiently small to ignore the possibility of more than one migration event in a given generation. The probability of seeing an *immigration event* when following the ancestry of a sample of n one generation back is therefore nm. Only a coalescent event

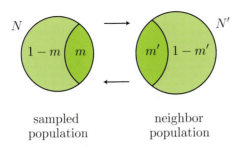

sampled   neighbor
population   population

Figure 5.4: Two populations with migration.

can disturb this description, and the probability that a coalescent event or an immigration event happens when going back to the parents of these n genes is

$$p_n + nm = \binom{n}{2}\frac{1}{2N} + nm,$$

when both the coalescent probability p_n and the immigration probability are sufficiently small to disregard the possibility of concurrent events. So far, the argument is the same as the one we used in describing the effect of mutation (pages 87–89 in Section 4.4), and the probabilities of the two possible events ending the wait are

$$\text{Prob(coalescent event occurs first)} \approx \frac{n-1}{n-1+4Nm} \quad \text{and}$$

$$\text{Prob(migration event occurs first)} \approx \frac{4Nm}{n-1+4Nm}.$$

In order to include the description of variation we ought to allow for mutation, and the event probability then becomes $p_n + nm + n\mu$. The probability of either of the events is again given by the relative probabilities. For instance, the probability that immigration occurs first is

$$\frac{4Nm}{n-1+4Nm+4N\mu}.$$

Mutation, however, does not influence the coalescence of the sample when the reference is observations of nucleotide sequences. The mutations may even be ignored during the description of the genealogy and subsequently tossed at random onto the genealogy. Thus, we initially concentrate on the study of the effects of migration without reference to sequence variation and mutation. We shall see that they have a more profound influence on the genealogy than those of mutation.

If the first event is a coalescence, say, at τ_1 in Figure 5.5, then we just keep waiting, and the only difference is that we have a sample of $n-1$ genes from a population. The situation therefore resembles the wait after the first coalescent

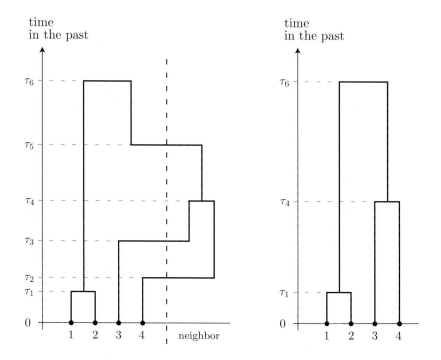

Figure 5.5: The coalescent of a sample of four genes in a population where rare events of migration occur to and from a neighbor population. The coalescent tree is shown to the right, and the left diagram shows the genealogy of the sample in time and space.

event in the mutation process, and the subsequent probability of a coalescence or immigration event is

$$p_{n-1} + (n-1)m = \binom{n-1}{2}\frac{1}{2N} + (n-1)m.$$

If the first event to take place after this wait is immigration, say, the event at τ_2 in Figure 5.5, then the ancestral material of the sample is split between the two populations, with $n-2$ genes in the sampled population and one gene in the neighbor. The population structure thus becomes apparent in the genealogy of the sample. When comparing these genes to their parental genes three events may be disclosed. Analysis of the $n-2$ genes may detect coalescence or immigration just as before, but now with the probabilities p_{n-2} and $(n-2)m$, respectively. The single gene in the neighbor population may descend from a parental gene either in that population or in the sampled population, that is, its ancestry shows immigration from the sampled population. Thus, the probability of an event is

$$p_{n-2} + (n-2)m + m' = \binom{n-2}{2}\frac{1}{2N} + (n-2)m + m',$$

where m' is the probability that a gene in the neighbor has a parent in the sampled population. We assumed, of course, that m' is small enough to allow us to ignore multiple events within one transition from offspring to parents. The probability that one of the three events occurs determines the distribution of the waiting time until the next event, and the probability that the next event is a particular one equals the relative probability of that event. For instance, the probability of an immigration event in the sampled population is

$$\frac{(n-2)m}{p_{n-2} + (n-2)m + m'}.$$

One additional kind of event must be considered when describing the genealogy of our sample. If the third event is also an immigration event in the sampled population, then the ancestral material of the sample is made up of $n-3$ genes in the sampled population and two genes in the neighbor (as after the event at τ_3 in Figure 5.5). Thus, in the following generations a coalescent event is possible in the neighbor population (like the event at τ_4 in Figure 5.5), and the per generation probability of an event becomes

$$\binom{n-3}{2}\frac{1}{2N} + (n-3)m + 2m' + \binom{2}{2}\frac{1}{2N'}.$$

The probability that any one of these four events happens is given by the relative event probabilities. For instance, the probability of a coalescent event in the neighbor population is

$$\frac{p_2'}{p_{n-3} + (n-3)m + m' + p_2'}.$$

This process continues. In the example in Figure 5.5 the final coalescent event occurs at τ_6 in the sampled population, but if the event at τ_5 had instead been an immigration event in the neighbor population, then the sample could have found its most recent ancestor in the neighbor.

The coalescent events in Figure 5.5 are τ_1, τ_4, and τ_6, and the corresponding coalescent tree is shown on the right side of the figure. If the sequences of the four genes are known, then the pattern of mutations may reveal aspects of the coalescent through an analysis like the one we used in Section 4.3 on page 84. The only effect of the migration events τ_2, τ_3, and τ_5 is to alter the distribution of the waiting times between coalescent events, and to delimit the number of genes into which a gene can coalesce. Thus, the further analysis of the coalescent in a structured population should focus on a comparison of the coalescent in a structured population and Kingman's coalescent.

5.1.2 The final waiting time

Few theoretical results exist for the coalescent in a structured population, and most of our knowledge of the process comes from computer simulations. Some very illustrative approximate results exist, however, but before discussing this, a simple exact result for a sample of two genes will be developed. Takahata (1988) and Nath and Griffiths (1993) consider such a sample in a simple model of a population with a neighbor, and their closer scrutiny of the final waiting time in the coalescent of the sample of n genes discussed above provides a good illustration of migration effects on the coalescent. In the following we will address the "sampled" population as population number 1 and the "neighbor" population as number 2.

Depending on the location of the two sampled genes, one of three waiting times is relevant. These are the stochastic variables T_{20}, T_{11}, and T_{02} that 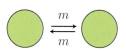 describe the wait when both genes are sampled in population 1, one in each population, and both in population 2. The simplification of the general two-population model above is that the populations are of equal size ($N' = N$), and that migration is symmetric ($m' = m$). Due to this symmetry of the populations we expect that $\mathrm{E}\,T_{02} = \mathrm{E}\,T_{20}$.

To calculate the mean $\mathrm{E}\,T_{20}$ we assume that both sampled genes initially are in population 1. The event probability is therefore equal to

$$\binom{2}{2}\frac{1}{2N} + 2m = \frac{1}{2N}(1 + 4Nm).$$

The probabilities that the first event is a coalescent or an immigration are the relative event probabilities

$$\frac{1}{1 + 4Nm} \quad \text{and} \quad \frac{4Nm}{1 + 4Nm}.$$

If we describe the waiting time to the first event by X, we can consider the average waiting time to a coalescent given the waiting time to the first event. This is

$$\mathrm{E}(T_{20}\,|\,X) = X + \left(\frac{1}{1 + 4Nm} \times 0 + \frac{4Nm}{1 + 4Nm} \times \mathrm{E}\,T_{11} \right).$$

The argument is straightforward. Wait the time X to the first event. If the first event is a coalescent, we need not wait any longer, but if the first event is an immigration, we shall expect a further wait equal to the time we would have waited if we initially had sampled one gene in each population. The average coalescence time is $\mathrm{E}\,T_{20} = \mathrm{E}\big(\mathrm{E}(T_{20}\,|\,X)\big)$ (equation (A.7) in Appendix A on page 349), and therefore we get

$$\mathrm{E}\,T_{20} = \frac{2N}{1 + 4Nm} + \frac{4Nm}{1 + 4Nm}\,\mathrm{E}\,T_{11}. \tag{5.5}$$

The mean $\mathrm{E}\,T_{11}$ is calculated by noting that the total event probability when one gene is sampled in each population equals $m + m$, because immigration may happen in either population. The probability that the first event is either of these equals $\frac{1}{2}$, and therefore

$$\mathrm{E}(T_{11}\,|\,X) = X + \tfrac{1}{2}\mathrm{E}\,T_{20} + \tfrac{1}{2}\mathrm{E}\,T_{02}.$$

Taking the average we get

$$\mathrm{E}\, T_{11} \;=\; \frac{1}{2m} + \tfrac{1}{2}\mathrm{E}\, T_{20} + \tfrac{1}{2}\mathrm{E}\, T_{02}.$$

Because of the symmetry of the two populations, the average coalescence time is

$$\mathrm{E}\, T_{11} \;=\; \frac{2N}{4Nm} + \mathrm{E}\, T_{20}, \tag{5.6}$$

where the numerator and denominator of the first term were multiplied by $2N$.

Equations (5.5) and (5.6) comprise a system of two linear equations with two unknowns, $\mathrm{E}\, T_{20}$ and $\mathrm{E}\, T_{11}$, and their solution is

$$\mathrm{E}\, T_{20} = 4N \quad \text{and} \quad \mathrm{E}\, T_{11} = 2N\frac{1+8Nm}{4Nm}. \tag{5.7}$$

This is a simple and straightforward result, but it is nonetheless surprising. The coalescence time of a sample of two genes from one population is independent of the immigration probability and exactly twice the coalescence time of Kingman's coalescent in a population of size N. The value makes sense if $4Nm$ is very large because migration is so frequent that the two populations fuse to one of size $2N$. Conversely, when $4Nm$ is very small there is a high probability that the two genes coalesce before an immigration event is detected. Thus, we would expect the waiting time to be very close to that in Kingman's coalescent in a population of size N. Indeed, Nath and Griffiths (1993) showed that the distribution of T_{20} converges to the waiting time distribution in Kingman's coalescent when $4Nm \to 0$, but the average coalescence time is still twice that of Kingman's coalescent.

What is wrong? Nothing, except perhaps our intuition that the expected value of a stochastic variable is a good but rough predictor of its typical value. Here we have a biologically relevant counterexample to this intuitive notion. This may be seen from equation (5.5). When $4Nm$ is very small the first term is $2N$, which is the expected value for Kingman's coalescent. The second term has a very small coefficient. Shouldn't we ignore that? The answer is no, because $\mathrm{E}\, T_{11}$ is inversely proportional to $4Nm$ and therefore very large, with the result that the second term becomes a bit *larger* than the first (see equation (5.7)).

The reason for the counterintuitive result is that for $4Nm$ very small, the distribution of T_{20} is very close to the distribution of T_2 in Kingman's coalescent, although there is a tiny probability that T_{20} becomes very large. Thus, the genetic variation in a population that we consider to be completely isolated would be described well by Kingman's coalescent, but we will get the occasional surprise of a very long coalescent time when inferring the genealogy of a sample on the basis of the observed variation.

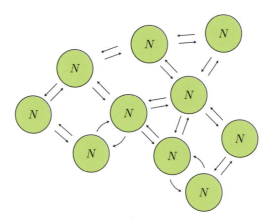

Figure 5.6: An island model with geographical structure.

5.2 Population Structure

Individuals of a population are spread over a habitat, maybe even in patches within a habitat, and though mating is independent of genotype, and therefore random, the probability that two randomly chosen individuals meet may depend on their location. Mating is therefore not strictly random in that it is not pan-mictic, and the geographical structure of the environment needs to be specified in the description of breeding in the population. The relevant structure is defined by the movement of individuals, and the genetic result is deduced from the composition of the local breeding population in terms of the geographic origin of its individual members (Figure 5.6). The effects of migration are in this sense very different from the disturbance caused by rare immigrants, even though the two views are only quantitative extremes on a continuum. Migration becomes the cohesive force in a population spread over a geographical range that is too wide to be covered by the range of individual movements.

The simplest model for investigating population structure partitions the population into d randomly mating subpopulations, and migration among them occurs every generation. This *d-island model* is a standard model of population genetics and was introduced in the original work of Haldane (1930) and Wright (1931). The model used in Section 5.1.1 is therefore a two-island model, and we will later investigate the finite version of Wright's island model called Wright's d-island model.

Geographical structure is described by the immigration to the local breeding population. The fraction of immigrants in subpopulation i that originated from population j is m_{ij}, $i, j = 1, 2, \ldots, d$, in that the fraction m_{ii} originates from subpopulation i. By this definition $m_{i1} + m_{i2} + \cdots + m_{id} = 1$. These migration frequencies describe where the individuals in the local population came from, and they are therefore referred to as the backward migration frequencies (Malécot 1948, Bodmer and Cavalli-Sforza 1968). The description of migration is collected in the matrix

$$M = \left\{ \begin{array}{cccc} m_{11} & m_{12} & \cdots & m_{1d} \\ m_{21} & m_{22} & \cdots & m_{2d} \\ \vdots & \vdots & \ddots & \vdots \\ m_{d1} & m_{d2} & \cdots & m_{dd} \end{array} \right\},$$

called the backward migration matrix, or in genetic contexts just the *migration matrix*. When applied to sexual organisms this description assumes that the result of migration is the same for the two sexes. This assumption is not restrictive when the focus is on neutral genetic variation (Box 16).

The backward migration matrix describes the result of migration. In ecology migration is usually described with reference to the movements of individuals as observed by, for instance, capture–recapture methods. This generates information on the forward migration frequencies $\tilde{m}_{ij}$, $i, j = 1, 2, \ldots, d$, where $\tilde{m}_{ij}$ is the frequency of individuals in population i that moved to population j ($\tilde{m}_{i1} + \tilde{m}_{i2} + \cdots + \tilde{m}_{id} = 1$). $\tilde{m}_{ii}$ is the fraction that did not migrate, or migrated but returned home to breed. The backward migration frequencies are simple functions of these when the relative sizes $\tilde{c}_i$, $i = 1, 2, \ldots, d$, of the subpopulations before migration are known ($\tilde{c}_1 + \tilde{c}_2 + \cdots + \tilde{c}_d = 1$). The relative size of subpopulation i at the time of breeding is c_i

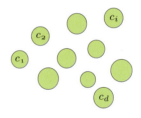

($c_1 + c_2 + \cdots + c_d = 1$), and this may be determined by the immigration in population i. A fraction $\tilde{m}_{ji}$ of the individuals in population j emigrates to population i, and therefore a fraction $\tilde{m}_{ji}\tilde{c}_j$ of the individuals in the entire population moves from subpopulation j to subpopulation i, that is, a fraction

$$c_i = \sum_{k=1}^{d} \tilde{m}_{ki}\tilde{c}_k$$

of the individuals in the population ends up in subpopulation i after migration. Among the breeding individuals in subpopulation i a fraction

$$m_{ij} = \frac{\tilde{m}_{ji}\tilde{c}_j}{c_i}$$

therefore came from subpopulation j. The forward and backward migration matrices are thus different in general, except in special circumstances where migration does not change the relative sizes of the populations, that is, when $\tilde{c}_i = c_i$ for all $i = 1, 2, \ldots, d$. This is the special case of *conservative migration*.

A finite version of Wright's island model may be formulated within this framework, but we present a slightly more general d-island model due to Deakin (1966). A fraction m of each subpopulation migrates, and a fraction $1 - m$ stays at home. The migrants enter a migrant pool and are returned at random to the subpopulations to replace the emigrants. Each migrant thus has the probability

---

Box 16: Sex-dependent migration

In higher organisms the migration patterns of the two sexes may be very different. For instance, in many parasitic wasps one sex is unable to fly: in the fig wasp *Blastophaga psenes* and in the parasitic wasp *Nasonia vitripennis* only females are able to fly. In many mammals one sex, often the male, is the predominant rover.

For autosomal genes in a diploid organism, males and females give the exact same contribution to the next generation: for every pair of homologous genes in every individual, one is of maternal and one of paternal origin. Thus, if the contribution to the gene pool is $m_{ij}^{\female}$ and $m_{ij}^{\male}$, then

$$m_{ij} = \tfrac{1}{2}\left(m_{ij}^{\female} + m_{ij}^{\male}\right)$$

is the contribution of immigrants from population j to the offspring population in population i.

Because of this simple mixing, sex heterogeneity in migration leaves no trace in evolution. Heterogeneity may only produce short-term sex differences for genes on the X chromosome. However, Y chromosome variation is only influenced by male migration because of its patrilinear transmission. Variation in mitochondrial genes show matrilinear transmission in mammals, and $\boldsymbol{M}^{\female}$ and $\boldsymbol{M}^{\male}$ may therefore be separated by studying variation on the Y chromosome and in the mitochondria (Seielstad et al. 1998, Stoneking 1998).

$\tilde{c}_i$ of ending up in population i. In terms of the forward migration matrix, this model is given by

$$\tilde{m}_{ii} = (1 - m) + m\tilde{c}_i \quad \text{and} \quad \tilde{m}_{ij} = m\tilde{c}_j, \; j \neq i. \tag{5.8}$$

This model, which we refer to as *Deakin's model*, has conservative migration. The backward migration matrix is therefore also given by an equation like equation (5.8).

Wright's d-island model is obtained by assuming equal population sizes $c_1 = c_2 = \cdots = c_d = \frac{1}{d}$, and the fraction of immigrants from population j in population i is then

$$m_{ij} = \frac{m}{d} \qquad \text{for } i, j = 1, 2, \ldots, d, \; i \neq j, \tag{5.9}$$

$$m_{ii} = 1 - m + \frac{m}{d} \quad \text{for } i = 1, 2, \ldots, d. \tag{5.10}$$

This is in agreement with the original Wright's island model, where a fraction of the emigrants return to their place of origin.

The effect of migration on the population dynamics in the local population is given in terms of forward migration, that is, in terms of the local population sizes $c_i N_T$, $i = 1, 2, \ldots, d$, after migration. We cannot assume similar simplifications in ecological models that correspond to our population genetic models with discrete nonoverlapping generations. Demographic and ecological processes occur continuously throughout the life of the organisms, and the relevant stages for the regulation of the population size need not be the stages that are relevant for the genetic description of the local population (see Christiansen 2004). Thus, the difference between c_i and $\tilde{c}_i$ need not reflect any ecological properties of the local habitats. Rather, their ratios calibrate the evolutionary effects of the movements of individuals.

5.2.1 The simple two-island model

The simplest migration structure of the two-island model is given by the migration matrix

$$\left\{ \begin{array}{cc} 1-m & m \\ m & 1-m \end{array} \right\}$$

(used in Section 5.1.2). We include the possibility of mutation, assume the infinite alleles model, and only describe the development of gene identity. Two variables describing the relationship among genes chosen within and between the two populations will be used. F_t is the probability of choosing two identical genes from the gamete pool of either population in generation t, and G_t is the probability of choosing two identical genes from different gamete pools. If we assume 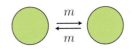 that no genes are identical at the outset, $F_0 = G_0 = 0$, then the subpopulations behave in the same way, and F_t and G_t suffice to describe the development of identity in the entire population. This model is formulated in the recurrence equations in Box 17.

When steady state is reached in the model, the ratio of the between-population gene identity and the within-population gene identity is given by

$$\hat{I} = \frac{\hat{G}}{\hat{F}} \approx \frac{m}{m+\mu} \tag{5.11}$$

when $\mu < m \ll 1$ (Nei and Feldman 1972). Thus, the amount of divergence depends on the relative sizes of the mutation rate and the migration rate. The ratio between $\hat{G}$ and $\hat{F}$ is always less than one, and when $m \gg \mu$ it is close to one and the two populations are very similar. When m is of the order μ, divergence is apparent between the populations, and for a very low migration rate similarities between the populations are hard to detect. Nei suggests using I as a measure of genetic similarity between populations (see Nei 1987) and the alternative of using $-\log I$ as a measure of genetic distance.

The convergence to the steady state can be illustrated by neglecting mutation (the rate is given in Box 17). This provides the rate of convergence to homozygosity, which may be compared to that in a randomly mating population with $2N$ individuals (the combined size of the two islands) and hence expressed as the effective population size (Figure 5.7). For $m = \frac{1}{2}$ the population is like one random mating population, and indeed we get $N_e = 2N$. This

---

Box 17: Gene identity structure of the two-island model

Two genes chosen from the gamete pool in one of the populations in generation t originate from the same gamete pool in generation $t-1$ with the probability $(1-m)^2 + m^2$; that is, both are either residents or immigrants in the population. The probability is $2m(1-m)$ that one is an immigrant and the other is a resident, and that they originate from different gamete pools. Thus, the two genes are identical by descent with the probability

$$F_t = \left((1-m)^2 + m^2\right)\left(\frac{1}{2N} + \left(1 - \frac{1}{2N}\right)F_{t-1}\right) + 2m(1-m)G_{t-1}.$$

Two genes chosen from different gamete pools in generation t have the probability $2m(1-m)$ of originating from the same gamete pool in generation $t-1$ and the probability $(1-m)^2 + m^2$ of originating from different gamete pools. Thus, the two genes are identical by descent with the probability

$$G_t = 2m(1-m)\left(\frac{1}{2N} + \left(1 - \frac{1}{2N}\right)F_{t-1}\right) + \left((1-m)^2 + m^2\right)G_{t-1}.$$

This model splits the population in the Wright–Fisher model without adding a source of variation. The result is therefore that $F_t \to 1$ and $G_t \to 1$ as $t \to \infty$. We can find the ultimate rate of convergence to homozygosity by using the method in the model on page 55 and express it in terms of the effective population size, N_e (Figure 5.7). For a low number of migrants ($4Nm < \frac{1}{5}$) we get a good approximation by

$$N_e \approx \frac{1}{4m(1 - 4Nm)}.$$

Thus, for very low migration rates, the rate of convergence to homozygosity is dominated by the migration frequency m and not the number of migrants $4Nm$.

---

rate is almost unchanged until m becomes rather small, but for very low immigration probabilities the effective population size increases without bounds, and the effective population size is certainly larger than the actual population size whenever $m < \frac{1}{2}$.

This result emphasizes that the effective population size is a reference to the dynamics in the Wright–Fisher model and not an expression of a virtual count of individuals. The convergence to homozygosity in a two-island model with low migration corresponds to the convergence in a very large gamete pool model. The populations do not behave as if they had more individuals. In this case, the interpretation in terms of a virtual number of individuals is obviously not productive for gaining biological insight into the process. Even in models

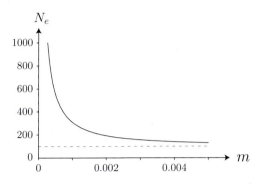

Figure 5.7: Convergence to homozygosity in the two-island model for $N = 50$. The dashed line shows the effective population size for $m = \frac{1}{2}$.

where an interpretation of the effective population size in terms of an effective number of individuals in the population seems straightforward, this example warns against a too naive interpretation of N_e.

The results are not restricted to two populations, but extend to Wright's d-island model (5.9–5.10), as shown by Latter (1973). When $\mu < m \ll 1$, Nei's genetic similarity becomes

$$\frac{\hat{G}}{\hat{F}} \approx \frac{(d-1)m}{(d-1)m + d\mu}. \tag{5.12}$$

The dependence on the number of subpopulations is weak. The genetic similarity increases as d increases. The extreme values are

$$\frac{m}{m+2\mu} \quad \text{for } d = 2 \quad \text{and} \quad \frac{m}{m+\mu} \quad \text{for } d \to \infty.$$

The first expression equals the genetic similarity in the simple two-island model because the migration parameter in that model equals twice the m in Wright's d-island model. Equation (5.11) therefore provides a description that is fairly robust with respect to the number of populations.

For low migration, each population behaves almost like a population in the Wright–Fisher model. Intermittently and separated by long time intervals, however, immigrants introduce extinct lineages. The convergence result therefore parallels the results that we reached for the coalescence process in Section 5.1.2. For low immigration probabilities, a sample of genes in one population usually coalesces within that population without an immigration event happening. A long wait is expected, however, before the ancestral material of a sample containing genes from both populations is present in one of the populations, allowing the final coalescence.

5.2.2   Coalescence in a widespread population

The analysis of the coalescent structure of a sample from a structured population is simple if we assume that the population consists of a lot of subpopulations with limited exchange of migrants. A simple example of such a model is obtained from Wright's d-island model,

$$m_{ii} = 1 - m + \frac{m}{d} \quad \text{and} \quad m_{ij} = \frac{m}{d}, \; j \neq i,$$

where d is very large, that is, $n \ll d$. Each population has the size N.

Suppose initially that the sample consists of one gene from each of n randomly chosen subpopulations—a common situation at a time when sequencing was expensive. When comparing the sample to its parental genes, the probability of detecting an immigration event is nm under the usual assumption that m is small enough to allow us to ignore the possibility of more than one event. However, most immigration events have no impact on the appearance of our sample because they trace the origin of the n genes to ancestral material, which is still n genes distributed in n populations. The only interesting events are therefore those that reveal an immigration event from a population that already contains another ancestral gene. Thus, the probability of an interesting immigration event is

$$nm \times \frac{n-1}{d-1} = \binom{n}{2} \frac{2m}{d-1} \, .$$

An interesting event is therefore like a coalescence, not of two sampled genes but of two subpopulations containing ancestral material. Given two genes in a population, three kinds of events can happen: a coalescent, immigration of one of the two genes, or an interesting immigration into the population. The probability of the last event is inversely proportional to the number d of subpopulations. As we assumed d to be very large, the process may be approximated by disregarding this possibility until after one of the first events has occurred. That is, we assume that a coalescent event or an immigration event occurs very fast compared to how long it takes for an interesting immigration event in the population to occur.

In a subpopulation with two ancestral genes the probabilities of a coalescent event and of an immigration event are approximately

$$\frac{1}{1 + 4Nm} \quad \text{and} \quad \frac{4Nm}{1 + 4Nm}$$

if we ignore terms of the order of $\frac{1}{d}$. These have the same characteristics that we found for the final waiting time in Section 5.1.2. If an immigration event occurs, the ancestral material of the sample is again distributed as n genes in n populations. Here we ignore the possibility that the immigration event creates another population with two ancestral genes, because the probability of such an event is of the order of $\frac{1}{d}$. If a coalescent does happen, then the process reaches a new state where the sample is represented by $n - 1$ ancestral genes in $n - 1$

populations. To this level of approximation the probability that an immigration event results in a coalescent is therefore

$$\frac{1}{1 + 4Nm},$$

and the length of time with two ancestral genes in one population is on average

$$\frac{2N}{1 + 4Nm}$$

generations.

If this waiting time is short compared to that to an interesting immigration event in the population, then the probability of a coalescent event within a time interval is given approximately by

$$\binom{n}{2} \frac{2m}{d-1} \frac{1}{1 + 4Nm} = \binom{n}{2} \left(\frac{1}{2N(d-1)} \frac{4Nm}{1 + 4Nm} \right).$$

In Kingman's coalescent the corresponding probability is

$$\binom{n}{2} \frac{1}{2N}.$$

We can therefore describe the coalescent process as a Kingman coalescent in a population of size N_e with

$$\frac{1}{2N_e} \approx \frac{1}{2N(d-1)} \frac{4Nm}{1 + 4Nm}.$$

We then get the result

$$N_e \approx Nd \left(1 + \frac{1}{4Nm} \right) \tag{5.13}$$

(Wakeley (1999) showed this equation by an argument that involves a so-called compound distribution, see Box 18). Thus, the coalescent in a sample of at most one gene from each subpopulation in Wright's d-island model for d large is indistinguishable from the coalescent in a Wright–Fisher model of size N_e.

The above analysis is due to Wakeley (1999). He carried the argument further to describe a sample where several genes may come from a single population, such that coalescence can occur without an initial wait for an immigration event. This is the general coalescence in Wright's d-island model. The waiting time to an immigration event is still very large compared to the waiting time to a coalescent event. In the beginning coalescence is therefore abundant and immigration events very rare. After a while, the result is that the ancestral genes of the sample are scattered as single genes in subpopulations—the situation we discussed above. Wakeley calls the initial stage the *scattering phase* because its end result is scattered genes. Few, if any, immigration events are seen in this phase, and Wakeley provides a description of the distribution of the number

---

Box 18: * A compound exponential distribution

The waiting time to an interesting immigration event in the Wakeley (1999) model is geometrically distributed with a small event probability. We can therefore consider it as exponentially distributed with parameter, say, λ. The probability of a coalescent after such an event is π, say, and the probability that we get the first coalescent after i such events is therefore $\pi(1-\pi)^{i-1}$. Given that the coalescence occurs after i events, the waiting time to coalescence is distributed as the sum of i exponential waiting times. This sum is Gamma distributed $\Gamma(i,\lambda)$ (see page 361). The unconditioned density of the waiting time to the first coalescent event is then

$$
\begin{aligned}
\sum_{i=1}^{\infty} f(x\,|\,i)\,\pi(1-\pi)^{i-1} &= \sum_{i=1}^{\infty} \frac{\lambda^i}{\Gamma(i)}\,x^{i-1}\,e^{-\lambda x}\,\pi(1-\pi)^{i-1} \\
&= \lambda\pi e^{-\lambda x}\sum_{i=1}^{\infty} \frac{\left(\lambda(1-\pi)x\right)^{i-1}}{(i-1)!} \\
&= \lambda\pi e^{-\lambda x}e^{\lambda(1-\pi)x}\sum_{i=1}^{\infty} \frac{\left(\lambda(1-\pi)x\right)^{i-1}}{(i-1)!}e^{-\lambda(1-\pi)x} \\
&= \lambda\pi e^{-\lambda\pi x},
\end{aligned}
$$

because the sum is a sum of the probabilities in a Poisson distribution. The waiting time to a coalescent is thus exponentially distributed with event parameter $\lambda\pi$. In our application we have

$$
\lambda = \binom{n}{2}\frac{m}{d-1} \quad \text{and} \quad \pi = \frac{1}{1+4Nm}
$$

and the exponential distribution of the waiting time therefore has the event parameter

$$
\binom{n}{2}\frac{2m}{d-1}\frac{1}{1+4Nm} = \binom{n}{2}\left(\frac{1}{2N(d-1)}\frac{4Nm}{1+4Nm}\right).
$$

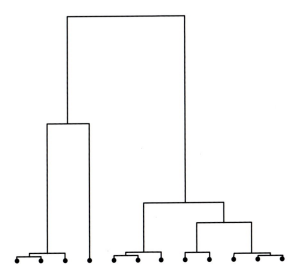

Figure 5.8: A coalescent in Wright's d-island model. The example of a coalescent tree in Figure 2.7 is modified to illustrate the scattering and collecting phases.

of genes entering the final stage of the coalescent process called the *collecting phase*.

The requirement that the number of subpopulations d is large does not seem to be very serious. Simulations by Wakeley (1998) suggest that $d > 4n$ is quite sufficient for the results to be applicable in the collecting phase.

The description of sequence variation in a species with a widespread distribution may be discussed in terms not only of the number n of sequences sampled, but also of how the samples should be distributed geographically. Thus we need to toss mutations onto the coalescent in Wright's d-island model. For argument's sake consider Figure 5.8, which shows a modification of the coalescent tree in Figure 2.7 (page 35) to reflect Wakeley's results (a modification causing $\hat{\theta}_T < \hat{\theta}_W$, see page 103). The scattering phase is short compared to the collecting phase, and even though the number of ancestors of our sampled sequences during the scattering phase is large, the total length of the coalescence tree (the gene×time sum) is dominated by the contribution in the collecting phase. Sampling that omits the scattering phase is therefore preferable, and sampling one sequence from each geographically distinct region of the species distribution therefore appears to be optimal. The remaining question is then to pinpoint such geographically distinct regions—a question our theoretical population genetic results cannot resolve. For instance, to squeeze the human population into a Wright's d-island model to design an investigation of variation surely seems ludicrous—the criterion is the distribution of the population, not the structure of models we can handle. The experience from considering Wright's d-island model, however, is a guide to the properties that should be emphasized, namely mutually isolated parts of the population, that is, parts that rarely exchange migrants. In the human population those isolated parts correspond to the classical anthropological ethnic groups—self-contained human populations with distinct morphological and cultural characteristics occupying a well-defined geographical area. These should clearly form the basis of a description of genetic variation in the human population (Cavalli-Sforza et al. 1994), because a sample of DNA sequences within such a population is often likely to coalesce within the population. Of course, the historical and prehistorical movements and expansions of human populations complicate the designation of original ethnic populations, but ignoring these and spreading the sampling efforts evenly over inhabited areas would overemphasize the contribution of genetic variants in recently spreading populations, such as the Bantu in Africa. Geographical foci of divergence are African and New Guinean tribes, while large populations of Europe and Asia are quite homogeneous.

5.2.3   Strong migration

When immigration is frequent, the effect of population structure disappears. This is already indicated in equation (5.13), where the effective population size during the collecting phase becomes equal to the total population size Nd for the immigration frequency m large. In two populations with symmetric migration, for instance, the waiting times for a sample of two genes to coalesce are independent of the sample configuration in that $ET_{11} \approx ET_{20} = 4N$ when $4Nm$ is large (see equation (5.7) on page 117). The coalescence time is thus about the same whether the sample contains genes from just one or from two populations.

Similar results hold for island models in general, as long as the models describe connected populations. An assembly of populations is connected when a gene can have ancestors in any subpopulation regardless of where it is sampled. The local population sizes are N_i, the total population size is

$$N_T = \sum_{i=1}^{d} N_i,$$

and the migration matrix is $\boldsymbol{M}$. The total number of immigrant genes in subpopulation i is expected to be $2N_i(1 - m_{ii})$, $i = 1, 2, \ldots, d$, each generation. If these numbers are large, we can assume that a lot of immigration events happen between coalescent events in a small sample of n genes. The result is that the ancestors of every gene in the sample have become dispersed and occur in a randomly chosen subpopulation, and we only have to specify what we mean by "randomly chosen subpopulation." This is the so-called strong migration limit (Nagylaki 1980).

In Box 19 it is argued that the probability distribution of the location of a gene after many migration events is given by a stationary distribution characteristic for the migration matrix independently of where the gene started. The stationary distribution is $\hat{\boldsymbol{\pi}} = (\hat{\pi}_1, \hat{\pi}_2, \ldots, \hat{\pi}_d)$, where $\hat{\pi}_i$ is the probability of finding the gene in population i. Thus, the probability of a coalescent in

---

Box 19: Stationary distribution for migration

Consider a single gene in the population at a time where the probability of finding it in population i is π_i. Given that the gene is in population i, the probability of finding its ancestor in population j in the previous generation is m_{ij}. The unconditional probability is therefore

$$\pi'_j = \sum_{i=1}^{d} \pi_i m_{ij}.$$

This may be written as $\boldsymbol{\pi}' = \boldsymbol{\pi M}$, where $\boldsymbol{\pi} = (\pi_1, \pi_2, \ldots, \pi_d)$. The stationary distribution $\hat{\boldsymbol{\pi}}$ is unchanged from generation to generation, that is, $\hat{\boldsymbol{\pi}}' = \hat{\boldsymbol{\pi}}$, and therefore

$$\hat{\boldsymbol{\pi}} = \hat{\boldsymbol{\pi}} \boldsymbol{M}.$$

The stationary distribution is hence a left eigenvector of the eigenvalue 1 of the migration matrix. The convergence to the stationary distribution follows from the basic limit theorem for Markov chains (see Appendix A.2 on page 355).

---

population i is

$$\binom{n}{2} \frac{\hat{\pi}_i^2}{2N_i},$$

and the total probability of a coalescent event during one transition from off-spring to parents is therefore

$$\binom{n}{2} \sum_{i=1}^{d} \frac{\hat{\pi}_i^2}{2N_i}.$$

The coalescence probability is the same as in Kingman's coalescent in a population of size

$$N_e = \left(\sum_{i=1}^{d} \frac{\hat{\pi}_i^2}{N_i} \right)^{-1}, \tag{5.14}$$

and the coalescent process is therefore indistinguishable from the process in a randomly mating population of size N_e.

In Wright's d-island model $\hat{\boldsymbol{\pi}} = (\frac{1}{d}, \frac{1}{d}, \ldots, \frac{1}{d})$, and for $4Nm$ large the population acts like a single population of size $N_e = dN = N_T$. Any other conservative migration matrix (see page 119) with all subpopulations of equal size also has $\hat{\boldsymbol{\pi}} = (\frac{1}{d}, \frac{1}{d}, \ldots, \frac{1}{d})$, and therefore $N_e = dN = N_T$. In general, $N_e = N_T$ when the size of the local population is proportional to the probability of finding a specified gene in the population, that is, $N_i = \hat{\pi}_i N_T$.

5.2.4 The stepping-stone model

Wright's island model gives a first impression of the effect of geographical substructure in a population, but it is only adequate in special circumstances to describe the effects of structure in natural populations. The model is inherently without spacial structure because any population is subject to the same level of immigration from all other populations. It may, for instance, represent the exchange of migrants among breeding populations in species with long-distance seasonal migration. Geographical structure introduces the possibility of low genetic exchange between distant populations, the *isolation by distance* effect also studied by Wright (1943). An extreme example is when only close neighbors exchange migrants (Figure 5.6 on page 118).

A simple population model for studying the effects of isolation by distance is the stepping-stone model, so named by Motoo Kimura, who was inspired by the stepping stones in Japanese gardens. Each subpopulation is of size N and

it exchanges migrants only with its two immediate neighbors. The chain of populations is infinitely long, $i = 0, \pm 1, \pm 2, \ldots$, and in this respect it resembles Wright's island model. This assumption is made to avoid the complication of populations at the ends of the chain. The infinity assumption allows all populations to be treated equally (the same way as we treat the populations in Wright's d-island model). A finite version of the stepping-stone model with this property is obtained by placing the d islands on a circle so that the two populations at the "ends" exchange migrants as though they were neighbors. Theoretical considerations of the genealogy of a sample of genes refer to this circular stepping-stone model, but even then the model is considerably more complicated than Wright's d-island model. For instance, analysis of a sample of at most one gene from each population requires information about which populations in the chain are sampled. Immigration events that are trivial in the island model influence the future probability of coalescence in the stepping-stone model. A more practical application of the finite circular version of the stepping-stone model is in numerical studies using computer simulations.

The steady state in the stepping-stone model was expressed in terms of consanguinity coefficients by Malécot in the 1950s (see Malécot 1969), and independently by Kimura and Weiss (1964), who used a different method more reminiscent of our calculation of equilibrium variances in gene frequencies. The two methods describe the same characteristics of the populations, and we will only expand the simpler description used by Malécot. We may therefore view it as referring to the infinite alleles model (Section 4.5 on page 95).

The probability of drawing two identical gametes from the gamete pools in populations h steps apart is described by F_h, $h = 0, 1, 2, \ldots$, and F_0 is thus the probability of forming an identical homozygote from gametes drawn from within a population. The equilibrium in these coefficients is approximately

$$\hat{F}_h = \frac{1}{1 + 4N\sqrt{2m\mu}} \, \exp\left(-h\sqrt{\frac{2\mu}{m}}\right) \qquad (5.15)$$

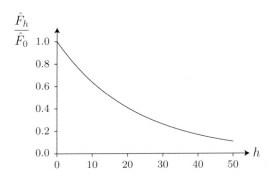

Figure 5.9: Variation in genetic similarity as a function of distance in the stepping-stone model ($m = 0.1$, $\mu = 10^{-4}$).

for a small migration fraction m and an even smaller mutation rate $\mu \ll m$ (see Box 20 on page 134).

The amount F_0 of variation in a population is determined by $4N\sqrt{2m\mu}$, and it may therefore be much larger than that in an isolated population determined by $4N\mu$. The reason is that variation is shared along the stepping-stone chain of populations, and the local effective population size is therefore larger than the size of a subpopulation.

The genetic coherence of the populations in the stepping-stone chain is illustrated in Figure 5.9. The figure shows the genetic similarity between populations, namely the ratio between $\hat{F}_h$ and $\hat{F}_0$, as a function of distance. The genetic similarity between close populations is very high. This will show up as very similar gene frequencies in closely situated populations. Populations far apart, however, are expected to show very little resemblance, and they will therefore show very different gene frequencies. This kind of variation, called a *cline*, is common in nature. Figure 5.10 shows an example of a gene frequency cline from Kattegat to the Baltic Sea in populations of a shallow-water marine fish. Natural habitats rarely have a geographical structure that may be described by simple theoretical models like the stepping-stone model. Whether or not the cline in Figure 5.10 may be described as a result of random genetic drift and migration is therefore hard to judge. Drift and migration, however, have the same influence on the pattern of variation at all loci, so any difference in gene frequencies between *Z. viviparus* populations in the North Sea and the Baltic Sea will cause similar clines through the Danish Belts (Christiansen and Frydenberg 1974), that is, the local gene frequencies are determined by the gene frequencies in the two seas and the migrational distances from the local population to either of them. A cline is also observed in the gene frequencies at a hemoglobin locus, and those at the two loci show the linear relationship expected when drift and migration caused the local variation (Hjorth and Simonsen 1975).

Finite versions of the stepping-stone model are used in studies applying simulations (see, e.g., Schierup and Christiansen 1996). In a finite stepping-stone model the populations at the ends of the chain clearly evolve differently

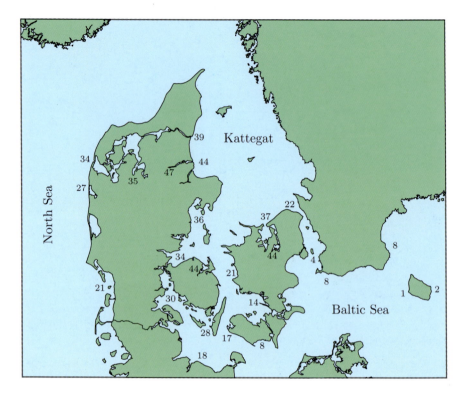

Figure 5.10: Variation in frequencies of the *EstIII*[1] allele in the eelpout *Zoarces viviparus* around Denmark. Data from Frydenberg et al. (1973).

from those inside the chain. Realizing this we see that the properties of any population depends on its position in the chain. This complicates computations as well as the interpretation of results. This obstacle is overcome by joining

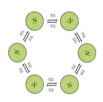

the ends to produce a circular stepping-stone model, where each population is positioned in the same way relative to the other populations in the chain. The model is therefore highly symmetric and much preferred over the linear model.

The stepping-stone model easily generalizes to a model that covers populations distributed over a two-dimensional area. A finite version of this model corresponding to the above circular model is formed from a square of populations, that is, first rolling them into a pipe by connecting populations on two opposite edges, and then bending this pipe into a doughnut by connecting the populations in the circles at the opposite ends of the pipe.

Analysis of the genealogy of a sample of genes cannot be made in the stepping-stone model because of its infinite population size. We have to refer to finite versions of the model. Matsen and Wakeley (2006) generalized the

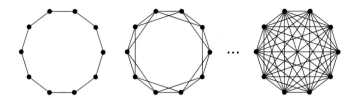

Figure 5.11: Examples of migration models in the isotropic class. Populations connected by lines are those with links of positive immigration frequency.

results of the coalescent in a widespread Wright's island model (Section 5.2.2) to a class of migration models that comprise the circular and the doughnut-shaped stepping-stone models. This class is the *isotropic* models, where "the pattern of migration for each subpopulation is identical to the migration pattern of any other subpopulation" (Strobeck 1987), and each population has the same size. After a scattering phase these models enter into a collecting phase where the sample genealogy is given by Kingman's coalescent, corresponding to a population of size

$$N_e \approx Nd\left(1 + \frac{1}{4Nm}\right),$$

where m is the immigration rate in the local population $\big($as in equation (5.13)$\big)$. As before, the result holds for a large number of subpopulations. The isotropic class contains models with varying amounts of local isolation, for instance ranging from the circular stepping-stone model all the way to Wright's island model with migration (Figure 5.11). Isotropic migration results in the models shown when migration is equal over the same distance. The rightmost graph is Wright's island model if the immigration frequencies are equal along all links, and it approaches the circular stepping-stone model when immigration becomes small along links between populations that are not neighbors.

5.3 Genotypic Proportions

A population that receives immigrants will appear mixed like the population of *Zoarces viviparus* at Hadsund in Mariager Fjord (Figure 5.1 on page 108). Mixing two randomly mating populations also produces characteristic deviations from the Hardy–Weinberg proportions, the so-called *Wahlund effect* (Wahlund 1928). The effect is an excess of homozygotes compared to the Hardy–Weinberg proportions, corresponding to the gene frequency in the mixed population. In a two-allele polymorphism the genotype frequencies in the mixed population are

$$\mathsf{AA}:\ \bar{p}^2 + \mathrm{Var}(p), \quad \mathsf{Aa}:\ 2\bar{p}\bar{q} - 2\mathrm{Var}(p), \quad \mathsf{aa}:\ \bar{q}^2 + \mathrm{Var}(p), \tag{5.16}$$

Box 20: * Gene identity structure in the stepping-stone model

Sampling within a population we get

$$F_0' = \left((1 - 2m)^2 + 2m^2\right)\left(\frac{1}{2N} + \left(1 - \frac{1}{2N}\right)F_0\right) + 4m(1 - 2m)F_1 + 2m^2 F_2.$$

Sampling genes in neighboring populations gives

$$\begin{aligned} F_1' &= 2m(1 - 2m)\left(\frac{1}{2N} + \left(1 - \frac{1}{2N}\right)F_0\right) \\ &+ \left((1 - 2m)^2 + 3m^2\right)F_1 + 2m(1 - 2m)F_2 + m^2 F_3. \end{aligned}$$

Sampling in populations two steps away gives

$$\begin{aligned} F_2' &= m^2\left(\frac{1}{2N} + \left(1 - \frac{1}{2N}\right)F_0\right) \\ &+ 2m(1 - 2m)F_1 + \left((1 - 2m)^2 + 2m^2\right)F_2 + 2m(1 - 2m)F_3 + m^2 F_4. \end{aligned}$$

Sampling in populations more than two steps away excludes the possibility that the two genes sampled are copies of the same gene in the previous generation, and we therefore get

$$\begin{aligned} F_h' &= m^2 F_{h-2} + 2m(1 - 2m)F_{h-1} + \left((1 - 2m)^2 + 2m^2\right)F_h \\ &+ 2m(1 - 2m)F_{h+1} + m^2 F_{h+2}, \quad h = 3, 4, 5, \ldots. \end{aligned}$$

These equations only describe the effect of random genetic drift, and we therefore expect the population to converge to homozygosity. This never occurs in an infinite chain, as populations far apart are virtually isolated from each other. The aim of the model, however, is not to describe the convergence to homozygosity, but rather the coherence of the population, that is, the genetic similarity among populations—the effect of isolation by distance. Introduction of mutation (rate μ) provides a characteristic level of variation, giving some coherence even between distant populations. Another possibility is to introduce gene frequency conservation by a low frequency of immigration in any subpopulation, as in Wright's island model where the collection of subpopulations provides the constant gene frequency—a kind of "long-distance migration." The use of the model as a biological description will obviously always refer to a finite number of populations. The assumption of coherence between distant populations diminishes their weight in the analysis and in the results. The equilibrium solution is approximately given by equation (5.15) for m small and $\mu \ll m$.

where

$$\bar{p} = \sum_{i=1}^{d} c_i p_i, \qquad \mathrm{Var}(p) = \sum_{i=1}^{d} c_i (p_i - \bar{p})^2, \qquad (5.17)$$

p_i is the gene frequency in component i, $i = 1, 2, \ldots, d$, and c_i is the relative contribution of subpopulation i to the mixed population.[2] This well-known signature of population mixing disappears among the offspring upon breeding by random mating. The genotype frequencies among offspring produced after random mating will be in Hardy–Weinberg proportions corresponding to the gene frequencies in the mixed population.

The mixing of individuals, however, need not have an immediate effect on the amount of mixing of genes among neighboring populations. Genetic mixing requires a breeding population consisting of individuals of mixed origin. The populations of *Zoarces viviparus* in Mariager Fjord (Figure 5.1 on page 108) were investigated again in 1977 (Christiansen et al. 1988), and the variation in vertebra counts showed a remarkable similarity to that observed by Schmidt (1917). This constancy over time seems at variance with regular interbreeding, corresponding to the observed mixing at Hadssund, and Schmidt's observations could therefore reflect a mixing of two isolated populations, which occasionally happens outside the mating season. Observation of population mixing is therefore not necessarily a sign of genetic exchange of migrants between neighboring populations. Immigration of genes occurs at the time of breeding and is manifested in the formation of zygotes by the union of gametes from individuals of different origin. The coherence of a population in genetic terms is defined in this way and is referred to as a *Mendelian population* (Dobzhansky 1950), in contrast to the aggregation of individuals observed when investigating a natural population.

5.3.1 Description of population structure

Genotypic data from a geographically structured population are described in terms of deviation from the Hardy–Weinberg proportions. The genotype frequencies z_{AA}, z_{Aa}, and z_{aa} ($z_{\mathsf{AA}} + z_{\mathsf{Aa}} + z_{\mathsf{aa}} = 1$) can be expressed as

$$z_{\mathsf{AA}} = p^2 + pqF_{IS}, \quad z_{\mathsf{Aa}} = 2pq(1 - F_{IS}), \quad z_{\mathsf{aa}} = q^2 + pqF_{IS}, \qquad (5.18)$$

where $p = z_{\mathsf{AA}} + \frac{1}{2}z_{\mathsf{Aa}}$, $q = \frac{1}{2}z_{\mathsf{Aa}} + z_{\mathsf{aa}}$, and

$$F_{IS} = \frac{z_{\mathsf{AA}} - p^2}{pq} = \frac{z_{\mathsf{AA}} z_{\mathsf{aa}} - \frac{1}{4}z_{\mathsf{Aa}}^2}{p^2 q^2}.$$

The index F_{IS} is positive when the population shows an excess of homozygotes with respect to the Hardy–Weinberg proportions, and negative when the heterozygotes are in excess.

---

[2]The variance in the heterozygote is really $\mathrm{Cov}(p, q)$. Realizing this, the multiple-allele version of equation (5.17) is readily constructed.

Suppose we know the genotype frequencies $z_{\mathsf{AA}i}$, $z_{\mathsf{Aa}i}$, and $z_{\mathsf{aa}i}$ in d subpopulations, $i = 1, 2, \ldots, d$. The gene frequency in the total population is then

$$\bar{p} = \sum_{i=1}^{d} c_i p_i,$$

where c_i, $i = 1, 2, \ldots, d$, are the relative sizes of the d subpopulations and $p_i = z_{\mathsf{AA}i} + \frac{1}{2} z_{\mathsf{Aa}i}$. The frequency of the homozygote AA is

$$\bar{z}_{\mathsf{AA}} = \sum_{i=1}^{d} c_i z_{\mathsf{AA}i}.$$

Expanding the expression for $\bar{z}_{\mathsf{AA}}$ by using the description (5.18) of the genotype frequencies within each population gives

$$\bar{z}_{\mathsf{AA}} = \sum_{i=1}^{d} c_i(p_i^2 + p_i q_i F_{ISi}) = \sum_{i=1}^{d} c_i p_i^2 + F_{IS} \sum_{i=1}^{d} c_i p_i q_i,$$

where

$$F_{IS} = \frac{\displaystyle\sum_{i=1}^{d} c_i p_i q_i F_{ISi}}{\displaystyle\sum_{i=1}^{d} c_i p_i q_i} = \frac{\displaystyle\sum_{i=1}^{d} c_i(z_{\mathsf{AA}i} - p_i^2)}{\displaystyle\sum_{i=1}^{d} c_i p_i q_i}$$

is the average deviation from Hardy–Weinberg proportions in the local population. The sums may be expressed in terms of $\bar{p}^2$ and $\bar{p}\bar{q}$ and the Wahlund variance, in that equation (5.16) gives $\bar{z}_{\mathsf{AA}} = \bar{p}^2 + \bar{p}\bar{q}F_{ST} + F_{IS}\bar{p}\bar{q}(1 - F_{ST})$, and F_{ST} is given by

$$F_{ST} = \frac{\mathrm{Var}(p_i)}{\bar{p}\bar{q}}.$$

Thus, F_{ST} corresponds to the level of gene identity between populations in Wright's island model.

The frequency $\bar{z}_{\mathsf{AA}}$ may also be described in terms of $\bar{p}$ and $\bar{q}$ by equation (5.18), and this we write as $\bar{z}_{\mathsf{AA}} = \bar{p}^2 + \bar{p}\bar{q}F_{IT}$, where

$$F_{IT} = \frac{\bar{z}_{\mathsf{AA}} - \bar{p}^2}{\bar{p}\bar{q}}$$

describes the level of gene identity in the total population. The two expressions for $\bar{z}_{\mathsf{AA}}$ therefore provide the formula

$$(1 - F_{IT}) = (1 - F_{IS})(1 - F_{ST}), \tag{5.19}$$

linking inbreeding within a population and consanguinity between populations to the level of consanguinity in the total population (Wright 1943). The formula is easily extended to a deeper hierarchical structure, and it is possible to define sensible measures for multiple alleles that obey formula (5.19) (Nei 1977).

When deviation from Hardy–Weinberg proportions is due to the same amount of inbreeding F in each population, then $F_{IS} = F$. In the infinite alleles model, formula (5.19) describes the effect of the population structure in Wright's island model. The statistical analysis of genetic variation in this model or in models with a more hierarchical population structure is discussed by Weir (1996).

Exercises

Exercise 5.2.1 Show that Deakin's model is conservative, and that $M = \tilde{M}$.

Exercise 5.2.2 Find the migration matrix for Wright's two-island model and compare it to the migration matrix for the simple two-island model given on page 121.

Exercise 5.2.3 Imagine a sample of n DNA sequences from a large distributed population that has been carefully collected to get as comprehensive a coverage of the population as possible. Such a sample is not inconceivable in investigations of human populations. Upon analysis of this sample we see a gene tree with a nice coherence to the coalescent, and using the method outlined in Section 4.3 we get an estimate of $\theta = 4N_e\mu$.

Discuss such observations. If we have a good idea of the value of μ, what would an estimate of θ tell us about the population?

Exercise 5.2.4 Consider two populations of equal size N with $m_{12} = \frac{1}{2}m$ and $m_{21} = m$. Show that $\hat{\pi} = (\frac{2}{3}, \frac{1}{3})$, and show that $N_e = \frac{9}{5}N$ describes the coalescence in the population when m is large.

Exercise 5.2.5 Consider three populations of equal size N with

$$
\left\{
\begin{array}{ccc}
m_{11} & m_{12} & m_{13} \\
m_{21} & m_{22} & m_{23} \\
m_{31} & m_{32} & m_{33}
\end{array}
\right\}
=
\left\{
\begin{array}{ccc}
1 - \frac{1}{2}m & \frac{1}{2}m & 0 \\
\frac{1}{2}m & 1 - m & \frac{1}{2}m \\
0 & \frac{1}{2}m & 1 - \frac{1}{2}m
\end{array}
\right\}.
$$

Find the population size that describes the coalescence in the population when m is large.

Exercise 5.2.6 Consider three populations of equal size N with

$$
\left\{
\begin{array}{ccc}
m_{11} & m_{12} & m_{13} \\
m_{21} & m_{22} & m_{23} \\
m_{31} & m_{32} & m_{33}
\end{array}
\right\}
=
\left\{
\begin{array}{ccc}
1 - m & m & 0 \\
\frac{1}{2}m & 1 - m & \frac{1}{2}m \\
0 & m & 1 - m
\end{array}
\right\}.
$$

Find the population size that describes the coalescence in the population when m is large.

Exercise 5.2.7 Discuss the result for strong migration in Wright's two-island model. How does it relate to Figure 5.7 on page 123 and equation (5.7) on page 117?

Exercise 5.3.1 Assume d populations with gene frequencies p_i, $i = 1, 2, \ldots, d$, of allele A and reproducing by random mating. A population forms in a new area by immigration from these populations, in that the fraction c_i arrives from population i, $i = 1, 2, \ldots, d$. Show that the frequency of genotype AA among the immigrants is $\bar{p}^2 + \mathrm{Var}(p)$, where $\bar{p}$ and $\mathrm{Var}(p)$ are given by equation (5.17).

Exercise 5.3.2 Consider two populations of equal size, and suppose the gene frequencies of A are $p + \delta$ and $p - \delta$. (1) Find the mean and the variance of the gene frequencies, and write down the genotype frequencies in the population formed by a mixture of the two populations. (2) What is the maximum excess of homozygotes for $p = \frac{1}{2}$? (3) What is the maximum excess of homozygotes for a general p?

Exercise 5.3.3 The cod, *Gadus morhua*, is polymorphic for two alleles, say, A_1 and A_2, at a hemoglobin gene. The populations in the inner Danish waters are rather homogeneous, and so are the populations in the northern Baltic Sea. The genotypic composition of populations in these areas are shown in Table 5.1. Show that these observed genotypic counts are consistent with the

Table 5.1: Population samples of cod (data from Sick 1965).

	A_1A_1	A_1A_2	A_2A_2	Total
Denmark	258	324	106	688
Baltic	0	10	230	240

assumption that the genotypic frequencies in the sampled populations are in Hardy–Weinberg proportions.

Samples from the western Baltic Sea around the island of Bornholm are rather heterogeneous. The genotypic counts of two samples in this area are shown in Table 5.2. Are these observed genotypic counts consistent with the

Table 5.2: Samples of cod in the western Baltic Sea (data from Sick 1965).

Sample	A_1A_1	A_1A_2	A_2A_2	Total
1	67	118	81	266
2	49	73	128	250

assumption that the genotypic frequencies in the sampled populations are in Hardy–Weinberg proportions?

Try to give a biological interpretation of the results of the analysis, and give a theoretically based argument for its plausibility.

Chapter 6

Linkage

The fundamental rules of population genetics that describe the transmission of genes from a parent population to its offspring population are a consequence of Mendel's first law. Mendel also studied the simultaneous inheritance of several characters that by themselves obeyed the law of inheritance. In one experiment he analyzed variants of pea color and form (Table 2.2 on page 23) and crossed plants grown from true-breeding yellow, round peas with plants grown from true-breeding green, wrinkled peas. As expected, all the F_1 peas were yellow and round, and in the 556 F_2 peas all four combinations of the traits were observed (Table 6.1). Forming the marginal sums confirms that the traits of both

Table 6.1: F_2 segregation in a two-character cross

	round	wrinkled	$\sum$
yellow	315	101	416
green	108	32	140
$\sum$	423	133	556

characters segregate $3 : 1$ among F_2 peas. In addition, this segregation ratio is observed for pea color whether the peas are round or wrinkled, and likewise for the segregation of pea form. Thus, the segregation in one character is independent of that in the other character. Mendel made similar observations in all the pairwise combinations he investigated, and this led him to conclude (in more modern words) that alleles of different Mendelian genes are transmitted independently to the offspring. These observations were formulated into a second law of inheritance that describes the transmission of different genes. It states that the versions of different genes are chosen independently.

The genotypes of the two parental lines can be described as AA BB and aa bb. Thus, the F_1 genotype is Aa Bb, and Mendel's conclusion stipulates that a gamete from such a plant is formed by taking an allele of each gene at random. The four possible gametes are therefore produced equally often. For instance, the

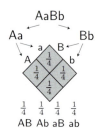

probability that a gamete carries allele A is $\frac{1}{2}$, the probability that it carries allele B is $\frac{1}{2}$, and according to the independence assumption, the probability of carrying A and B is $\frac{1}{2} \times \frac{1}{2} = \frac{1}{4}$. The two gametes AB and ab present in the two original lines therefore occur with the same probability as the recombined gametes Ab and aB. This phenomenon is also described as *independent reassortment* of parental alleles.

Combination of gametes into genotypes again occurs at random, such that the 16 different gametic combinations are formed equally often. If in addition the trait of the character in an individual is determined by the corresponding genotype independently of the other genotype, the trait combination of the genotypes

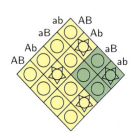

is given by the figure to the left. The presence of an A allele makes the pea yellow □ and a B allele makes it round ●. Absence of an A allele gives a green pea ■, and absence of a B gives a wrinkled pea ♣. The parental type of green, wrinkled peas (aa bb) therefore occurs with the probability $\frac{1}{16}$ among the F_2 peas. In Mendel's experiment this produces the expectation of 34.75, which is in reasonable agreement with the observed number in Table 6.1.

In Mendel's cross, color and shape were known to segregate according to Mendel's law. Without such knowledge, we would observe the character "pea appearance" segregate into four classes in F_2. From our analysis we should observe a segregation of $9:3:3:1$ expressed in Mendelian terms.

The Mendelian expectation is thus that an individual transmits the genes of maternal and paternal origin independently. The genes are said to *recombine*. Shortly after the acceptance of Mendel's first law in 1900, the second law of reassortment of parental alleles of different genes was found not to hold in general. Bateson et al. (1905) observed in experiments with sweet peas (*Lathyrus odoratus*) that for some pairs of genes the parental gametes were produced more often than the recombinant gametes. The two genes thus segregated as if they were partially *linked*.

An experiment that showed linkage studied the two characters of flower color (P- purple and pp red)[1] and pollen shape (L- long and ll round). The cross of PPLL and ppll plants yielded F_1 plants with purple flowers and long pollen. Selfing of these plants yielded segregation among the F_2 plants, where

the numbers of the four types in F_2 were 177, 15, 15, and 49, respectively. Compared to Mendel's expectation ($\frac{1}{16} \times 256 = 16$) the observed number of the

---

[1]The genotype designation P- is shorthand notation for the two genotypes PP and Pp.

double recessive individuals is elevated. They made more such crosses of PPLL and ppll plants and obtained a total of 6952 offspring with the segregation shown in Table 6.2. In this experiment the double recessive type occurred in a frequency

Table 6.2: F_2 segregation in a two-character cross

	long	round
purple	4831	390
red	393	1338

of 0.19 and the Mendelian expectation is only 0.06 (or 435 individuals). The two classifications in the table are clearly not independent. Recombination certainly does occur, but recombinant gametes are formed less often than the parental gametes. The recombinant gametes are Pl and pL, and the recombinant phenotypes are formed as Pl/Pl or Pl/pl and pL/pL or pL/pl. The observations in Table 6.2 thus suggest that when recombination occurs the two kinds of gametes are equally likely to result.

Mendel's first law is universal, but his second law is obeyed only by some genes, while others exhibit the phenomenon of *linkage*. Mendelian genes may interact in their transmission.

6.1   The Karyotype

It is tempting to link the acceptance and "rediscovery" of Mendelian inheritance to the parallel behavior of Mendelian factors and chromosomes in meiosis. The proliferation of cells in an individual happens by mitosis, where a prominent feature of the cell-division process is the duplication of chromosomes. The chromosomes appear as recognizable features in the cell nucleus when division begins (the prophase). They each show as a pair of *chromatids* glued together at a characteristic place called the *centromere*. As division proceeds the chromosomes collect in the equatorial plane of the cell, and the centromeres split and the sister chromatids are pulled to opposite poles. Finally, the cell divides. Either a cell membrane forms in the equatorial plane (plants), or the cell constricts at the central plane (animals). The cell division is complete and the chromosomes disappear from view.

The number of chromosomes and their morphology vary among species, and these differences may even be used as taxonomic characters in the definition and recognition of species. Each species thus has its own characteristic *kary-otype*, and sometimes even differences between individuals occur. Humans have 46 chromosomes varying in both size and morphology. The great apes have 48. The parallel to Mendel's law is apparent in the haploid–diploid life cycle of sexually reproducing higher organisms. The diploid phase is initiated by fertilization in which two haploid cells fuse, and meiosis of a diploid cell gives rise to haploid offspring (see page 8). The meiotic prophase initially resembles the mitotic prophase, but as the chromosomes appear they form pairs, usually of morphologically identical chromosomes, dubbed *homologous* pairs. As the homologous chromosomes pair, align, and form tight pairs, they are stabilized by the formation of *chiasmata*, where a pair of chromatids cross, break, and reunite (Figure 6.1; we return to this process in Section 6.4.2). These pairs collect in the division plane, and during the first meiotic division, the homologous chromosomes segregate to opposite poles. Each chromosome still appears two-stranded with the chromatids joined at the centromere. The first division thus reduces the cell from diploidy to two haploid daughter cells. The second division is just a mitotic division initiated immediately after the first.

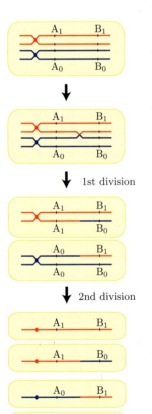

Figure 6.1: Simplified meiotic cell divisions of a diploid cell with one pair of chromosomes.

In humans (and many higher organisms) this description of the karyotype and meiosis is a bit too simple. The female meiotic prophase shows 23 homologous pairs of chromosomes, whereas male meiosis shows only 22 homologous pairs. The remaining two chromosomes in the male form a heterologous pair and segregate in the first meiotic division. The chromosomes in the heterologous pair are called the X and Y chromosomes, while the corresponding homologous pair in females is XX. Males produce two kinds of sperm: X bearing and Y bearing. These determine the sex of the zygote. The male sex is thus dubbed heterogametic, and females are homogametic.

Chromosomal sex determination is widespread in the animal kingdom and also occurs among plants. The most common is the XY system, where the male is heterogametic. The alternative is the WZ system, where the female is heterogametic WZ and the male is homogametic ZZ. This is the case in birds and Lepidoptera (butterflies and moths).

The chromosomes always present in homologous pairs are referred to as the *autosomes*, and the chromosomes of the heterologous pair are the *sex chromosomes*.

6.1.1 Sex linkage

A convincing parallel between hereditary transmission and chromosome segregation emerged with the discovery of sex-linked inheritance in the fruit fly, *Drosophila melanogaster*, by Thomas Hunt Morgan in 1911. He studied the inheritance of the recessive trait of white eye color (allele w), and found sex-asymmetric transmission. Crossing white-eyed males and wild type female flies always yielded red-eyed F_1 offspring. The complications began in F_2, where segregation was observed only in males. All females were red-eyed. In addition, the segregation among males was $\frac{1}{2}$ white to $\frac{1}{2}$ red eye color. The reciprocal cross of white-eyed females and wild type males gave red-eyed females and white-eyed males in the F_1 generation. The F_2 offspring from the cross between the F_1 males and females yielded segregation of $\frac{1}{2}$ white to $\frac{1}{2}$ red in both sexes.

Heterogametic meiosis

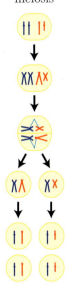

These peculiar results become understandable if the gene is placed on one of the sex chromosomes of *D. melanogaster*. The cells of female fruit flies carry eight chromosomes in four pairs, as shown in the drawing below. The autosomes are two pairs of large metacentric chromosomes (centromere in the middle) and one pair of very small telocentric chromosomes (centromere at one end). Males have only these three homologous pairs, one large telocentric chromosome, the X chromosome, and one acrocentric chromosome (centromere close to one end of the chromosome), the Y chromosome. Females have four pairs of homologous chromosomes.

Morgan suggested that the alleles w and the corresponding wild type allele, $+$, were placed on the X chromosome.[2] Morgan's first cross is then between a $++$ ♀ and a w ♂ (the male is *hemizygous*—he has only one X chromosome and one gene, and so he always appears like a homozygote for traits determined by allelic variation on the X chromosome). All F_1 offspring receive an X chromosome with a $+$ allele from the mother, and they all thus become red-eyed. The females receive a w allele from the male and are heterozygotes $w+$. A

cross of F_1 individuals is therefore between a $w+$ ♀ and a $+$ ♂. Female F_2 offspring receive a $+$ allele from the father, and all have red eyes. The observed segregation in the male F_2 offspring reflects the segregation of the alleles in the mother. The reciprocal cross of a ww ♀ and a $+$ ♂ of course gives white-eyed males and red-eyed females in F_1, and the F_2 segregation resembles a backcross in Mendelian analysis (see page 8).

Associating the transmission of the white gene with that of the X chromosome thus fully explains the aberrant inheritance of the white eye trait. This added considerable credibility to the gene–chromosome theory, which places the genes in the chromosomes. The genes on the X chromosome all show sex-linked

---

[2]The wild type alleles are referred to as $+$ alleles in the *Drosophila* literature. If ambiguous the wild type allele corresponding to the mutant w is designated w^+.

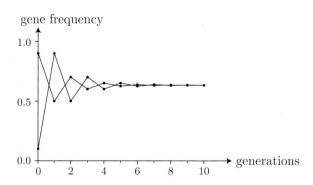

Figure 6.2: Changes in gene frequency at a sex-linked gene (see Box 21). The initial gene frequencies are $p_{\female} = 0.9$ and $p_{\male} = 0.1$, and therefore $p = 0.633$.

inheritance, and to emphasize that Mendelian rules of segregation pertain to genes on the autosomes, they are said to show autosomal inheritance.

Genes on the X chromosome segregate according to Mendel's law in females, whereas in males the X chromosome just passes through to female offspring. The result is that the frequency of genes of a given allele type is conserved among the X chromosomes in the population of eggs and sperm that unite to form the individuals in the population (Box 21). Initial difference in the gene frequencies in males and females causes these to vary in a seesaw manner from generation to generation (Figure 6.2). The gene frequencies, however, equilibrate quite rapidly at equality in females and males.

6.1.2   Linkage map

The placement of genes on the chromosomes soon connected recombination of linked genes to chiasmata where chromatids cross. Genetic recombination of linked genes became synonymous with chromatid crossing and interchange by breakage and reunion and is described as *crossover*.

Further genetic analysis of the segregation of two or more genes in crosses developed the theory of linked genes, especially the use of the backcross of F_1 individuals to a line carrying only the recessive alleles of the investigated genes. Sex-linked genes are particularly easy to analyze in this way. When the recessive alleles are introduced through the female, then the F_1 offspring are female heterozygotes and male recessive hemizygotes. For autosomal markers the F_1 females need to be collected as virgins and then crossed to males from recessive homozygotes. The first thorough linkage analysis describing the relationships among sex-linked mutants of *D. melanogaster* was the work of Thomas Hunt Morgan and his students. During this work Mendelian genes were firmly located on the chromosomes, and the degree of linkage was shown to be a measure of the distance between two genes. A gene was thus carried at a particular location on a chromosome, the gene *locus*, and soon the gene and its locus were used synonymously. A *linkage map* of chromosome number 1 (the X chromosome) of

---

Box 21: Population genetics of sex linkage

Conservation of genetic variation is of course valid for sex-linked genes. The only problem is how to count the frequency of the various alleles. Consider two alleles with gene frequencies $p_\female$ and $p_\male$ in females and males. Female offspring get a gene from both parents, whereas males get their gene from the mother, and the gene frequencies in the offspring population are therefore

$$p'_\female = \tfrac{1}{2}(p_\female + p_\male) \quad \text{and} \quad p'_\male = p_\female.$$

The difference in gene frequency between the sexes is diminished in the offspring population, in that $p'_\female - p'_\male = -\tfrac{1}{2}(p_\female - p_\male)$ and its sign changes. The frequencies in the two sexes thus oscillate from generation to generation and converge to an equilibrium where the gene frequencies are equal (Figure 6.2). The two sexes contribute equally to the next generation, so the females contribute twice as many X chromosomes as the males to the offspring population. The gene frequency in the population of transmitted X chromosomes is then $p = \tfrac{1}{3}(2p_\female + p_\male)$, and the population of X chromosomes transmitted by the offspring is $p' = \tfrac{1}{3}(2p'_\female + p'_\male) = \tfrac{2}{3}p_\female + \tfrac{1}{3}p_\male = p$. The gene frequency among transmitted X chromosomes is therefore conserved from generation to generation, and by defining the population gene frequency as

$$p = \tfrac{2}{3}p_\female + \tfrac{1}{3}p_\male,$$

we maintain the law of the conservation of gene frequencies.

Genetic drift in a finite population may be viewed in terms of inbreeding in the gamete pool model. Let F_t be the probability of sampling two identical genes in generation t with one from the female pool and one from the male pool. Thus, F is the inbreeding coefficient of females. The probability of sampling two identical genes from the female pool in generation t is G_t, and that of sampling two identical genes from the male pool is H_t. The recurrence equations of these are then

$$
\begin{aligned}
F_t &= \frac{1}{2}\left(\frac{1}{2N_\female} + \left(1 - \frac{1}{2N_\female}\right)G_{t-1}\right) + \tfrac{1}{2}F_{t-1}, \\
G_t &= \tfrac{1}{2}F_{t-1} + \tfrac{1}{4}G_{t-1} + \tfrac{1}{4}H_{t-1}, \\
H_t &= \frac{1}{N_\male} + \left(1 - \frac{1}{N_\male}\right)F_{t-1}.
\end{aligned}
$$

Numerical calculations suggest that the resulting inbreeding effective population size may be approximated by

$$\frac{1}{N_{e(F)}} = \frac{1}{5.50N_\male} + \frac{1}{3.67N_\female}.$$

When $N_\female = N_\male = N$ we get $N_{e(F)} \approx 2.2N$.

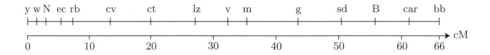

Figure 6.3: Linkage map of linkage group I (the X chromosome) in *Drosophila melanogaster*. The positions of a selection of genes are shown. The chromosome is telocentric (centromere placed at one end of the chromosome), and the centromere is to the right.

D. melanogaster was published by the group in 1914 (Figure 6.3). The unit of the map (cM, centiMorgan) is the distance where one crossover is expected per hundred gametes. One Morgan (M) is the distance of 100cM on a linkage map; the reason for such a backward definition is that the possibility of two or more crossovers can be neglected within a map distance of 1cM.

The recombination frequency between two linked loci is always less than $\frac{1}{2}$, the recombination frequency between two genes on different chromosomes. The reason is apparent from Figure 6.1 on page 142, as crossover occurs between two of the four chromatids in the meiotic prophase, and in half the gametes the parental chromosomes are therefore intact in the neighborhood of the crossover point.

A linkage map is constructed from observations of fairly closely linked markers. This minimizes the disturbance of double crossovers between two markers, which will be hidden among the observations of parental gametes. Thus, the recombination frequency between the loci for yellow (body) and vermilion (eyes) (y and v in Figure 6.3) is not 0.33, but less, because an even number of crossovers cannot be observed. To observe this, a typical experiment crosses wild type flies (brownish body, red eyes) and yellow flies with vermilion eyes (y and v are both recessive):

$$\frac{yv}{yv} \quad \times \quad ++$$

$$\downarrow$$

$$F_1: \quad \frac{++}{yv} \quad \times \quad yv$$

$$F_2: \quad yv, y+, +v, ++ \quad \text{in } \male\male$$

Cross of F_1 females and males yields offspring segregating in four phenotypic classes: (yellow,vermilion), (yellow,+), (+,vermilion), and (+,+), where + designates the corresponding wild type trait. Counting the F_2 phenotypes gives an estimate of the frequency of recombination.

To build a linkage map, loci must be placed in order. Backcrosses involving alleles at three loci accomplish this. An example is shown in Table 6.3, which is extracted from data assembled by Bridges and Olbrycht (1926). Flies with cut wings and vermilion and garnet eyes (corresponding to *ct*, *v*, and *g*, which are recessive alleles at three sex-linked loci) were crossed to wild type. The

Table 6.3: Backcross involving three sex-linked characters in *D. melanogaster*

cut	+	cut	+	+	cut	cut	+
vermilion	+	+	vermilion	+	vermilion	+	vermilion
garnet	+	+	garnet	garnet	+	garnet	+
1015	1370	249	254	185	159	8	9

eight possible combinations of the phenotypes are observed among the 3249 flies in F_2. The phenotypes are observed in complementary pairs with similar frequencies[3]—complementary in the sense that a parental trait of each character is only present in one of the phenotypes. Being a backcross, the phenotype reflects the composition of the gamete contributed by the F_1 parent, and the observation of complementary phenotypic pairs indicates even segregation of *complementary gametes*. The parental phenotypes are the most numerous, and among the recombinant phenotypes one pair of complementary types is rare, with a total frequency of about 0.005. This suggests the linkage relationship

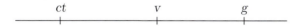

because the rare types are then the products of two crossovers, one between *ct* and *v*, and one between *v* and *g*. Accepting that the recombination frequencies in the two intervals are the observed values 0.160 and 0.111, we would expect to observe about 58 double crossovers—quite a lot more than the 17 observed. This is a common observation, and is interpreted as being due to interference in chiasmata formation, where the formation of one hampers the formation of another close by. The result we observe is *interference* in recombination in adjacent regions of the linkage map. The usual measure used to quantify this effect is the *coefficient of coincidence*, which is the observed frequency of double crossovers divided by the frequency expected if the recombination in the two regions was independent. In our example the coefficient of coincidence is $\frac{17}{58} = 0.3$. For closely linked loci the coefficient approaches zero, absolute interference, and for very loosely linked loci it may be as high as one, signifying no interference. For very closely linked markers the observed coincidence may actually be observed as larger than one due to a phenomenon called gene conversion (see Section 6.4.2).

Morgan and his students continued this work to include the large autosomes (chromosomes II and III) of *D. melanogaster*. Linkage group I contains all the sex-linked genes known to Morgan, and it was the first to be analyzed thoroughly. Work on the autosomes is more tedious, but the crossing scheme is very similar: F_1 females are backcrossed to males from the recessive stock, because *Drosophila* male meioses do not produce crossovers. In the 1930s linkage maps were joined with morphological maps of the giant chromosomes of the salivary glands of larvae (see Figure 6.5 on page 177).

---

[3]The recessive traits are caused by dysfunctional alleles, and they may therefore cause lower viability. The observed deficit of the recessive parental type in particular is often seen in such experiments.

6.2 Population Genetics of Linkage

Gametes transmit genomes and not single genes from parents to offspring, and in their formation recombination redistributes the genetic information in the two genomes that formed the diploid individual—reassortment of chromosomes and meiotic crossover mediate this redistribution. On the population level the effect of recombination depends on which genomes are found together in diploid individuals. Mating is important, and we therefore have to consider bona fide diploid models, not just haploid models with a diploid superstructure.

The basic population genetic study of recombination is thus founded on the simplest diploid breeding assumption, which is that of random mating. We shall study the simultaneous transmission of variation at two autosomal loci between generations in a two-allele model that parallels the experiments of Mendel and of Bateson and coworkers. The loci are named A and B, and the alleles are A_0, A_1, B_0, and B_1. Conservation of the gene frequencies at the two loci again follows from Mendel's first law, and our focus is now on the joint transmission of the alleles. Four types of gametes exist, namely A_0B_0, A_0B_1, A_1B_0, and A_1B_1. They are usually numbered 1, 2, 3, and 4, with the frequencies x_1, x_2, x_3, and x_4 in the population ($x_1 + x_2 + x_3 + x_4 = 1$). The gene frequency of allele A_0 is then $p_A = x_1 + x_2$, and that of allele B_0 is $p_B = x_1 + x_3$. The alternative alleles A_1 and B_1 have the frequencies $q_A = x_3 + x_4$ and $q_B = x_2 + x_4$.

The four gametes form ten two-locus genotypes, that is, four double homozygotes, four that are homozygote at one locus and heterozygote at the other, and two double heterozygotes. The two double heterozygotes are the genotypes in the F_1 of the two Mendelian crosses $A_0A_0,B_0B_0 \times A_1A_1,B_1B_1$ and $A_0A_0,B_1B_1 \times A_1A_1,B_0B_0$. In the first of these the alleles at the two loci are said to be introduced in coupling, and in the second in repulsion, with the arbitrary parallel of 0 and 1 alleles at the two loci. In Mendelian crosses the obvious classification of the alleles is into dominant and recessive. In the corresponding heterozygotes the alleles are said to be in coupling and repulsion phase. An alternative nomenclature stems from linked genes in that the first of the heterozygotes

$$\frac{A_0B_0}{A_1B_1} \quad \text{and} \quad \frac{A_0B_1}{A_1B_0}$$

are said to be the in the *cis* phase (next to each other on a chromosome), and the second are in the *trans* phase (Figure 6.1 on page 142 shows the meiosis in a *cis* heterozygote).

The double heterozygotes segregate the parental gamete in the frequency $1 - r$ and the recombinant gamete in the frequency r (Table 6.4), where r is the recombination frequency between the loci ($0 \leq r \leq \frac{1}{2}$). The gamete frequencies among female and male gametes need not be the same, because the frequency of recombination may differ between female and male meiosis. In *Drosophila*, for instance, crossovers do not occur in male meiosis. Humans are less extreme and recombination happens in both sexes, but the total length of the linkage map for the autosomes is 43M in females and 27M in males, and in specific regions of the chromosomes the sex difference may be even greater.

Table 6.4: Segregation from double heterozygotes

Genotype	A_0B_0	A_0B_1	A_1B_0	A_1B_1
A_0B_0/A_1B_1	$\frac{1}{2}(1-r)$	$\frac{1}{2}r$	$\frac{1}{2}r$	$\frac{1}{2}(1-r)$
A_0B_1/A_1B_0	$\frac{1}{2}r$	$\frac{1}{2}(1-r)$	$\frac{1}{2}(1-r)$	$\frac{1}{2}r$

For the remaining eight genotypes, the constituent gametes can be deduced by knowing only the one-locus genotypes. This is not possible in the double heterozygote, making observation of two-locus gametic frequencies in diploid organisms difficult. The single heterozygotes segregate their two gamete types in even Mendelian proportions, and the homozygotes only produce one kind of gamete.

The frequency of the genotype formed by joining a female gamete i and a male gamete j is z_{ij}, i, $j = 1, 2, 3, 4$ (Table 6.5). This way of defining geno-

Table 6.5: Two-locus genotypic proportions

Maternal gamete	Paternal gamete			
	A_0B_0	A_0B_1	A_1B_0	A_1B_1
A_0B_0	z_{11}	z_{12}	z_{13}	z_{14}
A_0B_1	z_{21}	z_{22}	z_{23}	z_{24}
A_1B_0	z_{31}	z_{32}	z_{33}	z_{34}
A_1B_1	z_{41}	z_{42}	z_{43}	z_{44}

typic frequencies is very convenient for calculations. For instance, the genotype frequencies sum to one, which may simply be expressed as

$$\sum_{i=1}^{4}\sum_{j=1}^{4} z_{ij} = 1.$$

We see that in this sum the frequency of the heterozygote formed by gametes i and j, $i \neq j$, enters as $z_{ij} + z_{ji}$. We consider only autosomal loci, and to simplify the discussion we further assume that the recombination frequency is the same in the two sexes, thus giving equal proportions among the gamete types produced by the two sexes. We may therefore assume that $z_{ij} = z_{ji}$, and gamete frequencies may be calculated as

$$x_i = \sum_{j=1}^{4} z_{ij}, \ i = 1, 2, 3, 4 \quad \text{and} \quad \sum_{i=1}^{4} x_i = 1.$$

That is, our formulation has a built-in factor of $\frac{1}{2}$ in the heterozygote frequencies. The genotype A_0B_0/A_0B_1 occurs in the frequency $z_{12} + z_{21}$ and A_0B_0/A_0B_0 in the frequency z_{11}.

The effect of recombination on the production of gamete types in a population evidently depends on the genotypic frequencies, because segregation is influenced by linkage only in the double heterozygotes. Segregation is therefore dependent on the mating pattern in the population. We assume *random mating*, such that the genotypic frequencies are in *Hardy–Weinberg proportions*, that is, $z_{ij} = x_i x_j$ for all i, $j = 1, 2, 3, 4$ (Table 6.6). Initially we assume a population

Table 6.6: Two-locus Hardy–Weinberg proportions

	A_0B_0	A_0B_1	A_1B_0	A_1B_1
A_0B_0	x_1^2	$x_1 x_2$	$x_1 x_3$	$x_1 x_4$
A_0B_1	$x_2 x_1$	x_2^2	$x_2 x_3$	$x_2 x_4$
A_1B_0	$x_3 x_1$	$x_3 x_2$	x_3^2	$x_3 x_4$
A_1B_1	$x_4 x_1$	$x_4 x_2$	$x_4 x_3$	x_4^2

large enough to allow us to disregard the effects of random genetic drift. We thus explore the population genetic effects of linkage in a deterministic setting.

6.3   Linkage Equilibrium

In a large population with genotypes in Hardy–Weinberg proportions the gametic recurrence equations may be formed by a calculation of the gamete frequencies in the population based on a general description of the process of segregation and recombination. Chiasmata are formed and recombination occurs in any meiosis whether or not we can detect it.

Thus, a fraction $1 - r$ of all gametes produced in the population is formed after a meiosis in which recombination did not occur between locus A and locus B. In this part of the gamete pool the frequency of gamete A_0B_0 is x_1. The remaining fraction of r contains recombinant gametes. In these the allele at each of the two loci originates as one from each of the two different gametes that formed the individual producing the gamete. These two gametes were chosen independently (the assumption of random mating) in the gamete pool that formed the parent. A gamete carrying allele A_0 is therefore chosen with probability p_A and one with allele B_0 with probability p_B. The frequency of gamete A_0B_0 among recombinant gametes is therefore $p_A p_B$, and in total we get the gamete frequency in the offspring population as

$$x_1' = (1 - r)x_1 + r p_A p_B. \tag{6.1}$$

The similar recurrence equations for the gamete frequencies x_2, x_3, and x_4 may be obtained by using similar arguments, but because of gene frequency conservation at the two loci they follow from $x_2 = p_A - x_1$, $x_3 = p_B - x_1$, and $x_4 = 1 - p_A - p_B + x_1$. The recurrence equation (6.1) is therefore quite sufficient to describe the evolution in gamete frequencies in the population—given, of course, that we may ignore stochastic effects, that is, in each generation the expected number of recombinant gametes in the population is large, $2Nr \gg 1$.

The gamete frequencies remain unchanged from generation to generation when $r = 0$, or when $x_1 = p_A p_B$. In the first instance the gamete frequencies are conserved, because the gametes in the model are transmitted unchanged between generations, like the alleles in a one-locus four-allele model. The second instance corresponds, for any recombination frequency, to an equilibrium in the population. This population state is called *linkage equilibrium*.

The effect of the recombination fraction r can be revealed by collecting terms with r in equation (6.1):

$$x'_1 = x_1 - rD, \quad \text{where} \quad D = x_1 - p_A p_B. \tag{6.2}$$

The gamete frequencies therefore change when recombination occurs ($r > 0$) and $D \neq 0$. The variable D measures the deviation of the gamete frequencies from linkage equilibrium, and it is called the linkage disequilibrium measure, or the *linkage disequilibrium* for short. The gene frequencies are conserved, and therefore the linkage disequilibrium in the offspring generation can be immediately calculated as $D' = x'_1 - p_A p_B$. Inserting x'_1 from equation (6.1), we get

$$D' = x_1(1 - r) + r p_A p_B - p_A p_B = (1 - r)(x_1 - p_A p_B).$$

The recurrence equation describing the evolution of the linkage disequilibrium is therefore

$$D' = (1 - r)D. \tag{6.3}$$

After t generations of recombination and random mating, this equation iterates to

$$D^{(t)} = (1 - r)^t D.$$

When recombination occurs between the two loci, the factor $1 - r$ is less than one, causing a diminished deviation from linkage equilibrium every generation: $D^{(t)} \to 0$ for $t \to \infty$. Linkage equilibrium is approached from any initial gamete frequency in the population.

At linkage equilibrium the frequency of gamete $A_0 B_0$ is given by $p_A p_B$, the product of the gene frequencies of the alleles at the two loci. The gametic frequencies at linkage equilibrium are therefore

$$x_1 = p_A p_B, \quad x_2 = p_A q_B, \quad x_3 = q_A p_B, \quad x_4 = q_A q_B.$$

(This may be shown either by symmetry arguments or by using the expressions for the gene frequencies in terms of the gamete frequencies.) These are called the *Robbins proportions* after R. B. Robbins, who showed this result in 1918. At linkage equilibrium the alleles at the two loci therefore occur independently in the gametes of the population. That is, the probability of collecting an $A_0 B_0$ gamete by sampling at random is the same as the probability of first picking an A_0 allele and subsequently a B_0 allele. The Robbins proportions are shown as a surface in the gamete frequency domain in Figure 6.4.

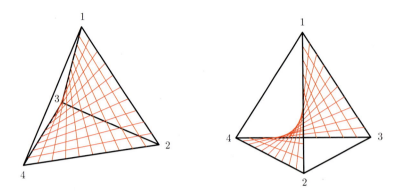

Figure 6.4: The surface of Robbins proportions in the gamete frequency domain, see Box 22. The four monomorphic equilibria $x_1 = 1$, $x_2 = 1$, $x_3 = 1$, and $x_4 = 1$ are at the vertices. The linkage equilibrium surface is indicated by the red lines. Notice that it includes the four edges where one of the alleles is fixed. At the edge 1–4 only the *cis* gametes are present in the population, and the edge 2–3 is that of *trans* gametes. For given gene frequencies, the linkage disequilibrium is maximal on these two edges.

The gamete frequencies are completely described by the gene frequencies p_A, q_A, p_B, and q_B and the linkage disequilibrium measure D. This induces the parameterization of the gamete frequencies shown in Table 6.7. This useful way

Table 6.7: Two-locus gamete frequencies

	B_0	B_1
A_0	$p_A p_B + D$	$p_A q_B - D$
A_1	$q_A p_B - D$	$q_A q_B + D$

of expressing the gamete frequencies allows us to describe the range of values that the linkage disequilibrium might attain, and because no gamete frequency can be negative, the linkage disequilibrium is bounded in the interval

$$- \min(p_A p_B, q_A q_B) \leq D \leq \min(p_A q_B, q_A p_B).$$

The gene frequencies are, of course, free to vary between 0 and 1. This property of the linkage disequilibrium makes its numerical value difficult to gauge, and therefore alternative measures have been suggested. A simplification is to pass D relative to its maximal value:

$$\acute{D} = \begin{cases} \dfrac{D}{\min(p_A p_B, q_A q_B)} & \text{when} \quad D \leq 0, \\[2mm] \dfrac{D}{\min(p_A q_B, q_A p_B)} & \text{when} \quad D \geq 0 \end{cases}$$

Box 22: Illustrations

Four gamete frequencies x_1, x_2, x_3, x_4 are needed to describe two loci with two alleles each. We really only need x_1, x_2, and x_3, as $x_1 + x_2 + x_3 + x_4 = 1$. For illustrations the use of x_1, x_2, and x_3 is awkward, as we always implicitly have to calculate x_4 in order to view all aspects of the variation.

One locus with the alleles A and a is described by their frequencies p and q, $p + q = 1$, which may be illustrated by splitting a line of length one:

For genotypic proportions z_{AA}, z_{Aa}, and z_{aa} at one locus, $z_{AA} + z_{Aa} + z_{aa} = 1$, we can do the same trick by using an equilateral triangle with altitude one:

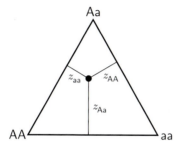

in that the distances to the three sides from a point in the triangle always sum to one. The corresponding figure in three dimensions is a tetrahedron with altitude one that can depict the two-locus two-allele gamete frequencies:

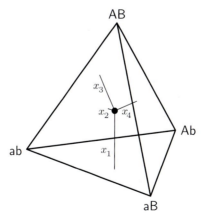

The distances to the four faces from a point in the tetrahedron always sum to one. The vertices correspond to populations monomorphic for gametes AB, Ab, aB, and ab, respectively.

(Lewontin 1964). $\acute{D}$ evidently assumes values between -1 and 1. A widely used measure is the *gametic correlation*:

$$\rho = \frac{D}{\sqrt{p_A q_A p_B q_B}} \tag{6.4}$$

(Wright 1933). This is calculated as the correlation between the indices of the two alleles in the gametes of the population.[4]

A randomly mating population in linkage equilibrium may be viewed as having attained the two-locus counterpart of Hardy–Weinberg equilibrium in that the genotypic frequencies are in the same Hardy–Weinberg proportions generation after generation. This state is called a *Robbins equilibrium* (Edwards 1977). Here the frequency of a genotype is found by multiplying the gene frequencies of all alleles in the genotype, and then multiplying by 2 for each locus where the individual is heterozygote.

The rate of convergence to Robbins equilibrium is given by the factor $1-r$ in the recurrence equation (6.3) for D. Free recombination, $r = \frac{1}{2}$, thus gives the fastest convergence, halving the linkage disequilibrium each generation. Even linkage disequilibrium between loci that show independent reassortment in Mendelian crosses exhibits a gradual convergence to zero. It is fast, but alleles at freely recombining loci may nonetheless show association in their population occurrence for several generations. For $0 < r < \frac{1}{2}$ the half-life of the linkage disequilibrium is on the order of magnitude of the reciprocal of the recombination frequency. The population is eventually expected to reach linkage equilibrium, where the loci are independent in their variation in the population while genetic linkage is apparent in the segregation of gametes in the individual. With absolute linkage, $r = 0$, the linkage disequilibrium remains unchanged from generation to generation.

The generalization of these results to cover an arbitrary number of loci was obtained by Geiringer (1944), Bennett (1954), and Reiersøl (1962): *"If recombination can occur between any pair of loci then, in time, the gametic frequencies will converge to linkage equilibrium where the alleles at the different loci are distributed independently in the gametes"* (Christiansen 2000). This multilocus principle extends to the general case of differences in recombination frequencies between sexes and to sex-linked loci (see Section 6.1.1 on page 143). It is also valid in populations of plants where seeds may be produced by self-fertilization in addition to random outcrossing (see Section 3.1.2 on page 59).

Recombination in a population with an element of random mating thus causes the gametic frequencies to converge to Robbins proportions in which the probability of picking at random a given gamete from the population equals the probability of picking at random each of the alleles carried in the gamete. Knowing that the gamete carries a given allele therefore gives no information about alleles at other loci in the gamete. The offspring of an individual, however, show segregation that exhibits linkage among loci in which the individual is heterozygote.

---

[4]. . . and therefore has nicer statistical properties.

Box 23: The forward calculations

A_0B_0 gametes are produced by individuals of a genotype formed by at least one A_0B_0 gamete. The homozygote only produces gametes of type A_0B_0. The gametes of the single heterozygotes segregate into two equally frequent classes, one of which is A_0B_0. Finally, the *cis* double heterozygote produces A_0B_0 in the frequency $\frac{1}{2}(1-r)$ (Table 6.4). Only one additional genotype may produce A_0B_0, namely the *trans* double heterozygote A_0B_1/A_1B_0, and it produces A_0B_0 in the frequency $\frac{1}{2}r$. In total, we get

$$x_1' = z_{11} + \tfrac{1}{2}(z_{12} + z_{21} + z_{13} + z_{31}) + \tfrac{1}{2}(1-r)(z_{14} + z_{41}) + \tfrac{1}{2}r(z_{23} + z_{32}).$$

Under the assumption that $z_{ij} = z_{ji}$, the equation simplifies to

$$x_1' = z_{11} + z_{12} + z_{13} + (1-r)z_{14} + rz_{23}.$$

Collecting the terms that involve the recombination fraction produces the effect of recombination as dependent only on the gametic phase disequilibrium, $\Delta = z_{14} - z_{23}$, and we get $x_1' = x_1 - r\Delta$. Segregation of gametes depends only on the recombination frequency in the double heterozygotes. The population-level dependence is expressed by the difference Δ of their frequencies. If the frequencies of the two double heterozygous genotypes are equal, then the differences in their population-level segregation cancel out. The forward recurrence equation can be turned into a recurrence equation of the gamete frequencies by describing the rules for gamete union.

6.3.1   The forward equations

The assumption of Hardy–Weinberg proportions allows a *backward* reference to gametic origin, and this is the key to the above arguments. Different recombination frequencies in the sexes may cause different gamete frequencies in eggs and sperm in the population. With random mating, the genotypic frequencies are in random union of gamete proportions. These retain an element of independence, allowing a backward argument that yields results similar to those obtained above (Christiansen 2000). Without random mating the gamete frequencies in the population need to be calculated based on a description of the process of segregation and recombination as it occurs in the individual. In this sense it is a forward calculation, and the resulting equations are *forward recurrence equations*. Still, the assumption of no difference between meiosis in the two sexes implies $z_{ij} = z_{ji}$ for all i, $j = 1, 2, 3, 4$; the resulting forward argument is given in Box 23.

   The forward gametic equations provide the recurrence equations in terms of *gametic phase disequilibrium*, $\Delta = z_{14} - z_{23}$.. The assumption of random mating produces genotypic frequencies in Hardy–Weinberg proportions (Table 6.6), and the gametic phase disequilibrium is therefore $\Delta = D = x_1x_4 - x_2x_3$.

6.3.2   Population mixing

The Wahlund effect in a mixture of previously isolated randomly mating populations describes an excess of homozygotes compared to the Hardy–Weinberg proportions, corresponding to the mean gene frequency in the mixed population (Section 5.3 on page 133). Thus, for two or many loci the Wahlund effect generates an excess of homogametic types when the genotypic frequencies are compared to the two- or multilocus Hardy–Weinberg proportions. The deviations from the Hardy–Weinberg proportions disappear after reproduction by random mating in the mixed population.

Hardy–Weinberg proportions reflect independence of gametes in a genotype, and a similar effect of mixing is therefore expected when mixing gametes in Robbins proportions. For two or more loci, linkage disequilibrium will therefore emerge in a mixed population (Sinnock and Sing 1972, Mitton et al. 1973, Nei and Li 1973, Feldman and Christiansen 1975). When mixing populations in linkage equilibrium, the difference between the gamete frequencies and the Robbins proportions in the mixed population is

$$\sum_{i=1}^{d} c_i p_{\mathsf{A}i} p_{\mathsf{B}i} - \bar{p}_{\mathsf{A}} \bar{p}_{\mathsf{B}} = \sum_{i=1}^{d} c_i (p_{\mathsf{A}i} - \bar{p}_{\mathsf{A}})(p_{\mathsf{B}i} - \bar{p}_{\mathsf{B}}) = \mathrm{Cov}(p_{\mathsf{A}i}, p_{\mathsf{B}i}),$$

where c_i is the relative size of the ith component of the mixture. $p_{\mathsf{A}i}$ and $p_{\mathsf{B}i}$ are the gene frequencies in that component. Their averages are $\bar{p}_{\mathsf{A}}$ and $\bar{p}_{\mathsf{B}}$ (see equation (5.17) on page 135). The linkage disequilibrium in the mixed population is therefore expected to be related to the covariance in gene frequencies among the components of the mixture.

This expectation of deviation from Robbins proportions alludes to the linkage disequilibrium. The genotypic frequencies in a mixed population, however, are not in Hardy–Weinberg proportions, because the genotypic frequencies already bear the signature of mixing, and so the gametic phase disequilibrium is describing the effect of breeding on the gametic frequencies (see Christiansen 2000). The offspring population is formed by random mating, so to describe the longer range effect of mixing we perform the simpler task of calculating the linkage disequilibrium in the offspring population. This is really a lot easier because recombination occurs within individuals, and the gamete production of individuals that originate from subpopulation i produces the same gamete frequencies as if mixing had never occurred. The frequency of gamete $\mathsf{A}_0\mathsf{B}_0$ produced by subpopulation i individuals is therefore given by the backward equation

$$x'_{1i} = (1 - r)x_{1i} + r p_{\mathsf{A}i} p_{\mathsf{B}i} \quad \text{or} \quad x'_{1i} = p_{\mathsf{A}i} p_{\mathsf{B}i} + (1 - r)D_i$$

(see equation (6.1) on page 150), where D_i is the linkage disequilibrium in subpopulation i at the time of mixing. The total production of gamete $\mathsf{A}_0\mathsf{B}_0$ in the mixed population therefore becomes

$$x'_1 = \sum_{i=1}^{d} c_i x'_{1i} = \sum_{i=1}^{d} c_i \big(p_{\mathsf{A}i} p_{\mathsf{B}i} + (1 - r)D_i \big) = (1 - r)\bar{D} + \sum_{i=1}^{d} c_i p_{\mathsf{A}i} p_{\mathsf{B}i},$$

where $\bar{D}$ is the average linkage disequilibrium in the components of the mixture. The linkage disequilibrium in the offspring population is $D' = x_1' - \bar{p}_A\bar{p}_B$, and the covariance in gene frequencies reemerges as the linkage disequilibrium created by the mixing:

$$D' = (1 - r)\bar{D} + \text{Cov}(p_{Ai}, p_{Bi}). \tag{6.5}$$

The average linkage disequilibrium in the subpopulations after breeding is expected to be $(1 - r)\bar{D}$ whether interbreeding occurs or not, so this term is contributed by the linkage disequilibrium in the subpopulations before mixing occurs. The amount of linkage disequilibrium created by the mixing therefore contributes to the linkage disequilibrium in the population of offspring of the mixed population.

The discussion of the effect of mixing is facilitated by assuming that the mixing subpopulations are in linkage equilibrium. The signature of mixing is then the emergence of linkage disequilibrium equal to the covariance in gene frequencies. This deviation from Robbins proportions is degraded at a finite rate determined by the frequency of recombination between the loci, and the signature of mixing can therefore be traced through many generations.

6.3.3 Recurrent mixing and migration

The finite rate of degradation of linkage disequilibrium causes recurrent migration among the subpopulations in a geographically structured population to accumulate high levels of linkage disequilibrium between polymorphic loci with geographical gene frequency variation. Feldman and Christiansen (1975) studied this phenomenon for two loci in the stepping-stone cline model of geographical structure. This model is inspired by the stepping-stone model (Section 5.2.4

on page 129) and consists of a finite number of islands connecting two mainlands, which are large populations that are not disturbed by immigrants from the islands (as in the classical island model).

We assume that the two mainlands differ in gene frequencies at both loci. Migration along the stepping-stone chain of islands produces linear gene frequency clines (see Box 24). When these are established migration will result in the same covariance in gene frequencies in each population after immigration, and the production of linkage disequilibrium is the same in all the island populations. The variation in linkage disequilibrium through the cline is therefore determined by the linkage disequilibrium on the mainlands, the frequency of immigration, and the recombination frequency. It is rather complicated to determine the equilibrium in the linkage disequilibrium values, but simplifications

---

Box 24: * The stepping-stone cline model

The recurrence equations in the gene frequencies at locus A in the various populations in the stepping-stone cline model are

$$p'_{\mathsf{A}\,i} = m p_{\mathsf{A}\,i-1} + (1 - 2m) p_{\mathsf{A}\,i} + m p_{\mathsf{A}\,i+1}, \quad i = 1, 2, \ldots, \ell.$$

The equilibrium in these equations is

$$\hat{p}_{\mathsf{A}\,i} = \frac{\ell - i + 1}{\ell + 1} p_{\mathsf{A}\,0} + \frac{i}{\ell + 1} p_{\mathsf{A}\,\ell+1}, \quad i = 1, 2, \ldots, \ell,$$

and the populations converge to this equilibrium. Thus at equilibrium the gene frequencies exhibit a linear cline. The gene frequencies at locus B settle at a similar cline.

At equilibrium the mean gene frequency among immigrants and residents in population i is $p_{\mathsf{A}\,i}$. The covariance in gene frequency among the components in population i after migration is therefore at equilibrium equal to

$$\begin{aligned}
\mathrm{Cov}_i(p_{\mathsf{A}}, p_{\mathsf{B}}) = \ & m(p_{\mathsf{A}\,i-1} - p_{\mathsf{A}i})(p_{\mathsf{B}\,i-1} - p_{\mathsf{B}i}) \\
& + m(p_{\mathsf{A}\,i+1} - p_{\mathsf{A}i})(p_{\mathsf{B}\,i+1} - p_{\mathsf{B}i})
\end{aligned}$$

because the $(1 - 2m)$ term is zero. At equilibrium

$$\hat{p}_{\mathsf{A}\,i-1} - \hat{p}_{\mathsf{A}i} = \frac{p_{\mathsf{A}\,0} - p_{\mathsf{A}\,\ell+1}}{\ell + 1}, \quad i = 1, 2, \ldots, \ell, \ell + 1,$$

and the covariance is therefore the same in all the island populations,

$$\mathrm{Cov}_i(p_{\mathsf{A}}, p_{\mathsf{B}}) = \mathrm{Cov}(p_{\mathsf{A}}, p_{\mathsf{B}}) = \frac{2m(p_{\mathsf{A}0} - p_{\mathsf{A}\,\ell+1})(p_{\mathsf{B}0} - p_{\mathsf{B}\,\ell+1})}{(\ell + 1)^2}.$$

---

can be obtained by reasonable approximations. We found the linkage disequilibrium in the middle of the cline to be approximately given by

$$\hat{D} \approx \frac{\mathrm{Cov}(p_{\mathsf{A}}, p_{\mathsf{B}})}{r}$$

when r is not too small and the mainlands are at linkage equilibrium. Thus, apart from being proportional to the covariance in gene frequencies between neighbor populations, the linkage disequilibrium is amplified to become inversely proportional to recombination frequency between the loci. To see why this is a plausible result, consider a gamete in any of the stepping-stone populations. The probability that it was produced by recombination in a parent is r, and the waiting time until it experienced recombination is therefore geometrically distributed with event probability r. An average gamete has therefore been transmitted unaltered through r^{-1} generations, and therefore experienced that number of mixing events. Thus, recurrent mixing due to immigration into the local population may build a level of linkage disequilibrium of much greater magnitude than the amount created by a single episode of mixing.

Szymura and Barton (1991) provided a beautiful example of the creation of linkage disequilibrium by population mixing in their study of a hybrid zone between the fire-bellied toads *Bombina bombina* and *B. variegata*. They studied electrophoretical variation determined by six unlinked loci in populations of toads in and around the hybrid zone. Far from the zone the two species have different alleles at the six loci, so their presence in both species close to the zone confirms interbreeding of the two species. Szymura and Barton observed the expected high values of pairwise linkage disequilibria close to the hybrid zone, whereas samples from populations further away were at linkage equilibrium. The hybrid zone between *B. bombina* and *B. variegata* is characterized by the presence of morphologically intermediate individuals, which seem to produce fewer offspring than the pure forms. The studied loci are therefore expected to be influenced by natural selection because of the buildup of linkage disequilibrium in the two populations. We return to this study in Chapter 10, where hybridization is discussed, but we mention it here because the authors apply the emergence and preservation of linkage disequilibrium in mixed populations in their analysis.

6.3.4 Stochastic effects

Random genetic drift of course disturbs gamete frequencies. In the Wright–Fisher model the numbers of the four gametes at two loci with two alleles are multinomially distributed with probability parameters equal to the gamete frequencies in the pool of gametes produced by the parents. The changes in gene frequencies are well known (Chapter 2). With reference to the backward equation we may use these results by viewing the dynamics of linkage disequilibrium as an effect of random genetic drift in the sampling of parental gametes and stochastic effects in the formation of recombinant gametes. In parallel with the effects of mixing, the effects of random genetic drift on the linkage disequilibrium are thus expected to be higher for closely linked loci.

Deviations from Robbins proportions are important in any finite population. A population of size N is formed from $2N$ gametes, and n polymorphic loci each with two alleles may form 2^n gametes. Many studies of human populations now use hundreds of SNPs and these can form at least 10^{30} gametes,[5] so at any time our population can only contain a tiny fraction of these gametes. However, recombination has the potential to form all possible gametes, but the waiting time to a recombination event may be very long for closely linked loci, and any locus is therefore expected to be associated with a stretch of chromosome surrounding it (Sved 1971). For unlinked loci the situation is only slightly better: one two-allele locus at each of our autosomes makes us investigate more than 10^7 types of gametes—indeed a rare, if not impossible, order of magnitude for

---

[5] 2^{10} is a bit larger than 10^3, so 2^{100} is larger than 10^{30}.

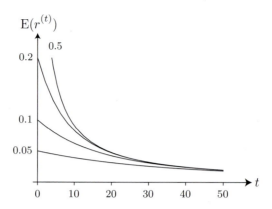

Figure 6.5: The decrease in the conserved piece of chromosome next to a marker locus during backcrosses for $r = 0.5, 0.2, 0.1$, and 0.05.

the sample size in population genetic investigations.

The anticipation of deviations from Robbins proportions turns the interest to the linkage disequilibrium measure. But its properties are hard to evaluate when gene frequencies are expected to change. Its expectation is zero, $E(D) = 0$, because of the expectation of linkage equilibrium, but this just reflects that positive and negative deviations have equal weight. We are actually interested in its variation, but it has proven difficult to find good approximations for the variance $E(D^2)$. A crude approximation of the expectation of the squared correlation in gene frequencies is

$$E\left(\frac{D^2}{p_A p_A q_B q_B}\right) \approx \frac{1}{4Nr + 1} \tag{6.6}$$

(Hill and Robertson 1968, Ohta and Kimura 1969, Hudson 1983). As expected a significant buildup of linkage disequilibrium occurs when the number of recombinant gametes produced in the population per generation is small.

6.3.5   Backcrosses

Animal and plant breeders are regularly in the situation that a mutant producing a desirable trait is available in a single individual. This situation applies equally to a new mutation as well as to the introduction of an entire gene into an individual. The interesting allele of the gene is A and the alternative allele is a (in case of a genetic transformation, think of the homologous nontransformed chromosomes as carrying allele a at the locus of the transformation). The goal of the transformation, is to introduce a transgene into livestock or a cultivar, and to accomplish this, backcrosses are performed between individuals of genotype Aa and individuals of genotype aa from the population. Among the offspring of the backcross we choose individuals of genotype Aa and again cross these to individuals from the population. This procedure is repeated every generation.

Initially the backcrosses multiply allele A, but in addition they allow recombination between the genome in which allele A appeared and the genomes in the population. As a first approach to the effects of recombination we consider a simpler experiment where only one individual of genotype Aa is chosen as parent every generation. At some generation allele A is associated with a contiguous stretch of chromosome identical by descent to the original chromosome marked by A. We need only consider the associated piece to one side, say, right,

and denote its length on the linkage map as r. In the next breeding event the right-hand piece of chromosome is conserved with probability $1 - r$, and in this case it remains size r. The right-hand piece is broken by recombination with probability r, and in this case the remaining right-hand piece from the original chromosome is on average of length $\frac{1}{2}r$, because any point within the piece is equally likely to be the point of recombination. The expected length of the new piece is therefore

$$\mathrm{E}(r'|r) = (1 - r) \times r + r \times \tfrac{1}{2}r = \left(1 - \tfrac{1}{2}r\right)r \,.$$

The conserved piece becomes smaller and smaller because it is, on average, decimated by a factor of $1 - \frac{1}{2}r$ every generation. This factor, however, approaches one as r becomes small, and the rate at which the conserved piece of chromosome shortens will be lower and lower.

The change in the mean of the conserved piece may be found by forming a recurrence equation for the distribution function of the length of the conserved piece, and the result is

$$\mathrm{E}(r^{(t)}) = \frac{1 - (1 - r)^{t+1}}{t + 1} \tag{6.7}$$

(Christiansen 1990*b*). Figure 6.5 shows that a long piece of associated chromosome decays quite rapidly, but as it becomes shorter, further decrease is slow.

6.4 The Ancestral Recombination Graph

A sequence covers a finite distance on the linkage map of the organism, and recombination can therefore happen within the stretch of chromosome it covers. Intragenic recombination disrupts the genealogy of a gene in a more fundamental way than mutation. It causes the gene to have two parental genes, and the genealogy of the gene therefore splits into ancestral material carried in two different genes. The first to describe this process as an extension of the coalescent was Hudson (1983, 1991).

Figure 6.6 shows an example of the genealogy of two sites, a red ● and a green ●, in a sample of four genes. The two sites find their ancestral material in four genes until τ_1 generations ago, when gene number 4 was transmitted

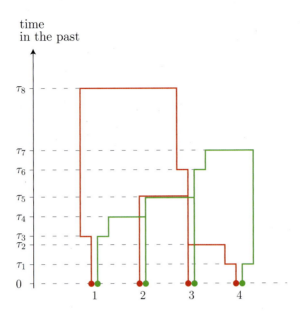

Figure 6.6: A genealogy of two sites in a sample of four genes.

through a meiosis in which recombination took place between the two sites. Thus, prior to τ_1 the ancestral sites were carried in two different genes, and the ancestral material of the sample is therefore carried in five genes. At time τ_2 the gene with the red site 4 coalesces with gene number 3, and the ancestral material of the sample is back to four genes. After yet another recombinational split at τ_3 and a coalescence of green sites at τ_4 we get what seems like an old-fashioned coalescent event of the gene at τ_5. However, we can only conclude that the event at τ_5 is a coalescent of the genes that carry the ancestral material of the red site in genes 2, 3, and 4, and of the green site in genes 1, 2, and 3. After a recombination event in the lineage of this gene, the green sites of the four genes find their most recent common ancestor at time τ_7 and the red site finds its most recent common ancestor at time τ_8.

The effect of recombination is that the genealogies of the two sites are different (Figure 6.7). The difference in coalescence times is already apparent in Figure 6.6, and Figure 6.7 shows that the topology of the trees also differ. For instance, the red sites of genes 1 and 2 are separated by the coalescence time, whereas the green sites are the first to coalesce. Thus, if we take mutation into account, genes 1 and 2 have a higher probability of differing in the red site than in the green site. Further illustrations of the buildup of the coalescent with recombination is obtained from the Hudson animator (Christensen et al. 2006).

The genealogy in Figure 6.6 has a common characteristic for genealogies with recombination, namely that the genealogies of the two sites coalesce in different genes. Thus, we have two site genealogies that must be fitted into one gene genealogy, and this can be accomplished by following the ancestral material of

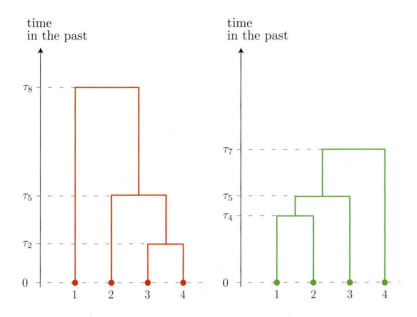

Figure 6.7: The genealogies of each of two sites in a sample of four genes.

the sampled gene sequences until it coalesces into one ancestral sequence. We thus consider observed sequences. To keep matters simple, Figure 6.8 shows the genealogy of sequences number 1 and 4 in the sample in Figure 6.6. The entire sequence is depicted as a bar; sequence 1 and its ancestral material are shaded yellow, ▭ and sequence 4 is shaded orange ▬. The two sites in Figure 6.6 are at the ends of the sequences. After a recombination event, like the events at τ_1 or τ_3, the parental sequences are shown with the nonancestral material unshaded ▭. The probability that a given sequence originates from a meiosis in which it experienced a recombination equals the probability r of recombination between the red and green sites. The stretch of DNA sequence is assumed to be short, r is small, and the possibility of double crossovers is disregarded. A single crossover point is thus assumed and the left and right parental contributions are shown to the left and to the right. After a coalescent event like the one at τ_5, the parental sequence is shown with the combined shading, and sites with coalesced ancestral material are shaded black ▬, like the event at τ_7 where the green site coalesces. This genealogy only shows the recombination events that break the association of ancestral material, that is, we neglect recombination events with a crossover within nonancestral material at the ends of the sequence. The full description is provided by the ancestral recombination graph, which includes all recombination events in the ancestral sequences (Griffiths and Marjoram 1997).

The sites at the ends of the sequence coalesce at times τ_7 and τ_8, and in the figure the sites in the middle of the sequence coalesce later, such that the time of coalescence is given by τ_{10} and τ_{11}. Again taking mutation into account, we expect the regions towards the ends of the sequences to be more similar than

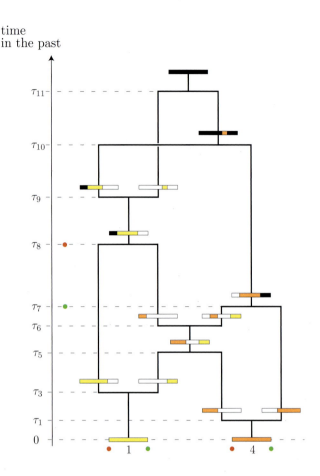

Figure 6.8: The genealogy of genes 1 and 4 in Figure 6.6. The two sites in that figure are assumed close to the ends of the gene. The figure is explained in the text.

those in the middle when sequences number 1 and 4 are compared. In general, a sequence coalescence will specify the coalescent tree for each site along the sequence. The sequence will thus be partitioned into segments in which the sites share a common coalescence. These trees will differ by recombination events, and their total length will determine the total variation observed within the corresponding segment.

τ_8	τ_{10}	τ_{11}	τ_7

The genealogy in Figure 6.8 is very simple, and even a small change can illustrate the complications that recombination can produce. Assume that everything is the same up to τ_{10}, but introduce a recombination event within the

time
in the past

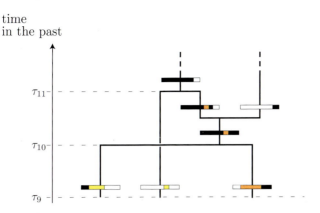

Figure 6.9: An alternative to the top of the genealogy in Figure 6.8.

coalesced material in the right part of the sequence between times τ_{10} and τ_{11} as in Figure 6.9. Then at time τ_{11} all sites have coalesced and the ancestral origin of all the genetic material in our sample has been found. The ancestral material, however, is disbursed in more than one sequence. Although the description of the genealogical process beyond the time when all sites have coalesced seems of purely academic interest, we can still ask the question whether the ancestral material will ever end up in one sequence. The answer is yes, but the waiting time until that happens may be so long that the possibility is irrelevant. In Figure 6.9 the probability that the two sequences left coalesce in a given generation is $\frac{1}{2N}$ and the probability that one of them experiences a recombination is r, the probability of recombination within the sequence. The probability of a coalescence before a recombination event is therefore equal to

$$\frac{1}{1 + 4Nr}.$$

If $4Nr$ is large, then this probability is small and we expect many recombination events before a coalescence happens. The typical situation is therefore that the ancestral material is spread into many sequences in the population. On the other hand, if $4Nr$ is small, then a genealogy like the one in Figure 6.8 is very likely. The average waiting time until a sample of two sequences coalesces onto a single ancestral sequence is not known, but Griffiths and Marjoram (1997) calculated the time until the ancestral recombination graph ends in one sequence as

$$4N \frac{e^\rho - 1 - \rho}{\rho^2}, \tag{6.8}$$

where $\rho = 4Nr$. This waiting time is longer than the time until the ancestral material is in one contiguous sequence. They require, for instance, that all the material in the three sequences after τ_9 in Figures 6.8 and 6.9 coalesce into one sequence before the ancestral recombination graph is terminated. For $\rho \rightarrow 0$

formula (6.8) returns the coalescence time 2N as in equation (2.16) on page 37, and for large ρ the waiting time increases exponentially as ρ increases. For comparison, the waiting time W_n until all sites in the sequence have coalesced in a sample of n sequences has a mean that is bounded to at most a linear increase in that

$$\mathrm{E}W_n \leq 4N\left(1 + \rho\,\frac{n^2 + n - 2}{4n(n+1)}\right) \leq 4N\left(1 + \tfrac{1}{4}\rho\right)$$

(Griffiths and Marjoram 1997). In both Figures 6.8 and 6.9 $W_2 = \tau_{11}$.

The situation in Figure 6.9 is still very simple because the ancestral material sits as one block in each of the two sequences. Repeated recombination can spread the material further (Wiuf and Hein 1997). A sequence like

$$\begin{array}{cccc} & & & \\ r_1 & r_2 & r_3 & r_4 \end{array} , \tag{6.9}$$

where $r_1 + r_2 + r_3 + r_4 = r$, has the probability $r_1 + r_2 + r_3$ of experiencing a recombination event that spreads the ancestral material. Thus, recombination within "holes" in the ancestral material also contributes to dissipation, and the probability of further spread of the ancestral material will therefore typically be larger than the length $r_1 + r_3$ of the ancestral material. This effect also contributes to the long waiting times to final coalescence of the whole sequence.

The description of the final coalescence of ancestral material into a single sequence gives a valuable impression of the properties of the ancestral recombination graph of a single sequence. The average waiting time until the sequence experiences a recombination is ρ^{-1}, and subsequently the average waiting time until it is again collected as a unit is approximately given by (6.8). These two waiting times are compared in Figure 6.10. For $\rho < 1$ the waiting time to recombination quickly becomes much longer than the collecting time and coalescence therefore predominates. For $\rho > 1$ splitting of ancestral material quickly becomes the dominant feature.

The average waiting time until recombination shown in Figure 6.10 refers to the length of the observed sequence on the linkage map. The split of the ancestral material is a more complicated process. The average waiting times until recombination of the ancestral material in the two sequences after time τ_{11} in Figure 6.9 differ by a factor of five, so *the smaller the piece, the longer the wait* (see Section 6.3.5 on page 160). The expectation is therefore that as we go further back in time the ρ value of a typical piece of ancestral material decreases, and eventually the rate of coalescence surpasses the recombination rate and the pieces start to collect again (Figure 6.10). Coalescence of atomized ancestral material is, however, likely to produce sequences like (6.9), where the distribution of the sizes of contiguous ancestral material remains unchanged. The long wait for the breakage of small pieces of ancestral material will therefore be important for the distribution of the lengths of intervals of contiguous ancestral material.

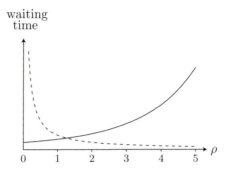

Figure 6.10: The average waiting time until a sequence experiences a recombination (dashed curve) compared to the average waiting time until the ancestral material associated with the observed sequences coalesces (solid curve).

6.4.1 Genealogy and recombination

In Section 4.3 we briefly discussed the analysis of sequence data and argued that mutations could aid in the reconstruction of the genealogy of the sequences. When recombination is taken into account, sequence analysis becomes vastly more complicated because we cannot assume that all sites have the same genealogy (Figure 6.7). Variation among the sequences is still the only source of information for our inference on genealogy, but mutation can in principle produce any pattern of variation. If in a sample of four sequences we observe the bases AA, AT, TA, and TT at two sites, we may explain it by three mutation events in the genealogy of the four sequences, with two events at one of the sites. If the ancestral type was AA, mutation events may produce AT and TA. To obtain TT at the two sites one of the mutants, AT or TA, must mutate again. When mutations are sufficiently rare to neglect the possibility of more than one per site (the infinite sites model), then the four sequences can only be explained if we postulate at least one recombination between the two sites. In this instance, the observation can be taken as evidence of recombination. This method for inferring recombination is Hudson and Kaplan's (1985) *four-gamete test*. The important pattern in the four-gamete test is, of course, that the four gametes are observed at two sites each with two bases. The diagram (4.3) on page 84 that describes the variation in the example analyzed in Section 4.3 shows no evidence of recombination even though two of the site pairs occur in four variants.

The four-gamete argument can be turned around in the infinite sites model. The only recombination events in the genealogy that can possibly be directly detected are those that result in four gametes with respect to two sites. To reflect this, two sites are considered *compatible* when the genealogy of their variation can be described by single mutations in a common coalescent tree (without recombination). This way of arguing, however, does not refer to the sequence as

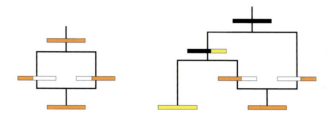

Figure 6.11: Two recombination events with minor effects on the genealogy.

an ordered row of bases, that is, to the phenomenon of genetic linkage. Without recombination the genealogical structure of a sample is inferred from the mutational differences among the observed sequences. The pattern of the mutations is neglected, and although a nonrandom distribution of the mutations over the sequence arouses our biological curiosity, it is without immediate consequence for coalescence analysis.

When recombination is allowed, the pattern of mutations does play a role in the analysis. The observation of three sequences with the ten variable sites within a long sequence (shown to the left) suggests a recent recombination between the variable sites 6 and 7, although it cannot pass the four-gamete test. Such observations are highly indicative of recombination, but they are expected only if recombination is very rare (Maynard Smith 1999). When recombination is more common it is unlikely for three sequences to have diverged at ten sites with only one recombination. The three sequences thus show a pattern that is expected as a rare deviation in cases where we can argue that recombination is sufficiently infrequent to be neglected. For a less extreme recombination rate the observation becomes likely following rare but massive immigrations.

GACCTGATTC
GACCTGGCAT
AGTACAGCAT

Only some of the recombination events in the ancestral material have the potential to leave a trace in the sample. In Figure 6.8 suppose that the coalescent event at τ_5 involved the two rightmost lineages. The impact of the coalescent event would then be described by the left graph in Figure 6.11. The recombination event at τ_1 would leave no trace in the sampled sequences, and the example shows that recombination events in the ancestral material of a sample may not be observable even in principle. On the other hand, if the recombination events at τ_3 and τ_6 had not occurred, the genealogy would look like the right graph in Figure 6.11, and the single recombination event would result in different coalescence times in different parts of the sequence. The two sequences would therefore show different amounts of divergence depending on site position. Coalescence time and topology were influenced by recombination in the example in Figure 6.6, but more than one recombination event was involved. In a sample of three sequences, one recombination event can, however, make the topology similarly heterogeneous. The left and right of the sampled sequences in Figure 6.12 have different coalescence times and mirrored topologies, like those in

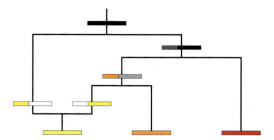

Figure 6.12: A simple recombination events with a topological effect on the genealogy.

Figure 6.7 on page 163. Moving the recombination event to the middle sequence produces what is essentially the genealogy in Figure 6.6.

A high frequency of recombination within the sequence will break associations among sites. The alternative bases found at variable sites will be independent—they are in linkage equilibrium, a state reached when many recombination events occur between adjacent mutation events.[6] If the sequences are samples from a population in which the variable sites of the sequence are in linkage equilibrium, then the genealogical information supplied by the variation at a site pertains only to that site. Thus, virtually nothing can be inferred about the genealogy of the sample.

Absolute linkage maintains an association among the sites of a sequence that can only be broken by mutational events, and such local breaks in the association were used in the sequence analysis in Section 4.3 (page 84). A break in the coherence within a piece of nucleic acid caused by mutation compares to the more profound breaks of association caused by recombination. In classical population genetic terms, mutation causes the gamete frequencies to converge to linkage equilibrium just as recombination does. Therefore, too much mutation will dissolve associations in the sequence and thereby hide the effects of linkage. The ancestors of a sequence can hence only be followed for a short while—in effect the phenomenon discussed in Section 4.4 (page 87). Too little variation weakens inference on genealogy. An adequate amount of mutation with absolute linkage engenders sufficient variation to resolve the genealogical structure and leaves enough constant sites in any comparison to allow us to trace ancestry back in time.

For a given mutation rate the requirement is thus that the sequence under study is sufficiently long, but this makes the likelihood of recombination increase. Thus, when planning observations for sexual organisms the investigator is caught between a rock and a hard place: long sequences provide the most useful variation and short sequences the most coherent analysis. Recall that sample size has not been mentioned in these arguments. These effects of sequence length are due to inherent limitations to the amount of genealogical

---

[6]This effect parallels the effect of strong migration discussed in Section 5.2.3 on page 128.

information that can be extracted from sequences. The genealogical structure is described by the parameters θ and ρ, and the estimates of these become better and better (lower and lower variance) the more sequences we sample. The effect of increasing the sample size is rather limited, however, and even for very large samples the estimates still have high variances (Fearnhead and Donnelly 2001). The mutation rate parameter θ is always determined much more accurately than the recombination rate parameter ρ.

Simulation tools to follow the effects of mutation and recombination on the genealogy of a sample are available at the website of Christensen et al. (2006). Recombination and coalescence both affect the topology of the ancestral recombination graph, and so the two processes have to combine in the real time of the simulation. From a given state in the process the waiting times till recombination and coalescence (and we might as well include mutation) are determined, the next event is thus determined by the shortest wait, and that event is applied to a randomly drawn sequence. Furthermore, in case of recombination or mutation the random position of the event has to be determined. This algorithm is then applied until we arrive at a specified ancestral state. If we are looking only for the topology of the ancestral recombination graph we do not need the waiting times. Relative event probabilities suffice.

What happens if we analyze the genealogy of a sample of sequences under the assumption that recombination is negligible even though it does occur? For two variable sites in a sample of four, Figure 6.7 demonstrates that we cannot hope to get a sensible answer. Schierup and Hein (2000a, 2000b) showed that the resulting genealogical tree deviates from Kingman's coalescent. The number of mutations expected in the various branches can no longer be expected to be proportional to branch length. The molecular clock seems to tick at different rates in the various branches. This may be illustrated by drawing the branch length as proportional to the expected number of mutations while keeping the total branch length of the genealogy the same as in Figures 6.6 and 6.7 (pages 162 and 163). Such a drawing is shown in the figure to the right. Because of the lack of a molecular clock, this tree can be drawn in different ways (it is unrooted), and in the figure the root of the tree is placed at $\frac{1}{2}(\tau_7 + \tau_8)$, the average of the coalescence times at the two variable sites.

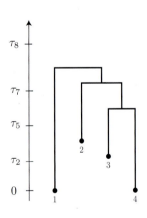

Apart from the loss of a common molecular clock, this tree also has long terminal branches, another characteristic found by Schierup and Hein. Given that we convinced ourselves that recombination can be neglected, such deviations from the expected pattern in Kingman's coalescent may be taken as evidence for interesting evolutionary phenomena (e.g., causing $\hat{\theta}_T > \hat{\theta}_W$, see page 103). Thus, unless the assumptions of the analysis are firmly supported, these phenomena may just be ghosts of the fundamental laws governing the transmission of genetic material from parents to offspring.

6.4.2 The mechanism of recombination

The molecular mechanisms of recombination between homologous chromosomes are of interest when sequence variation is studied. As indicated in Figure 6.1 on page 142, chromosome crossover (or chiasma formation) occurs while the chromosomes consist of two chromatids joined at the centromere. Consider the paired chromosomes in a chiasma and assume that they differ at three sites, namely where the two upper chromatids carry Ts and the two lower ones Gs (paired with As and Cs, respectively). Two of the chromatids interact in the

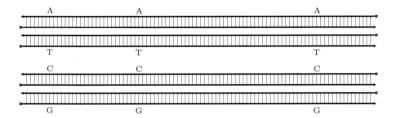

chiasma. In each, one of the DNA backbones is cut, roughly at homologous sites (the sketch below show only the two interacting chromatids). Each of

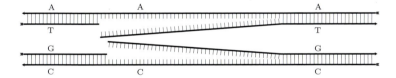

the two loose strands pair with the exposed strand in the other chromatid. This configuration, known as the *Holliday structure*, binds the two chromatids together. Note that the pairing of exposed strands creates a stretch of hybrid

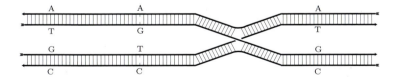

DNA, a so-called heteroduplex, in which mispairings occur at sites where the original strands differ.

The Holliday structure may be resolved in two ways. One is obvious in the way it is drawn here, namely to cut the two outer strands and synthesize replacements along the unbroken strands of DNA (top drawing in Figure 6.13). This resolution results in a crossover between the two chromatids involved in the Holliday structure in that the T–A pair to the left links up with the G–C pair to the right, and vice versa. The alternative resolution is to cut the two inner strands (bottom drawing in Figure 6.13). This may seem awkward, but remember that the Holliday structure in fact consists of two entangled double

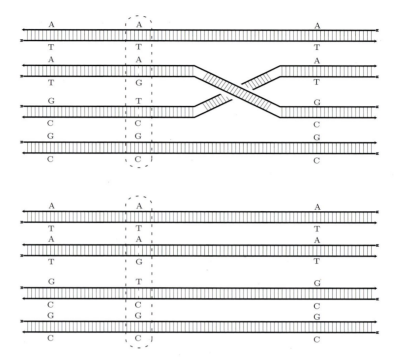

Figure 6.13: The alternative ways of resolving the Holliday structure. Top: crossover resolution. Bottom: resolution without crossover.

helices of DNA, so the two possibilities are actually equally likely. The result is not a crossover because the T–A pair to the left links up with the T–A pair to the right and the G–C pair to the left links up with the G–C pair to the right.

The second Holliday resolution causes an observable genetic effect even if it seems like nothing happened. The effect is also apparent in the first resolution, but its result is usually not as spectacular. The mismatches at the site of the middle marker (traced in the drawings) are repaired. If the strings running from left to right are used as templates, the upper chromatid will have the A changed to a C, and the lower will have C changed to A. Reading the strings running from left to right, this repair yields T, G, T, and G in the four chromatids, and the four chromatids become indistinguishable from the originals at the sites of the three markers. However, four possible repair configurations are determined by the choice of template strings. The resulting configurations at the middle marker are

$$
\begin{matrix}
\text{T} \\ \text{G} \\ \text{T} \\ \text{G}
\end{matrix}\text{,}\quad
\begin{matrix}
\text{T} \\ \text{T} \\ \text{T} \\ \text{G}
\end{matrix}\text{,}\quad
\begin{matrix}
\text{T} \\ \text{G} \\ \text{G} \\ \text{G}
\end{matrix}\text{,}\quad\text{and}\quad
\begin{matrix}
\text{T} \\ \text{T} \\ \text{G} \\ \text{G}
\end{matrix}\text{,}
$$

where the first corresponds to the example above. If each DNA strand is equally likely to become the repair template, then these four outcomes are equally likely.

The first Holliday resolution has the same possible outcomes of repair of the heteroduplex section. The first and the last of these do indeed follow the rule of balanced segregation of gametes after a meiosis, but the second segregates three Ts to one G and the third gives one T to three Gs. A random gamete still has the probability $\frac{1}{2}$ of carrying T and $\frac{1}{2}$ of carrying G, so the phenomenon does not disrupt Mendelian segregation. It was quite a surprise, however, when this skew segregation among the meiotic products was first observed in the 1950s in work with yeast, *Neurospora*, and other Ascomycetes, where segregation among the products of a single meiosis may be observed. It looked like rare changes to the alternative allele in the cross, and the phenomenon was dubbed *gene conversion*. It was soon shown to be associated with recombination, and in about half the observed incidences of gene conversion it was associated with recombination between closely linked markers on either side of the converting marker. This suggests that the two Holliday resolutions are equally likely. Gene conversion is a general phenomenon, but the observed frequency of the two resolutions seems to vary considerably.

If we observe the three markers in the above chiasma diagrams, the repair process in the first resolution may only move the apparent crossover point from the right to the left of the middle marker. In the second resolution, however, repair may produce a haplotype with respect to the three markers that reveal recombination simultaneously in both intervals between the markers—a double crossover. Such an observation is not expected in classical genetics because many observations have shown that double crossovers between closely linked markers are exceedingly rare, if not impossible.

This classical phenomenon is described as recombination interference and is caused by chiasmata interference in that chiasmata, and therefore crossovers, cannot be formed close together. Chiasmata interference is said to give rise to positive interference in recombination (fewer double crossovers than expected, assuming independence between recombination events along the chromosome). Gene conversion may then be viewed as extreme negative interference on short linkage distances (more double crossovers than expected, assuming independence). Recombination interference is thus positive between chiasmata and negative within a region typically covered by a chiasma. In the preceding section we discussed recombination of sequences assuming classical recombination, thus neglecting gene conversion, and in the next section we include the local effects of chiasma formation.

Recombination and gene conversion within an exon may cause changes with phenotypical effects that look like mutations. The process of recombination may, however, in itself cause bona fide mutations through the synthesis of DNA strands and the proofreading and repair of mismatches.

6.4.3 Gene conversion

The formation of a chiasma within the ancestral material of a sample may leave traces of the phenomena associated with the resolution of the Holliday structure and the ensuing repair of heteroduplex DNA. Let us in this discussion assume that a randomly chosen strand of the heteroduplex is the template for the repair. In the crossover resolution of the Holliday structure the result is then indistinguishable from the classical recombination description used in Figure 6.8 on

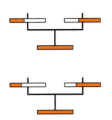

page 164. The alternative resolution produces no crossover, but a heteroduplex is still formed. Half of the heteroduplex repairs will leave no trace in the genealogy (when the template strand is contiguous with the rest of the ancestral material), and the other half will place the heteroduplex tract of ancestral material on a parental chromosome different from the one where the remainder of the ancestral material is placed (Figure 6.14). If the heteroduplex tract is contained within the ancestral material, then the ancestral material will appear as if it was subject to two simultaneous classical crossovers. On the other hand, when the heteroduplex tract partly overlaps the ancestral material, the repair will look like an ordinary crossover.

The double crossover phenomenon shown in Figure 6.14 is usually referred to as a *gene conversion*—a misnomer that arose because it is produced by a

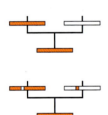

phenomenon related to that which causes the observation of gene conversion in Ascomycetes. The coalescent with mutation, recombination, and gene conversion is discussed by Wiuf and Hein (2000). Inclusion of gene conversion genuinely complicates the model in that the distribution of the length of the heteroduplex tract must be described. Wiuf and Hein assumed that this length is exponentially distributed—an assumption that neglects the experience that the length seems to be above a threshold. It does simplify the analysis, however, and it is unlikely to cause major bias in the analysis of data unless the sequences of the sample are highly variable. Wiuf (2000) relaxed that assumption.

6.4.4   The APOE risk factor in Alzheimer's disease

Alzheimer's disease is a common cause of dementia, mainly in elderly people. It has a weak tendency to aggregate in families, and some genes are known to influence individual susceptibility. One such gene is APOE, which modifies the probability of acquiring the disease. It is polymorphic in human populations. Allele $APOE_4$ increases the risk of late-onset Alzheimer's disease (age 60 years or above). Possession of allele $APOE_4$ is not enough to develop the disease, and individuals who do not carry the allele can still get it.

In a major study of the APOE risk factor Martin et al. (2000) analyzed the region around the APOE gene using SNPs. Table 6.8 shows the 13 SNPs closest (within 40,000 base pairs) to the locus of the nucleotide characteristic of allele $APOE_4$ (SNP528 shown in bold). The association between the presence of the disease and the presence of a SNP was tried by the standard 2×2 χ^2 test comparing the frequencies of the alleles among patients (cases) and unaffected persons (controls) in the study. The test probability appears as the last column

time
in the past

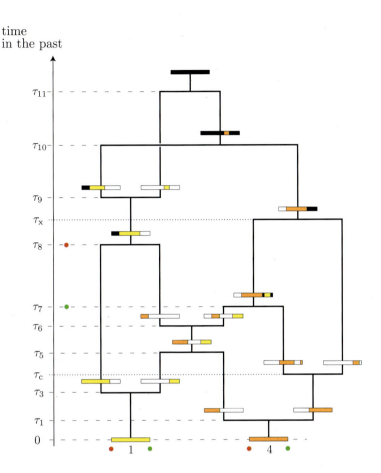

Figure 6.14: The genealogy of the sequences in Figure 6.8 including gene conversion. For simplicity only two extra events are assumed to affect the ancestral material, namely a conversion event occurring between τ_3 and τ_5 and a coalescent event between τ_8 and τ_9.

in the table. Five SNP alleles show significant association with disease presence, and of these, SNP528, SNP888, and SNP988 show strong association. One of the strong associations is, of course, exhibited by site 528 (characteristic of allele $APOE_4$).

The significance of the association is an ambiguous measure of its strength. The SNP877 alleles, for instance, were found in the frequencies 0.02 and 0.98 among controls. The association test thus has distinctly lower power than those for neighboring SNP888 and SNP988. The results therefore cannot conclude a major difference in association between $APOE_4$ and SNP887, SNP888, SNP877, and SNP988. Between this cluster of association and $APOE_4$, SNP952 and SNP873 show no indication of an association with the disease. The powers of the

Table 6.8: Association of SNP variation with Alzheimer's disease within a 40 kb distance[a] to the characteristic site of $APOE_4$. Data from Martin et al. (2000).

SNP	distance to APOE gene[a]		allele frequency	association between disease and allele
457	40	kb	0.47	0.15
464	35	kb	0.22	0.16
465	35	kb	0.41	0.006
874	21	kb	0.13	0.5
992	2.5	kb	0.36	0.005
528	0	kb	0.14	10^{-19}
952	2.8	kb	0.33	0.14
873	6.4	kb	0.36	0.37
887	8	kb	0.03	0.86
888	8.5	kb	0.46	10^{-5}
877	14.4	kb	0.02	0.21
988	16	kb	0.22	10^{-9}
533	40	kb	0.23	0.13

[a] 1kb = 1000 base pairs

analyses of these two SNPs are comparable to those of the analyses of SNP888 and SNP988, so we can conclude an island of genuine lack of high association between islands of high association around $APOE_4$ and around SNP888 and SNP988. This conclusion is outlined in the diagram below showing these seven SNPs with $APOE_4$ (528) as the leftmost SNP. The shades show the likelihood of association, with a high level of association as dark, and low or no association as light.

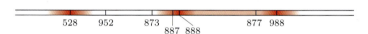

These associations between allele and trait are likely to be due to genetic associations between a SNP allele and $APOE_4$. The associations, in turn, are likely to be due to genealogical associations. From the analysis of the ancestral recombination graph we know that the ancestral material of a sample of genes may be noncontiguous in the chromosome under study. The dark areas of observed genealogical associations then indicate the current distribution of the genetic material associated with the ancestor of the $APOE_4$ alleles in the sample. Alternatively, the break in the association block leaving out SNP952 and SNP873 may be due to a chiasma that produced a gene conversion but no recombination event. The variation in SNP952 and SNP873 may also be due to mutations more recent than the $APOE_4$ ancestor. Such mutations may have siphoned these SNPs off the ancestral block surrounding the $APOE_4$ genes in the sample. Which of these explanations is the most reasonable depends on a juxtaposition of the local recombination frequency and the mutation rate in light of the estimated age of the $APOE_4$ allele.

6.5   Recombination and Physical Distance

As soon as linkage maps of Mendelian genes contributed to the genetic description of chromosomes, the search began for their relation to physical distances. Morphological variation of chromosomes was used early to affirm that recombination was produced by crossover, but the first placement of genes on a physical map of chromosomes was not achieved until the 1930s in *Drosophila melanogaster*. Dipteran insects (flies and mosquitoes) have giant chromosomes with a rich morphology in the salivary glands of larvae. Painter (1934) described variation in their morphology in *D. melanogaster* and used it for genetic analysis. A number of genes were mapped to locations on the giant chromosomes, and a comparison of such a map and the linkage map is shown for the X chromosome in Figure 6.5.[7] The physical map of the chromosome with its 20 morphologically

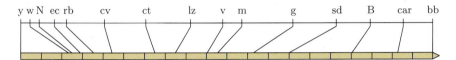

defined sections is symbolized by the lower bar, and the centromere is indicated to the right (the chromosome is telocentric). The 20 sections are roughly equal in size. The linkage map is shown above with the positions of a selection of genes with known location on the salivary gland chromosome map. Distances on the linkage map are not proportional to physical distances. For instance, the probability of a recombination within the first two morphologic sections is very low compared to that in the third section. The recombination distance between *y* and *w* is comparable to that between *w* and *N*, but the physical distance between the first two genes is more than two sections and that between the last two is a small fraction of section three.

In humans a similar heterogeneous distribution of recombination events is found along the chromosomes, and in addition, the distribution is different in females and males. Even when the different levels of recombination in female and male meioses are taken into account, local differences in recombination rate persist (Broman et al. 1998). The resolution in linkage maps is usually of the order of magnitude of 1cM, corresponding to a resolution on a sequence map on the order of a million bases, 1Mb. The APOE data discussed in Section 6.4.4 thus considers variation on a scale below that usually covered by a linkage map.

Fine-scale density of recombination along a chromosome is even more heterogeneous. Jeffreys et al. (2001) investigated recombination in a 216kb segment of the MHC II region on chromosome 6 in humans. They determined the SNP genotype of single sperm from six men with SNP genotypes that could provide information on recombination within the segment, producing so-called *haplotypes*. Most recombinant haplotypes corresponded to recombination events in

---

[7]The *Bar* gene, which was shown to be a duplication by Sturtevant and Morgan (1923) using Mendelian analysis, was then seen as a physical duplication pinpointing the *Bar* locus (Bridges 1936).

Box 25: The HapMap collection

International HapMap Project was launched by the International HapMap Consortium to describe "the common patterns of DNA sequence variation" in human populations (Gibbs et al. 2003). This is achieved by genotyping a sample of individuals at millions of SNPs to determine their gene frequencies and mutual linkage disequilibria. The reference sample was from four populations:

YRI Yoruba in Ibadan, Nigeria,

JPT Japanese in Tokyo, Japan,

CHB Han Chinese in Beijing, China,

CEU Utah residents with ancestry from northern and western Europe.

At present the data comprises haplotypes of 270 individuals (90 YRI, 45 JPT, 45 CHB, 90 CEU). The quality of data is checked by genotyping father–mother–offspring trios (30 YRI, 30 CEU). HapMap Public Release number 21 offer SNPs with the gene frequency distribution:

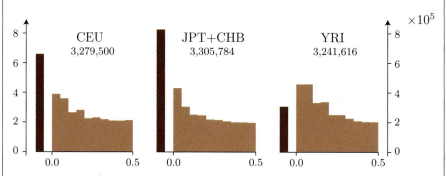

The two Asian samples have been pooled, and a SNP is counted in the histograms by the frequency of the rarest allele. The dark columns count monomorphic SNPs that are polymorphic in the total sample. The SNPs were carefully chosen to be informative and cover the human genome. These requirements are easily satisfied because 10 million SNPs with the rarest allele in a frequency above 0.01 are estimated to exist. That is, about one such SNP exists per 300b, and even though they cannot be expected to be evenly distributed over the genome, a dense map has indeed been produced. The International HapMap Consortium is committed to make the data publicly available to facilitate genetic mapping of heritable human traits, susceptibility to diseases in particular. The HapMap website (Thorisson et al. 2005) offers access to the data, to summaries of it, and to preliminary interpretations. For instance, the linkage disequilibrium structure is depicted in terms of presumed hotspots (Figure 6.15). In addition, the website offers access to analytical tools for comparing the HapMap data to your own.

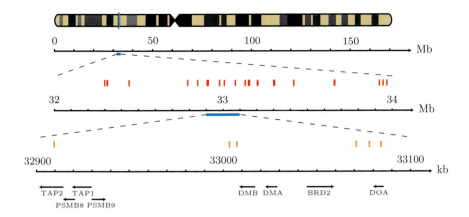

Figure 6.15: Hotspots (red bars) in the segment on chromosome 6 investigated by Jeffreys et al. (2001). Simplified after a HapMap screen (see Box 25). The hotspots found by Jeffreys et al. are marked by orange.

six subregions, called recombinational *hotspots*, with little recombination between these, and the hotspots occurred in three clusters (Figure 6.15, orange bars over bottom axis). Disregarding the low background level, recombination events centered on the hotspots, with deviations approximately corresponding to a normal distributed error with a variance of 0.09kb^2. The recombination at the strongest hotspot corresponded to 90–160mcM (millicentiMorgan: $1\text{mcM} = 10^{-3}\text{cM} = 10^{-5}\text{M}$) in six individuals, and 95 percent of the recombinations occurred within an interval of 1.2kb. Of the recombination in the 216kb segment, 95 percent occurred at hotspots covering a total of 10kb. These 10kb contributed about 150mcM on the linkage map, while the remaining more than 200kb gave about 10mcM. Thus, 5 percent of the sequence corresponds to more than 90 percent of the distance on the linkage map.

Heterogeneity of recombination and linkage disequilibrium in the MHC has been studied over a long period (see, e.g., Bodmer et al. 1986). Jeffreys et al. (2001) observed even stronger intermittent linkage disequilibrium on their fine-scale map. They determined the genotype at 179 SNPs of a sample of 50 individuals from the UK, and found very high levels stretching between the hotspot clusters and even between hotspots within clusters. The linkage disequilibrium between two SNPs on either side of a hotspot was very low. This close correspondence between hotspots and linkage disequilibrium led McVean et al. (2004) to a genome-wide study of association among SNPs using the HapMap collection of sequences (Box 25). The results of this investigation are now available as a HapMap resource (Myers et al. 2005).

Exercises

Exercise 6.1.1 Consider the segregation described in Figure 6.1. At which of the divisions do the two A genes segregate? At which of the divisions do the two B genes segregate?

Exercise 6.1.2 Consider three linked loci. The middle locus recombines with the others with frequency r_1 and r_2, respectively. If the coefficient of coincidence is ι, what is the frequency of recombination between the outer loci?

Exercise 6.2.1 An investigator reports an observation of a linkage disequilibrium value of 0.1, but he forgot to report the gene frequencies. Is this a large D? Evaluate the observation for 1) $p_A = p_B = 0.5$, 2) $p_A = p_B = 0.2$, and 3) $p_A = 0.5$ and $p_B = 0.2$.

Exercise 6.2.2 Show that the linkage disequilibrium equals half the difference between the frequencies of the *cis* and the *trans* heterozygotes. That is, it is given by $D = x_1 x_4 - x_2 x_3$.

Exercise 6.2.3 Show that D is the covariance between the indices of the two alleles in the gametes of the population, and show that ρ in equation (6.4) is the corresponding correlation.

Exercise 6.4.1 Might the conversion event in Figure 6.14 be detectable on the mismatches between the two observed sequences?

Chapter 7

Phenotypic Variation

The study of variation in genomes and changes in genetic composition of populations through time is a study of the record of evolution, the unbroken descent of genes through time. In any study of evolution inference is based on observations made today, and in terms of genetic observations we rely on the principle that extant genes trace an unbroken descent back through time to the first living organism. This states the axiom of evolution in terms of genetic material. For Charles Darwin and his predecessors Erasmus Darwin and Jean Baptiste de Lamarck this axiom was a simple expression of the unity of life. Expressing the common origin in terms of genes is somewhat more obscure. Current views of the origin of life suggest that the first organisms did not have genes made of DNA (the so-called RNA world, see page 17). The unity of life is, however, impressive on the molecular scale, suggesting that extant genes do indeed find common ancestors in an ancient DNA genome. Other aspects of the living cell, the cell membrane for instance, may trace their ancestory even further back.

The second pillar of Darwin's theory of evolution is the evolutionary mechanism. In genetic terms it could simply be stated as: heritable variation within species is by natural selection converted to variation between species. The evolutionary changes in neutral molecular traits described in Chapters 2, 4, 5, and 6 transform variation within species to variation between species. The arguments in those are solely founded on the Mendelian and other rules of transmission of genetic material. We have thus contemplated those aspects of evolutionary theory that were added to Darwin's theory of evolution in the neo-Darwinian theory that found its final form around 1930. This theory solved one of Darwin's fundamental problems with his theory, namely the maintenance of variation in natural populations.

Darwin based his theory on the biometric formulation of inheritance mainly due to Francis Galton. This theory was founded on the likeness in appearance between relatives, and the transmission of traits of a character from parents to offspring was a cornerstone (Figure 7.1). Describing inheritance on the basis of these relationships assumes direct transmission of the trait from parents to offspring, much like the passing of wealth by inheritance. The trait value of an

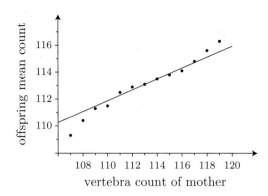

Figure 7.1: Inheritance of vertebra count in *Zoarces viviparus* (see Chapter 5, page 108) in mother–offspring samples taken 1914–19 in Isefjord, Denmark. The average vertebra count among offspring is shown as a function of the vertebra count in the mothers (ten offspring from each of the 857 mothers are counted). The slope of the fitted line is 0.40 (corresponding to a heritability of 0.80, see Section 7.2.2). Data from Smith (1921).

offspring would tend to be the mean of the parental values—assuming, of course, that the trait is equally apparent in both males and females. In every generation any individual thus tends to have a trait value corresponding to the average of the values of two individuals in the parental population. The result is that the variation in trait values in the offspring population is considerably lower than that in the parental population. Maintenance of variation in the population would therefore require the input of much novel variation. A key assumption in Darwin's theory was that the introduction of new heritable variation in the formation of offspring was random and related neither to the environment nor to the performance of the parents in that environment. The effect of natural selection would therefore be virtually nullified by the amount of new variation required in each generation to maintain the typical state of the population, thus blurring the effects of Darwin's mechanism for the process of evolution.

The solution to this problem turned out to be Mendelian inheritance, which is indirect in that genes are inherited, not traits. Gene frequency conservation (Chapter 2) is maintenance of variation in natural populations. Mendelian inheritance is observed as regular segregations of traits of a character like pea color in *Pisum sativum*, and such characters and traits are called Mendelian characters or traits. The vertebra count in an individual *Zoarces viviparus* (see Figure 5.1 on page 108) and individual height in humans do not show simple Mendelian inheritance. The traits of such morphological characters exhibit resemblance among family members. Such traits may be quantified, and characters such as vertebra count are called *quantitative characters*. Variation in heritable characters like this exists in virtually all natural populations. The fundamental theoretical reconciliation of Mendelian inheritance with the ob-

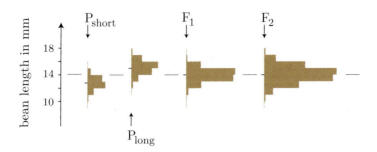

Figure 7.2: A cross of two pure lines of beans. Data from Johannsen (1903).

served inheritance of traits of quantitative characters (Figure 7.1) was obtained by R. A. Fisher (1918), and his work forms the basis of the deliberations in this chapter on quantitative genetics. Before Fisher's contribution much empirical work indicated that Mendelian genes adequately accounted for the phenomena observed in quantitative inheritance.

In the early twentieth century Mendelians proposed that inheritance of quantitative traits might be explained by assuming that the character was influenced by variation in many genes—hence the term polygenic character. Those of the Galton-based biometric inheritance school doubted this, and were unable to reconcile this hypothesis of multifactorial inheritance with observed patterns of quantitative inheritance. The first step in the Mendelian explanation of quantitative inheritance was taken when the Danish botanist Wilhelm Johannsen (1903) introduced an important distinction between the genotype and the phenotype of an individual—the phenotype is the appearance quality of the character of an individual, and the corresponding genotype is the entirety of the qualities of the genes in the individual that influence variation of the character. He performed an analysis of quantitative inheritance in the common bean (*Phaseolus vulgaris*) very similar to Mendel's analysis (Figure 7.2). Like Mendel's, his material consisted of pure-breeding lines, that is, lines of beans that had been maintained by self-fertilization of single individuals for many generations. He measured the size of the beans set by a self-fertilized plant from a pure-breeding line and found the measures to agree with the normal or *Gaussian distribution*. The mean and variance of the size of progeny beans were independent of the size of bean that gave rise to the parental plant. He posited that all the offspring beans were of the same genotype and attributed the differences to environmental influence. This phenomenon was seen in all of the 19 pure lines that he studied. Following Johannsen (1903), for a quantitative character we may thus formally write

$$\text{phenotype} = \text{genotype} + \text{environment}.$$

In Figure 7.2 the distribution of environmental effects is seen directly in F_1 as expressed in the phenotypic distribution. The variances of the measurements in the parental lines are about equal and also equal to the variance in F_1. The genotypic variability shows in F_2 as an increase in the variance of the measurements—seen in the figure as an increase in the range of variation of the phenotypes.

Herman Nilsson-Ehle (1909) described variation in oat (*Avena sativa*) and wheat (*Triticum aestivum*) that was due to the segregation of several genes. He crossed wheat strains that varied in the color of the pericarp, the maternal tissue enveloping the seed. He noted color variation in the red strains, and when he crossed red and white strains the F_1 was always red, but paler than the parental strain. Considering only the presence or absence of color, some crosses gave an F_2 segregation of $\frac{3}{4}$ to $\frac{1}{4}$; others gave a lower frequency of white. In one of these crosses he further analyzed the red F_2 plants. The 78 observed red F_2 plants were self-fertilized, and he interpreted some of the resulting segregations as 3:1 and 15:1, with the remaining sibgroups having none or a low number of white plants (Table 7.1). Nilsson-Ehle saw this as indicating the simultaneous

Table 7.1: Nilsson-Ehle's observations in F_3

F_2 plants	observed F_3 segregation	inferred F_3 segregation	total F_3 plants
8	307:97	3:1	404
15	727:53	15:1	780
5	324:6	63:1	330
50	803:0		803
78			2317

segregation of three Mendelian genes, each having two alleles, say, R_i and r_i, $i = 1, 2, 3$, with the R alleles dominant to the r alleles. The F_1 plants were thus $R_1R_2R_3/r_1r_2r_3$, and the presence of an R allele produced a red pericarp and the absence a white. Among the 78 F_2 plants in Table 7.1, 8 were thus heterozygote for one gene and homozygote rr for the others, and 15 were double heterozygotes with genotype rr for the remaining gene—with independent assortment, the expectations are 7.3 and 14.6, respectively. According to Mendel's second law we expect $\frac{1}{64}$ of the F_2 plants to be true-breeding white. About one white F_2 plant is thus expected, and observing none has the probability of 0.29. Of the F_2 plants, $\frac{1}{8}$ (or 9.8) are expected to be like the F_1 plants, that is, to be triple heterozygotes, and the F_3 offspring of these are expected to include $\frac{1}{64}$ white plants. Many of these F_2 plants are therefore expected to yield only red offspring because of the limited number observed, that is, they will be observed as if they were true-breeding red. True-breeding red is expected in $\frac{37}{64}$ of the F_2 plants (or 45.1 plants). The sum $9.8 + 45.1 = 54.9$ agrees well with the observed number of $5 + 50 = 55$ of these two genotypic classes in F_2. Genes with similar effect on the genotype exist, thus lending credibility to the hypothesis of multifactorial inheritance of quantitative characters.

This brief account of the reconciliation of phenotypic transmission with Mendelian inheritance does not pay the tribute to the intellectual process that it deserves. Provine (1971) gives a proper historical account of the development of thoughts and ideas in this field. The establishment of Mendelian inheritance as the prime source of biological inheritance was achieved by Fisher (1918) in a paper that founded the theoretical description of phenotypic transmission. We consider this theory in Section 7.2.

7.1 Quantitative Inheritance

For any given character we may represent the individual measurement by the stochastic variable P, its *phenotypic value*. Johannsen's (1903) distinction between the genotype and the phenotype of an individual may be expressed in terms of the *genotypic value* G, which is simply given by the conditional expectation of the phenotypic value given the genotype of the individual. Formally this may be written as

$$G = \mathrm{E}\big(P|\text{genotype}\big).$$

In Figure 7.2 the genotypic values of the parental pure lines are about 13mm and 15mm, and the hybrid value is about 14mm. The *environmental value* is the residual $E = P - G$. With these definitions we get Johannsen's partitioning as

$$P = G + E. \tag{7.1}$$

Johannsen's investigation of this equation was a population investigation. He fixed the genotype by using pure lines and he could then go on to investigate the influence of the environment by observing the variation in phenotypic value of the character for a given genotype. This is the characteristic approach used in investigations into quantitative inheritance: the transmission of traits is investigated in a population of individuals. Quantitative inheritance is therefore a population genetic phenomenon.

This statement may seem surprising. The transmission of physical traits from parents to offspring is a phenomenon that can be readily observed among relatives. However, this observation is based on a comparison with the population at large: relatives resemble each other, that is, they are more alike than randomly picked pairs of people. When staying among alien people you may find their appearance rather homogenous and thus find it difficult to perceive kin relationships. After a while, however, the population variation becomes apparent and families emerge from the background variation. In the same way, a Dane coming to Australia perceives gum trees as a fairly homogenous mass of blue-green foliage, but as experience grows the immense diversity materializes, and the beauty of the woodlands becomes even more striking.

The population variation is quantified by the *phenotypic variance* $V_P = \mathrm{Var}(P)$, and from the definition of the genotypic value we get the *genotypic variance* as $V_G = \mathrm{Var}\big(\mathrm{E}(P|\text{genotype})\big)$, and these are connected by the formula

$$\mathrm{Var}(P) = \mathrm{Var}\Big(\mathrm{E}(P|\text{genotype})\Big) + \mathrm{E}\Big(\mathrm{Var}(P|\text{genotype})\Big), \tag{7.2}$$

as seen from equation (A.8) on page 350. This equation defines the *environmental variance* as $V_E = \mathrm{E}\big(\mathrm{Var}(P|\text{genotype})\big)$, which describes the residual

---

Box 26: Genotype–environment interaction in quantitative traits

The environment influences the phenotype—a house plant getting little water and fertilizer surely looks different from one properly cared for. The figure shows a sketch of the effect of temperature on the expression of the genotype *vg vg* that gives vestigial wings in experiments with *Drosophila melanogaster*.

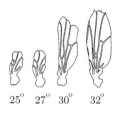

$25°$ $27°$ $30°$ $32°$

The genotype causes a rudimentary wing when its owner is raised at a temperature of 25°C, which is normal in genetic experiments with this organism. At higher temperatures the wing becomes substantially bigger without being quite normal (after Sturtevant and Beadle 1939). Students of quantitative inheritance of wing length in laboratory populations of *Drosophila* with genotype *vg vg* thus need to carefully consider the environment. In addition, the variance in the population also depends on temperature, and this will show as a temperature dependence of the environmental variance estimated in the population. Temperature also has an effect on the normal wild type *Drosophila*. The effect, however, is much smaller than for those of genotype *vg vg*. In analyses of data this will appear as a genotype–environment interaction.

---

variation as due to environmental influence. The environmental variance is thus the average over genotypes of the variance within genotype. In total we have

$$V_P = V_G + V_E, \tag{7.3}$$

a more practical expression of Johannsen's partitioning of the causes of phenotypic variation.

7.1.1 Conservation of variation

The neo-Darwinistic solution to the problem of hereditary variation is coupled with the conservation of gene frequencies. The coupling is not direct, however, because V_G is defined as a variance of phenotypic expectations. The conservation of variation is therefore discussed with reference to a population living in a *constant environment* to allow comparison of phenotypes between generations. This assumption is not true to nature, and the variation of the environment in time and space will influence the phenotypes of individuals (Box 26). The environmental value describes the difference of an individual from the mean corresponding to its genotype. It is therefore usually termed the *environmental deviation*. Indeed, from the definition we have

$$\mathrm{E}(E) = \mathrm{E}(P - G) = \mathrm{E}(P) - \mathrm{E}\big(\mathrm{E}(P|\text{genotype})\big) = \mathrm{E}(P) - \mathrm{E}(P) = 0,$$

which is consistent with the term deviation for E. The environmental deviation is of course limited by the assumption of a constant environment through time,

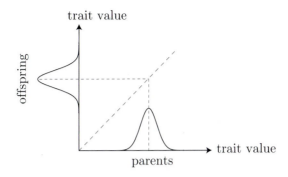

Figure 7.3: The population of parents and that of their offspring.

but this is not the most important aspect. Rather, we assume that the genotypic value may be treated as a constant from generation to generation. This is an important assumption because the relative magnitudes of genotypic values may even change their order depending on the environment. Nevertheless, for the theoretical analysis of the evolutionary conservation of genotypic variance we need the assumption of a constant environment to be able to address the inherent consequences of Mendelian genetics. The following analysis thus has a role corresponding to the analysis of random genetic drift in the Wright–Fisher model.

A further assumption is that we study variation of a *neutral character*, that is, a character where the genetic variation related to the variation is neutral. In a large population we may then assume that the gene frequencies are constant from generation to generation. This is not sufficient to bring the population into genetic equilibrium, so in addition we assume *linkage equilibrium*. Such a population is in *Hardy–Weinberg equilibrium* and *Robbins equilibrium*, where the genotypic frequencies are the same in every generation. The genotypic variance V_G is therefore constant through time. With these assumptions we get conservation of the phenotypic distribution from generation to generation (Figure 7.3).

7.1.2 Random genetic drift

In a small population we expect random genetic drift of the gene frequencies to influence a quantitative character, even though we would expect Hardy–Weinberg equilibrium and Robbins equilibrium. Rather than looking at the theory, let us consider an example. Rasmuson (1952) studied the effect of random genetic drift on an abdominal bristle number character in *Drosophila melanogaster*. This character shows heritable variation in populations of *D. melanogaster*, and Figure 7.4 shows the variation in bristle numbers in a population kept in experimental cages for a long time.

Rasmuson started ten populations from one pair of flies, and

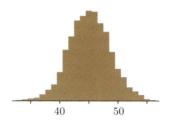

Figure 7.4: Distribution of abdominal bristle number in a laboratory population of *Drosophila melanogaster* based on 3264 females. The mean count is 45.15 and the standard deviation in the distribution is 3.51. Data from Sheridan et al. (1968).

in each following generation a random female and a random male were chosen as parents of the next generation. The results from generation 2 onward are illustrated in Figure 7.5. Special conditions at the start of the experiment minimized the variance among populations, so generation 1 shows some effects that are not interesting in the present context.

The figure shows a steady increase in the variance among the populations during the first five to ten generations, and after that the average of the character in each population hovers around a characteristic level (after generation 12 and again after generation 13 a population is lost, so during the final generations of the experiment only eight populations are monitored). This is what we should expect from the theory of random genetic drift. We can view the ten populations as a small version of Wright's island model with $N = 2$ (or less if sex-linked genes are involved). For a given varying gene the frequencies of the alleles should diverge among the populations due to genetic drift, and as the populations fix for one of the alleles this particular gene does not contribute to further divergence. The pattern in Figure 7.5 emerges from the joint effect of this stochastic process on many genes.

7.2   Kin Resemblance

The classical problem in quantitative genetics is that the genes and alleles that determine the studied variation are unknown. Molecular genetic markers (e.g., SNPs) allow localization of the loci of such genes, so-called *quantitative trait loci* or QTLs, and these loci are important in many applications (see Section 7.4). The classical problem remains, however, and the main inference on inheritance of quantitative traits still relies on the description of kin resemblance based on observations like those in Figure 7.1. The analysis of quantitative inheritance is therefore statistical in nature.

The nice bell-shaped distribution in Figure 7.4 and the nice linear relationship in Figure 7.1 suggest the use of linear normal models for the analysis of data. A normal-distribution assumption of phenotypic variation is usually reasonable

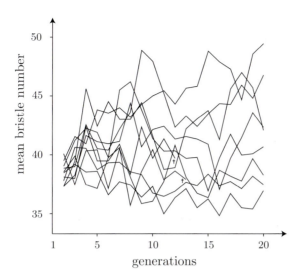

Figure 7.5: Average number of abdominal bristles observed in laboratory populations of *Drosophila melanogaster*. See text for explanation. Redrawn after Rasmuson (1952).

(maybe following transformation of data, e.g. by a logarithmic transformation), and the reference model for quantitative inheritance is therefore a Gaussian distribution of the phenotypic measurements.

Genetic models should refer to this observation, but for discrete loci with discrete variation this does not seem to be the most natural reference. We may, however, reach this aim via two paths. First, we may assume that the character is influenced by many variable genes. Their alleles necessarily show small differences in their effect on the character. The second path is that the environmental effects smooth the distribution. The environmental deviation is like an error (normally distributed with mean zero: $E \sim N(0, V_E)$) that smoothes the kinks in the distribution of genotypic effects.

Gaussian distribution:

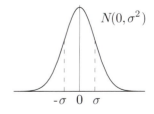

We will further make the simplifying assumption that the environmental deviation is independent of the genotypic value. A much more important assumption in the basic models of quantitative inheritance is that *the environmental deviations of family members are independent*. Without this basic assumption we cannot separate similarities due to genetic transmission and similarities engendered by correlations in environmental influences.

7.2.1   The additive model

The natural operation in the Gaussian distribution is addition of effects to the mean. In polygenic inheritance we consider a number of different loci that affect the trait of the individual, and as a first step we posit that the genotypic value is obtained by adding these effects. This *additive model* will be assumed as a simple initial model in our analysis. The first path is therefore an application of the central limit theorem,[1] provided we make a link between the effects of the alleles and the number of loci that ensure a given observed variance.

The properties of this model can be found from the properties of the contributions of each locus to the character. The assumed independence of the loci (Robbins equilibrium) allows us to consider just one locus. We therefore assume that the variation of the character is determined by the contributions of only one locus with alleles $A_1, A_2, \ldots, A_k$ in frequencies $p_1, p_2, \ldots p_k$, and we assume Hardy–Weinberg proportions. Write the genotypic value of an individual of genotype $A_i A_j$ as G_{ij}, that is,

$$G_{ij} = \mathrm{E}\big(P | A_i A_j\big),$$

and we consequently assume $G_{ij} = G_{ji}$. The phenotypic value of an individual of genotype $A_i A_j$ is thus given by $G_{ij} + E$, and the mean in the population is

$$m = \mathrm{E}(P) = \mathrm{E}(G) = \sum_{i=1}^{k} \sum_{j=1}^{k} G_{ij} p_i p_j, \qquad (7.4)$$

where the assumption of random mating allows us to use the Hardy–Weinberg proportions (G_{ij} and G_{ji}, for $i \neq j$, are included, both with a frequency of $p_i p_j$).

Due to gene frequency conservation the population mean will be conserved between generations, and its value therefore does not contribute to the description of quantitative inheritance. The phenotypic measurements may then be given as deviations from the mean, as *phenotypic deviations*, and we define the *genotypic deviation* in the same way as $g = G - m$. The genotypic variance is then simply

$$\mathrm{Var}(G) = \mathrm{Var}(g) = \sum_{i=1}^{k} \sum_{j=1}^{k} g_{ij}^2 p_i p_j, \qquad (7.5)$$

which is again conserved between generations.

In the more amenable model of L loci we may for each locus, say, ℓ, define the genotypic deviations as

$$g_{\ell,ij} = \mathrm{E}\big(P | A_{\ell,i} A_{\ell,j}\big) - m, \quad i, j = 1, 2, \ldots, k_\ell,$$

---

[1]In its presently most usable form it says: the distribution of the sum of n identically distributed independent random variables $X_1, X_2, \ldots, X_n$ with mean m and variance σ^2 / n approaches a normal distribution with mean m and σ^2 as n becomes large.

where k_ℓ is the number of alleles at this locus, and $\mathsf{A}_{\ell,i}$, $i = 1, 2, \ldots, k_\ell$, are its alleles. The genotypic variance due to the variation at locus ℓ is then

$$\mathrm{Var}(G_\ell) = \sum_{i=1}^{k_\ell} \sum_{j=1}^{k_\ell} g_{\ell,ij}^2 p_{\ell,i} p_{\ell,j}.$$

In the additive model the genotypic deviation of a multilocus genotype is the sum of these single-locus genotypic deviations. With independent loci (Robbins proportions) the genotypic deviations are independent random variables, and the genotypic variance is therefore given by

$$\mathrm{Var}(G) = \sum_{\ell=1}^{L} \mathrm{Var}(G_\ell). \tag{7.6}$$

The theory for the additive model with L independent loci is thus very easy, because variances and covariances may be calculated for single loci and summed.

7.2.2 Parent–offspring covariance

An example of the covariance between trait values of a parent and those of its offspring is given in Figure 7.1, and we may use our theory to learn about the genetic basis of such a relationship. The regression coefficient of the offspring phenotypic value on that of the mother (the slope of the line in the figure) is

$$\beta_{\mathcal{PO}} = \frac{\mathrm{Cov}(P_{\mathcal{P}}, P_{\mathcal{O}})}{V_P},$$

where $P_{\mathcal{P}}$ and $P_{\mathcal{O}}$ are the phenotypic values of mother and offspring, respectively. The slope in the figure is about 0.4, and to relate this to our genetic model we need to calculate the parent–offspring covariance in that model.

For the one-locus model the mother–offspring genotypic combinations are given in Table 7.2. From this we may immediately calculate the genotypic co-

Table 7.2: Mother–offspring combinations

mother genotype	offspring genotype $(i > 2)$				
	$\mathsf{A}_1\mathsf{A}_1$	$\mathsf{A}_1\mathsf{A}_2$	$\mathsf{A}_2\mathsf{A}_2$	$\mathsf{A}_1\mathsf{A}_i$	$\mathsf{A}_2\mathsf{A}_i$
$\mathsf{A}_1\mathsf{A}_1$	p_1	p_2	0	p_i	0
$\mathsf{A}_1\mathsf{A}_2$	$\frac{1}{2}p_1$	$\frac{1}{2}(p_1 + p_2)$	$\frac{1}{2}p_2$	$\frac{1}{2}p_i$	$\frac{1}{2}p_i$

variance of mother and offspring. Based on our assumption of independence of environmental deviations, this is equal to the phenotypic covariance of mother and offspring $C_{\mathcal{PO}} = \mathrm{Cov}(P_{\mathcal{P}}, P_{\mathcal{O}})$, so $C_{\mathcal{PO}} = \mathrm{Cov}(G_{\mathcal{P}}, G_{\mathcal{O}})$. In the calculation of the genotypic covariance we collect contributions from the homozygote mothers separately from those of the heterozygote mothers and get

$$\text{Cov}(G_{\mathcal{P}}, G_{\mathcal{O}}) = \sum_{i=1}^{k} g_{ii} p_i^2 \sum_{h=1}^{k} g_{ih} p_h + \sum_{i=1}^{k} \sum_{\substack{j=1 \\ j \neq i}}^{k} g_{ij} p_i p_j \sum_{h=1}^{k} (g_{ih} + g_{jh}) \tfrac{1}{2} p_h$$

$$= \sum_{i=1}^{k} \sum_{j=1}^{k} g_{ij} p_i p_j \sum_{h=1}^{k} g_{ih} p_h$$

(because i and j may be interchanged in the sums). The last sum in this expression,

$$a_i = \sum_{h=1}^{k} g_{ih} p_h, \quad i = 1, 2, \ldots, k, \tag{7.7}$$

may be understood as the *average effect* of allele A_i in the population. To see this, pick a random A_i allele. This allele occurs in genotype $\mathsf{A}_i \mathsf{A}_j$ with probability p_j (because of random union of gametes). Allele A_i is therefore in an individual with genotypic deviation g_{ij} with the probability p_j, and a_i is the average genotypic deviation of the allele in the population. The average effect of the alleles in the population is zero,

$$\sum_{i=1}^{k} a_i p_i = \sum_{i=1}^{k} \sum_{j=1}^{k} g_{ij} p_i p_j = 0,$$

as it should be.

Returning to the calculation of the covariance, we insert the average effects and get

$$\text{Cov}(G_{\mathcal{P}}, G_{\mathcal{O}}) = \sum_{i=1}^{k} \sum_{j=1}^{k} a_i g_{ij} p_i p_j = \sum_{i=1}^{k} a_i p_i \sum_{j=1}^{k} g_{ij} p_j = \sum_{i=1}^{k} a_i^2 p_i.$$

The parent–offspring covariance is thus the variance in average effects of the alleles. However, it is not related directly to the genotypic variance.

The resemblance between a mother and its offspring is entirely due to the one gene they share at each locus. The homologous gene in the mother is independent of the shared gene due to Hardy–Weinberg proportions. The shared gene in the offspring is independent of that of paternal origin due to random mating. The gene shared by mother and offspring is of allele type A_i with probability p_i, and it causes an effect of a_i in both individuals. This amounts to a genetic argument for the covariance expression above.

A natural question to ask in any statistical analysis is whether the gene effect revealed by the parent–offspring covariance suffices to explain the genotypic variation. We thus formulate the *additive model of allele effects* on genotypes as

$$g_{ij} \sim \gamma_{ij} = \begin{cases} 2\alpha_i & \text{for} \quad i = j, \\ \alpha_i + \alpha_j & \text{for} \quad i \neq j \end{cases} \quad \text{with} \quad \sum_{i=1}^{k} \alpha_i p_i = 0. \tag{7.8}$$

The least-square estimates[2] of the parameters $\alpha_1, \alpha_2, \ldots, \alpha_k$ are $\hat{\alpha}_i = a_i$, $i = 1, 2, \ldots, k$. The genotypic variance under this model is given by

$$\sum_{i=1}^{k} \sum_{j=1}^{k} (a_i + a_j)^2 p_i p_j.$$

This is known as the additive genotypic variance or just the *additive variance*,

$$V_A = 2 \sum_{i=1}^{k} a_i^2 p_i, \tag{7.9}$$

and the *parent–offspring* covariance is therefore

$$C_{\mathcal{PO}} = \tfrac{1}{2} V_A. \tag{7.10}$$

The resemblance of an offspring to a parent is thus fully described by the additive alleles model. The resemblance to both parents is described by covariance between the offspring value and the *midparent value*

$$P_{\mathcal{M}} = \tfrac{1}{2} \left(P_{\mathcal{P}\female} + P_{\mathcal{P}\male} \right).$$

The *midparent–offspring* covariance is therefore the same as the parent–offspring covariance:

$$C_{\mathcal{MO}} = \tfrac{1}{2} \left(\mathrm{Cov}(P_{\mathcal{P}\female}, P_{\mathcal{O}}) + \mathrm{Cov}(P_{\mathcal{P}\male}, P_{\mathcal{O}}) \right) = C_{\mathcal{PO}}. \tag{7.11}$$

Random mating stipulates independence between the parental values, so the variance of the midparent value is equal to $\tfrac{1}{2} V_P$. The midparent–offspring regression therefore has twice the slope of the parent–offspring regression.

The additive variance plays a key role in the description of quantitative inheritance because it describes the transmission of traits from parents to offspring. It may thus be seen as reflecting the transmissible part of the genotypic variation (Fisher 1918). Alternatively, it may be expressed as the transmissible part of the phenotypic variation as described by the *heritability*

$$h^2 = \frac{V_A}{V_P}. \tag{7.12}$$

This quantity is readily observable as twice the regression coefficient of offspring trait on parent trait. In Figure 7.1 the regression coefficient of the offspring phenotypic value on that of the mother is $\beta_{\mathcal{PO}} = \tfrac{1}{2} h^2$. The slope of the line is about 0.4, and the variation in vertebra count of *Zoarces viviparus* therefore has a heritability of about 0.8.

---

[2]The maximum-likelihood estimates of parameters are usually replaced by the least-square estimates in the analysis of linear normal models. This is because of their nicer distributional properties. The two kinds of estimates are proportional, they are very similar, and the least-square estimate converges to the maximum-likelihood estimate as the sample size increases.

We have no reason to believe that the additive model of allele effects gives an overall good description of the genotypic values. The genotypic residuals of the model are

$$d_{ij} = g_{ij} - (a_i + a_j), \tag{7.13}$$

and they are called *dominance effects*, and hence the genotypic deviations are specified as $g_{ij} = a_i + a_j + d_{ij}$. Being residuals, the dominance effects have no additive effects, and so

$$\sum_{j=1}^{k} d_{ij} p_j = \sum_{j=1}^{k} \big(g_{ij} - (a_i + a_j) \big) p_j = a_i - (a_i + 0) = 0,$$

using the definition (7.7) of the additive effects. The full one-locus model of genotypic deviations therefore extends the additive model (7.8) to

$$g_{ij} \sim \gamma_{ij} = \begin{cases} 2\alpha_i + \delta_{ii} & \text{for} \quad i = j, \\ \alpha_i + \alpha_j + \delta_{ij} & \text{for} \quad i \neq j, \end{cases} \quad \text{with} \quad \sum_{j=1}^{k} \delta_{ij} p_j = 0 \tag{7.14}$$

for $i = 1, 2, \ldots, k$. The variance of the dominance effects is the *dominance variance* V_D, and because it is a residual variance, we have

$$V_G = V_A + V_D \quad \text{and} \quad V_P = V_A + V_D + V_E. \tag{7.15}$$

This partitioning of the phenotypic variance holds for the additive model in a population in linkage equilibrium.[3] A parent–offspring regression is therefore not expected to estimate the full genetic variability in the character. To estimate the genotypic variance V_G we need additional information from other kinds of kin resemblance.

7.2.3   Covariance between sibs

One gene in the offspring is identical to one gene in the parent, and so the parent–offspring covariance depends on only the allelic type of this one gene. Sibs, on the other hand, share two, one, or no genes at any locus, and as sibs with two identical genes have the same genotype, the genotypic deviations must enter into the sib–sib covariance C_{SS}.

The probabilities that a pair of sibs share none, one, or two identical genes at a given locus are $\frac{1}{4}$, $\frac{1}{2}$, and $\frac{1}{4}$, respectively. If the sibs do not share genes their genotypes are independent. If they share one gene they have precisely the same relationship as a parent and an offspring. Finally, if they share two genes they have the same genotype. Thus, we have

$$\begin{aligned} C_{SS} &= \text{Cov}(P_{S_1}, P_{S_2}) \\ &= \tfrac{1}{4}\text{Cov}(P_{S_1}, P_{S_2} \,|\, \text{none shared}) + \tfrac{1}{2}\text{Cov}(P_{S_1}, P_{S_2} \,|\, \text{one shared}) \\ &\quad + \tfrac{1}{4}\text{Cov}(P_{S_1}, P_{S_2} \,|\, \text{two shared}) \end{aligned}$$

---

[3]Whether the additive alleles model is a sufficient description for given observations may be tested by comparing V_D to the error variance, which is a component of V_E.

$$= \tfrac{1}{4} \times 0 + \tfrac{1}{2} \times \tfrac{1}{2} V_A + \tfrac{1}{4} \sum_{i=1}^{k} \sum_{j=1}^{k} g_{ij}^2 p_i p_j = \tfrac{1}{4} V_A + \tfrac{1}{4} V_G.$$

The sib–sib covariance can therefore be written as

$$C_{\mathcal{SS}} = \tfrac{1}{2} V_A + \tfrac{1}{4} V_D \qquad (7.16)$$

by using the specification in equation (7.15). Data on parent–offspring and sib–sib covariance therefore allows the estimation of the components of the genotypic variance, and this determines the genotypic variance as $V_G = V_A + V_D$, assuming the additive model.

7.3 Inference on the Genotypic Variance

Observation of resemblance between parents and offspring and between sibs solves the estimation problem of finding the genotypic variance, but only in theory. We assumed linkage equilibrium, additive contributions of loci to the genotypic value, constant environment, random mating, and uncorrelated environment of relatives in addition to the more usual assumptions of a large isolated randomly mating population. To limit discussions, let us keep relying on the "usual assumptions" and discuss those that are particular to the simple model of quantitative inheritance discussed here. For a more thorough discussion, consult a textbook in genetic analysis of quantitative character variation (for instance, the excellent book by Falconer and Mackay (1996)).

Before raising too many warning signs, however, it is appropriate to recall the immense success of quantitative genetics as a guide in animal and plant breeding. Some of the problems to be discussed below, in particular those concerned with the influence of the environment, are pertinent in any situation where family resemblance is observed. The design of quantitative genetic experiments or observations is therefore an ongoing challenge. On the other hand, the simplicity of the genetic model has caused few problems in breeding applications, and for the lack of detailed knowledge about the genetic determination of trait variation, *the main criterion is that the models work.*

For two alleles the one-locus formulae become particularly simple. The mean of the average allele effects is zero, and therefore $a_A p = -a_a q$. The additive model of allele effects in equation (7.8) may therefore be given by one parameter

$$\alpha = \frac{\alpha_A}{q} = -\frac{\alpha_a}{p} \qquad (7.17)$$

(Table 7.3). The full model of genotypic deviations in equation (7.14) is in the same way given by

$$\delta = \frac{\delta_{AA}}{q^2} = \frac{-\delta_{Aa}}{pq} = \frac{\delta_{aa}}{p^2}.$$

The genotypic deviations fit the model in Table 7.3, and the estimates are

$$\hat{\alpha} = a = \frac{a_A}{q} = -\frac{a_a}{p} \quad \text{and} \quad \hat{\delta} = d = \frac{d_{AA}}{q^2} = -\frac{d_{Aa}}{pq} = \frac{d_{aa}}{p^2}.$$

Table 7.3: Two-allele genotypic model

effects	AA	Aa	aa	average
additive	$2q\alpha$	$-(p-q)\alpha$	$-2p\alpha$	0
dominance	$q^2\delta$	$-pq\delta$	$p^2\delta$	0
genotype	γ_{AA}	γ_{Aa}	γ_{aa}	0

These reparametrizations rely heavily on the assumption of two alleles per locus. The model in Table 7.3 is therefore a two-allele model that embraces all two-allele models. An impression of the complications for general multilocus determination of the genotypic values may be obtained by generalizing this model and considering the additive and dominance variances. For multiple loci the additive effects of the alleles at a particular locus may differ from the average effects when linkage disequilibrium is present. However, they depend only on the linkage disequilibrium between pairs of loci—not on higher order disequilibria (Box 27). The additive and dominance variances simplify when the locus effects are additive (Box 27), and for one locus with two alleles they become

$$V_A = 2a^2 pq \quad \text{and} \quad V_D = d^2 p^2 q^2. \tag{7.18}$$

Linkage disequilibrium produces correlations in genotypic, and thereby allelic, effects among loci, but its influence on the analysis of quantitative inheritance is difficult to gauge because linkage influences the genetic relationship among kin and linkage disequilibrium influences the genotypic proportions in the population.

Calculation of familial covariances is rather cumbersome, but in the additive model we again get $\text{Cov}_{\mathcal{PO}} = \frac{1}{2}V_A$ and $C_{\mathcal{SS}} = \frac{1}{2}V_A + \frac{1}{4}V_D$ (see Christiansen 2000). Linkage disequilibrium therefore influences the additive and the dominance variances, but not their determination in terms of the parent–offspring and the sib–sib covariances. These covariances therefore estimate V_A and V_D, and $V_G = V_A + V_D$. Linkage disequilibrium thus causes few problems as long as the genotypic values are determined by addition of contributions from each locus. Nevertheless, the variance components are expected to change as the population approaches Robbins equilibrium.

Nonadditive contributions of loci to the genotypic value pose more severe problems because the simple relationship between the single-locus variance components and the genotypic variance is lost, as is the correspondence between these and the familial covariances. When the additive and the dominance effects are specified, residual effects remain, the so-called interaction effects. In this sense we may specify the genotypic variance as $V_G = V_A + V_D + V_I$. The component V_I describes the variation due to interaction of the single-locus genotypic effects, and it is often referred to as the epistatic variance (for further discussions, see Bürger 2000).

Box 27: A quantitative genetic model with multiple loci

Assume L loci each with two alleles, say, 0 and 1, with gene frequencies p_ℓ and q_ℓ. The average effects $a_{\ell 0}$ and $a_{\ell 1}$ are calculated as before from the genotypic effects by the average over genotypes that includes alleles 0 and 1 at locus ℓ. As their average is zero their values can be summarized in one average effect a_ℓ per locus defined as in equation (7.17). We want to find the corresponding additive effect parameters α_ℓ, $\ell = 1, 2, \ldots, n$ (see Table 7.3). The additive effect parameters are found as the solutions to a set of equations of which the first is

$$a_1 = \alpha_1 + \sum_{\ell=2}^{L} \alpha_\ell \frac{D_{1,\ell}}{p_\ell q_\ell},$$

where $D_{1,\ell}$ is the linkage disequilibrium between loci numbers 1 and ℓ (see Christiansen 2000).

So much for the complications; now to the simple aspects. The equation above holds without the additive allele effect assumption of the simple two-locus model. In addition, the additive allele effects only depend on the pair-wise linkage disequilibria, so we do not have to bother with generalizing the two-locus linkage disequilibrium to more loci. The additive variance becomes

$$V_A = 2 \sum_{\ell=1}^{L} a_\ell \hat{\alpha}_\ell p_\ell q_\ell,$$

a fairly simple covariance between the average effects and the additive effects of the ℓ loci. The dominance effects and residuals of the additive allele model are related in a similar way and produce the dominance variance as

$$V_D = \sum_{\ell=1}^{L} d_\ell \hat{\delta}_\ell p_\ell^2 q_\ell^2.$$

Variation in the environment from generation to generation may wreak havoc on the analysis of quantitative inheritance (Box 26). Minor variations can be tolerated as variations causing changes in the population mean, but with only minor effects on deviations from the mean. Even in this case the comparison of the parent–offspring covariance and the sib–sib covariance may cause difficulties because two generations are involved in the analysis. A widely used way around this problem is to use the half-sib covariance instead of the parent–offspring covariance, because $C_{\mathcal{HS}} = \frac{1}{4} V_A = \frac{1}{2} C_{\mathcal{PO}}$.[4]

Deviations from random mating change the genotypic frequencies away from the Hardy–Weinberg proportions, and the entire analysis must be reformulated. We shall not persue this further, but just note that an important

---

[4]Half-sibs share a gene from the common parent with probability $\frac{1}{2}$.

problem solved by Fisher (1918) was that mating in which spouses are more similar than random pairs (assortative mating) may give slopes exceeding $\frac{1}{2}$ in the parent–offspring regression. The biometric school viewed this as a major problem for the Mendelian explanation of quantitative inheritance.

Correlated environment of relatives interferes strongly with the outlined methods of quantitative genetics because covariance among related individuals may have both genetic and environmental causes. This is a serious problem in mammals because the environment of an offspring early in its life is the mother. This may produce *maternal effects* that cause the offspring to resemble the mother more than would be expected from genetic influence alone. The immediate effect is that sibs and maternal half-sibs show a higher than expected resemblance. The effect might also show as a mother–offspring regression where the slope is steeper than $\frac{1}{2}$, and a higher resemblance to the mother than to the father. The father–offspring covariance and the covariance of paternal half-sibs are therefore generally expected to be more dependable.

In humans, environments of relatives are commonly correlated. Sibs are reared together by their parents, giving a sib–sib environmental covariance. In a broader sense the environments of parents and offspring are correlated due to inheritance of social status, wealth, trade, etc. This may cause nongenetic correlations in physical characters, skills, and behavioral characters. The study of quantitative genetics in humans is thus a difficult undertaking.

In addition to the influence of the immediately observable environment, nongenetic transmission occurs by teaching and learning. Teaching may be active (and for humans theoretical) in addition to the passive transmission of skills like, for instance, a fledgling following and watching an adult blackbird hunting earthworms.

These transmissions of environment and abilities are often embraced under the term *cultural inheritance* (Feldman and Lewontin 1975). It may be studied by augmenting the quantitative genetic models with biometric inheritance models describing cultural transmission (Cavalli-Sforza and Feldman 1971). Variation of the character IQ score[5] has been subject to much public debate since at least the 1960s, the question being whether "nature or nurture" is the cause of variation, that is, whether the cause is hereditary or environmental. A geneticist's answer to the "nature or nurture" question about variation is: "most probably both"—surely not an answer that would hit the evening news. A quantification is wanted, and for the IQ score this has been tried in terms of heritabilities calculated from familial data using various modifications of the simple quantitative inheritance model. The answers referred to in the public media are usually heritabilities of between 60 and 80 percent.

The IQ score is found from a knowledge and skills test, and it may therefore be subject to both cultural and genetic transmission. Experiments on humans for this character cannot be made, so the only source of information that can separate these two phenomena is relatives who have lived separate lives in mutual isolation. Similarity among sibs reared apart because of adoption points to

---

[5]Results of intelligence tests.

genetic transmission, and comparing these similarities to those among common sib pairs may contribute to a separation of cultural and genetic inheritance of the traits. Feldman et al. (2000) made such an analysis on a large previously published set of data concerning sib–sib covariances and other covariances of relatives. They concluded that the phenotypic variance could be partitioned into about $\frac{1}{3}$ genetically transmissible variance (genetic heritability $\frac{1}{3}$), a negligible dominance variance, $\frac{1}{3}$ culturally transmissible variance (cultural heritability $\frac{1}{3}$), and $\frac{1}{3}$ nontransmissible environmental variance—a surefire way to secure fierce criticism from both the "nature" and the "nurture" proponents.

Others have analyzed the same data and reached the conclusion that the cultural heritability is much smaller. The main difference is that Feldman et al. (2000) corrected for effects particular to twins, whereas others did not. Discussion of who is right or wrong is out of place here, but the example serves to illustrate the kinds of problems that arise in quantifying transmission of human traits.

7.3.1 Comparison of populations

The hereditary description of a quantitative character refers to the variation in a particular population at a particular time. The description of one population thus cannot be assumed to apply to another population. Both the genetic and the environmental influences on a character may cause serious deviations in the description of different populations. Genetic differences and genotype–environment interactions are particularly important.

Quantitative inheritance describes the transmission of the variants we see in a population. Two isolated populations may, however, differ in the loci that vary, that is to say, one population may be monomorphic for a locus where the other population is genetically polymorphic. The genetic basis of the variation is thus different in the two populations. The populations may also be fixed for different alleles at a locus, resulting in different genetic environments in the two populations. These effects are expected to be particularly important in small populations. Cattle breeds, for instance, where a few bulls sire the entire population of calves, are expected to diverge in this way.

For common polymorphisms the difference between populations is usually less drastic. A strictly additive character that shows variation determined by one autosomal locus with two alleles has the genotypic mean values $G_{AA} = \mathfrak{a}$, $G_{Aa} = 0$, and $G_{aa} = -\mathfrak{a}$, where we arbitrarily choose to measure the genotypic values as deviations from the midpoint between the values of the homozygotes. The population average is then $\mathfrak{a}(p-q)$, and the additive effect is $a = \mathfrak{a}$, independently of the gene frequencies. The additive variance, and thus the heritability, is therefore from equation (7.18) expected to be proportional to the frequency of heterozygotes in the populations. When dominance is allowed, that is, $G_{AA} = \mathfrak{a}$, $G_{Aa} = \mathfrak{d}$, and $G_{aa} = -\mathfrak{a}$, the population average is $\mathfrak{a}(p - q) + 2pq\mathfrak{d}$ and the additive effect depends on the gene frequency, in that, $a = \mathfrak{a} - 2(p - q)\mathfrak{d}$. The dominance effect of the two alleles is simply $d = 2\mathfrak{d}$. To keep the arguments as

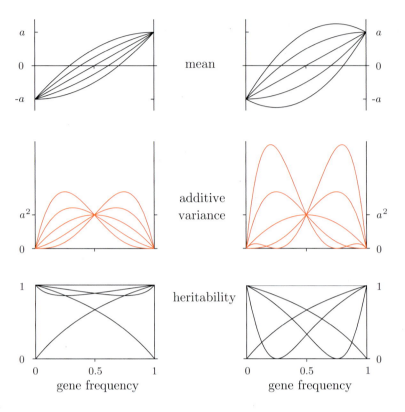

Figure 7.6: The population mean, additive variance, and heritability as functions of the gene frequency. The heritability is drawn assuming $V_E = 0$. The left panel shows five dominance levels corresponding to $\eth = -\mathfrak{a}$, $-\frac{1}{2}\mathfrak{a}$, 0, $\frac{1}{2}\mathfrak{a}$, and $\mathfrak{a}$. The right panel shows five dominance levels corresponding to $\eth = -2\mathfrak{a}$, $-\mathfrak{a}$, 0, $\mathfrak{a}$, and $2\mathfrak{a}$.

simple as possible we assume $V_E = 0$, that is,

$$h^2 = \frac{V_A}{V_A + V_D}.$$

Additivity gives a linear increase in the population mean with increasing frequency of A, symmetry in the variation of V_A, and a heritability of 1 ($V_G = V_A$; we assumed $V_E = 0$), as shown in Figure 7.6 ($\eth = 0$). Dominance gives asymmetry in the variation of V_A. For absolute dominance ($\eth = -\mathfrak{a}$ or $\mathfrak{a}$) the heritability is almost 1 for a rare dominant allele, and close to zero for a rare recessive. Intermediate dominance as exemplified by $\eth = -\frac{1}{2}\mathfrak{a}$ and $\frac{1}{2}\mathfrak{a}$ (left panel of Figure 7.6) produces a rather small effect on the heritability.

For $\eth > \mathfrak{a}$ or $\eth < -\mathfrak{a}$ the genotypic mean value of the heterozygote is outside the interval formed by those of the homozygotes (right panel of Figure 7.6). These cases are called over- and underdominance. The additive variance and the heritability are zero ($V_G = V_D$) for the gene frequency where the population mean value reaches an internal maximum or minimum.

7.3.2 Random genetic drift

The populations compared in studies of quantitative inheritance are often lines or breeds of domestic plants or animals. These are often viewed as isolated populations sharing a common origin, that is, as versions of Wright's island model, as illustrated by Rasmuson's (1952) experiment in Figure 7.5 (page 189). Genetic variation is thus present as variation within and between the islands, and we describe this by a partitioning of the variance in much the same way as we partitioned the phenotypic variance in equation (7.2) on page 185:

$$\mathrm{Var}(P) = \mathrm{Var}\Big(\mathrm{E}(P|\text{island})\Big) + \mathrm{E}\Big(\mathrm{Var}(P|\text{island})\Big),$$

where the first component is the between-population variance and the second is the within-population variance. We can partition the genotypic variance in the same way by assuming the environmental effects to have the same distribution on all islands.

The genetic divergence is modeled by the Wright–Fisher model with N individuals on each island (Section 2.2 on page 26); we assume one locus with two alleles to simplify arguments. A convenient measure of divergence is the expected identity coefficient F_t on an island after t generations in isolation. The average of the genotype frequencies in generation t are then given by equation (3.27) on page 69, and using these the population mean m_t at generation t in Wright's island model becomes

$$m_t = m_0 + pqF_t(g_{\mathsf{AA}} - 2g_{\mathsf{Aa}} + g_{\mathsf{aa}}).$$

In the additive allele model (7.8) the population mean therefore stays at the initial value as the identity coefficient increases. Specifying the genotypic effects in the original population by the full one-locus model (7.14) of genotypic effects, the population mean in generation t becomes

$$m_t - m_0 = pqF_t(\delta_{\mathsf{AA}} - 2\delta_{\mathsf{Aa}} + \delta_{\mathsf{aa}}) = 2pq(p - q)^2 \delta F_t \tag{7.19}$$

(Table 7.3).

The calculation of the change in the genotypic variance and its components therefore becomes a lot simpler if we restrict our attention to the additive model (7.8) when partitioning the genotypic variance into components within and between the islands.[6] The additive variance in the entire population becomes $V_{At} = (1 + F_t)V_{A0}$, the average of that within islands becomes $V_{Awt} = (1 - F_t)V_{A0}$, and that between islands becomes $V_{Abt} = 2F_tV_{A0}$ (Figure 7.7).

---

[6]The calculation is fairly straightforward, but alas, long and tedious.

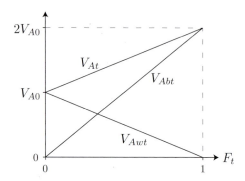

Figure 7.7: Partition of the genotypic variance as a function of the population identity coefficient in Wright's island model, assuming additive allele effects.

7.3.3 Applications of quantitative genetics

The preceding development of quantitative genetics may seem like a stroll through aspects of quantitative inheritance, with results casually mentioned here and there, and rightly so. The reason is that quantitative genetics is an ad hoc theory that, on the background of genetics, aims to describe observations. Practical criteria have therefore been very important in the historical development of the theory. As already stressed: the theory is and has been useful.

The main criterion of usefulness is the ability to predict the outcome of a breeding initiative involving artificial selection for increased performance in production. We return to this application of quantitative genetics in the next chapter; here it suffices to say that such attempts occur within a limited time. Breeding eelpouts with a high number of vertebrae produces offspring with a mean number of vertebrae that can be predicted from Figure 7.1, and this procedure of artificial selection may be repeated for a number of generations with the same slope of the line in the figure. This cannot go on forever because the present limited variation is eventually exhausted in the process, and further gains have to await the arrival of useful mutations.

The quantitative description of transmission relates to the population at the time of observation, and the usefulness of that description indicates that it is robust.

7.4 Mapping of Quantitative Trait Loci

In our simple one-locus quantitative genetics model with two alleles A and a, add a linked locus (a SNP) without effects on the character also with two alleles 0 and 1. The model is thus a two-locus two-allele model like the reference model in Chapter 6, and again we assume that the population is in Hardy–Weinberg equilibrium at both loci. The difference is that only the SNP and the quantitative character are observable, and rather than the linkage disequilibrium, we

Table 7.4: Two-locus mixed model

A0	A1	a0	a1
x_1	x_2	x_3	x_4

are here interested in finding the correlation between quantitative traits and genotypes at the marker locus. To do this we need to assign genotypic values to the SNP genotypes, and to keep matters simple we assume an additive alleles effect model with the genotypic values $\Gamma_{00} = 0$, $\Gamma_{01} = 1$, and $\Gamma_{11} = 2$ and no environmental deviations.[7] The assumption that environmental deviations are independent implies that the covariance between quantitative trait and SNP type equals the covariance between genotypic deviation and SNP value, that is, $\mathrm{Cov}(P, \Gamma) = \mathrm{Cov}(G, \Gamma) = \mathrm{Cov}(g, \Gamma)$. The mean of the genotypic deviation is zero, and therefore we have that $\mathrm{Cov}(g, \Gamma) = \mathrm{E}(g\Gamma)$, and so

$$
\begin{aligned}
\mathrm{Cov}(P, \Gamma) \quad = \quad & g_{\mathsf{AA}}(0 \times x_1^2 + 1 \times 2x_1x_2 + 2 \times x_2^2) \\
& + g_{\mathsf{Aa}}(0 \times 2x_1x_3 + 1 \times 2x_1x_4 + 1 \times 2x_2x_3 + 2 \times 2x_2x_4) \\
& + g_{\mathsf{aa}}(0 \times x_3^2 + 1 \times 2x_3x_4 + 2 \times x_4^2).
\end{aligned}
$$

Minor modifications of this yield

$$
\begin{aligned}
\mathrm{Cov}(P, \Gamma) \quad = \quad & g_{\mathsf{AA}}2x_2p_{\mathsf{A}} + g_{\mathsf{Aa}}2(p_{\mathsf{A}}x_4 + x_2q_{\mathsf{a}}) + g_{\mathsf{aa}}2x_4q_{\mathsf{a}} \\
= \quad & 2(p_{\mathsf{A}}q_1 - D)(g_{\mathsf{AA}}p_{\mathsf{A}} + g_{\mathsf{Aa}}q_{\mathsf{a}}) + 2(q_{\mathsf{a}}q_1 + D)(g_{\mathsf{Aa}}p_{\mathsf{A}} + g_{\mathsf{aa}}q_{\mathsf{a}}) \\
= \quad & 2q_1\big(p_{\mathsf{A}}(g_{\mathsf{AA}}p_{\mathsf{A}} + g_{\mathsf{Aa}}q_{\mathsf{a}}) + q_{\mathsf{a}}(g_{\mathsf{Aa}}p_{\mathsf{A}} + g_{\mathsf{aa}}q_{\mathsf{a}})\big) \\
& - 2D(g_{\mathsf{AA}}p_{\mathsf{A}} + g_{\mathsf{Aa}}q_{\mathsf{a}} - g_{\mathsf{Aa}}p_{\mathsf{A}} - g_{\mathsf{aa}}q_{\mathsf{a}}).
\end{aligned}
$$

The factor of q_1 in the first term is the mean of the genotypic deviation, which is zero, and the covariance therefore becomes

$$
\mathrm{Cov}(P, \Gamma) \quad = \quad -2D\big(g_{\mathsf{AA}}p_{\mathsf{A}} - g_{\mathsf{Aa}}(p_{\mathsf{A}} - q_{\mathsf{a}}) - g_{\mathsf{aa}}q_{\mathsf{a}}\big).
$$

Inserting the genotypic values according to the two-allele genotypic model in Table 7.3 on page 196 yields the very simple result $\mathrm{Cov}(P, \Gamma) = -2D\alpha$.

To find the correlation we need the genotypic variance at the SNP locus:

$$
V_{SNP} \quad = \quad (2p_0q_1 + 4q_1^2) - (2p_0q_1 + 2q_1^2)^2 \quad = \quad 2p_0q_1.
$$

The correlation is therefore given by

$$
\mathrm{Corr}(QTL, SNP) \quad = \quad \frac{-2D\alpha}{\sqrt{2p_0q_1V_P}}.
$$

From the definition (7.12) of heritability and using the expression for the additive variance in equation (7.18) we get

$$
\mathrm{Corr}(QTL, SNP) \quad = \quad \frac{-Dh}{\sqrt{p_0q_1p_{\mathsf{A}}q_{\mathsf{a}}}}. \tag{7.20}
$$

---

[7]We seek the correlation between the traits at the two loci, and hence the magnitudes of the values are irrelevant.

The correlation is observable and independent of the scale at which we measure the quantitative character. Knowing the heritability, we may on the basis of this formula infer the genetic correlation,

$$\frac{D}{\sqrt{p_0 q_1 p_A q_a}},$$

between the observable SNP and the putative QTL. Doing this for a number of SNPs will help localize the QTL. Formula (7.20) is therefore the basis for localizing genes of major effect by the method called *linkage disequilibrium mapping* (see, e.g., Hein et al. 2005). The magnitude of the correlation between the marker and the character, however, depends on the effect of the QTL and the linkage disequilibrium in such a way that both have to be appreciable for the association to be noted.

A gene causing disease has a major effect. The classical way of placing such a gene on a linkage map is to study segregation of the disease and genetic markers in families. Humans in particular have a small number of relatives, so the resolution of pedigree analysis is limited and not accurate enough to localize a gene in the DNA sequence. The extant disequilibrium is a relic of the unique chromosome where the disease mutation occurred, in addition to associations generated by random genetic drift and population admixture. Such associations of linked loci are utilized in linkage disequilibrium mapping, and the method of analysis is based on the coalescent with mutation and recombination. The linkage disequilibrium is expected to be more degraded the looser the linkage of the marker to the disease gene. Degradation is caused by crossovers in the meioses of the coalescent tree of the sample—usually many more than can be observed in a family pedigree.

The analysis of the APOE risk factor in Alzheimer's disease in Section 6.4.4 is linkage disequilibrium mapping of a genetic factor in a disease—a QTL. Even though the investigation is based on prior knowledge of the genetic factor, it illustrates the power of linkage disequilibrium mapping in QTL analysis on a fine genetic scale.

7.4.1   Diallel crosses

More extensive family investigations are possible in plants and animals. The method of analysis devised by Mendel (1866) and used for analysis of quantitative characters by Johannsen (1903) and others has developed into the diallel cross. Johannsen crossed two pure-breeding lines of *Phaseolus vulgaris* that differed by about 2mm in bean size and got an F_1 with an intermediate mean. Self-fertilization yielded an F_2 generation with a mean very close to that of the F_1 generation (Figure 7.2 on page 183). The variance in F_2, however, is distinctly higher than that in F_1, which in turn is comparable to the variances in the parental lines. The qualitative nature of these results is as expected from Mendelian segregation of genes like those described by Nilsson-Ehle (1909). All loci where the two lines differ are heterozygous in F_1, and they all segregate in F_2, suggesting a genotypic variance of about $\frac{1}{2}$mm^2, and an environmental

Table 7.5: A cross of two pure lines of beans

	P_{short}	P_{long}	F_1	F_2
number	671	1045	2009	3718
mean (mm)	12.8	15.1	14.0	14.1
Variance	0.81	1.06	0.91	1.52

variance of 1mm^2 (Table 7.5). The equality of the F_1 and F_2 means suggests that the character is additive, that is, $V_A = V_G$. Under this assumption the heritability is about $\frac{1}{3}$.

Sax (1923) took this argument further in a series of crosses between pure lines of *P. vulgaris*. He crossed large pigmented beans to small white beans and observed pigmented F_1 beans of intermediary size. This agrees with Johannsen's experiments and underscores the dominance of pigmented over nonpigmented. Pigmented beans are thus P- and the recessive pp gives white beans. In F_2 he found the pigmented beans to be larger than the white beans, and upon further analysis of these he concluded that the size difference between PP and pp beans was twice that of the difference between P- and pp beans. He thus described a QTL—or a series of closely linked QTLs—with additive allele effects on size linked to the P locus.

His analysis was in fact much more detailed, as his crosses were really between two lines of large "eyed" beans and several small white beans. Eyed beans are pigmented with partial pigmentation, giving lighter color around the scar— P-T- have eyes and P-tt give solid color, with the genotype pp concealing the eye character. He considered two lines with different T alleles giving large and small eyes, respectively. Seed weight QTLs were found to be associated with these T alleles. Segregation of a series of alleles for pigment color and density originally hidden in the white parental lines emerged in his crosses, and QTLs were also found associated with several of these.

Sax's (1923) investigation was by far the first QTL analysis, and he added localization of factors to the multiplicity of factors found by Nilsson-Ehle (1909). The accuracy of the QTL localization improved as the number of markers on the linkage map increased—with a local high in Robertson's (1966) analysis of bristle QTLs in *Drosophila melanogaster* taking advantage of the dense map of morphological markers in that species.

The systematic investigation of QTL localizations accelerated with the increasing availability of molecular markers. As in Sax's investigation the point of initiation is two pure or highly diverged lines with an interesting difference in one or more quantitative characters; the genetic markers are chosen among the molecular genetic differences between the lines. Cosegregation of morphology and markers may then place QTLs on the linkage map of the species—if such a map exists. If the diallel cross is an F_2 cross or a backcross the cross will in any event provide valuable information about the linkage map of the markers.

F_2 **cross**

The simple one-locus quantitative genetics model with a SNP marker locus is sufficient to describe the genetics of a diallel cross. As in Johannsen's experiment, we cross two parental populations, namely P_{A0} homozygous for gamete A0 and P_{a1} homozygous for gamete a1. The linkage disequilibrium in the F_2 population is then

$$D = \left(\tfrac{1}{2}(1-r)\right)^2 - \left(\tfrac{1}{2}r\right)^2 = \tfrac{1}{4}(1-2r),$$

using the gamete frequencies given by Table 6.4 on page 149. The genotypic covariance is then given by

$$\mathrm{Cov}(QTL, SNP) = -\tfrac{1}{4}(1-2r)\left(g_{AA} - g_{aa}\right).$$

For $r = \tfrac{1}{2}$ we get independent segregation, and the covariance is equal to zero, but it also becomes zero when $g_{AA} = g_{aa}$, that is, when $\alpha = 0$.

We may therefore write the covariance as

$$\mathrm{Cov}(QTL, SNP) = -2D\alpha,$$

where α is the additive effect of the alleles at the QTL locus in the F_2 population. The covariance and the correlation are thus given by the same formulae as in any other population. The F_2 population is not in equilibrium, however, so our previous interpretation of the additive variance should be handled with caution in this instance.

The inference in the diallel cross is simple compared to linkage disequilibrium mapping because the ancestry of every offspring is known. The additive effect and the dominance effect are determined solely by the QTL, that is, by its genotypic variation in F_2. A linked marker, however, will carry some information about the additive and the dominance effects so that

$$\alpha = \frac{g_{00} - g_{11}}{8D} \quad \text{and} \quad \delta = \frac{g_{00} - 2g_{01} + g_{11}}{16D^2},$$

where g_{00}, g_{01}, and g_{11} are the genotypic deviations of the marker genotypes given in the last column of Table 7.6.

The situation may be simplified further by assuming that the QTL effect can be described by the additive allele effect model ($\delta = 0$). From Table 7.6 we thus expect the observed traits to regress onto the genotypic values of the marker locus with a slope of $4D\alpha$. The comparison of the variances in the F_1 and $\bar{F}_2$ generations provides $\alpha = \sqrt{2V_G}$ and thereby D, and the QTL are placed at a distance of $r = \tfrac{1}{2} - 2|D|$ from the locus of the SNP.

This is but a sketch of the analysis of data. A proper statistical analysis provides an expected error on the estimate of r, for instance by depicting the profile likelihood function $L(r)$ as a function of r. This pinpoints the most likely place of the QTL and the amount of support to that projected by the data—given, of course, that only one QTL exists, and that its effect is strong enough to show in the experiment. Further discussion of this method is found in Falconer and Mackay (1996) and Lynch and Walsh (1998).

Table 7.6: Two-locus F_2 genotypic proportions

	AA	Aa	aa	Σ	g
00	$\frac{1}{4}(1-r)^2$	$\frac{1}{2}r(1-r)$	$\frac{1}{4}r^2$	$\frac{1}{4}$	$4D\alpha + 4D^2\delta$
01	$\frac{1}{2}r(1-r)$	$\frac{1}{2}(1-r)^2 + \frac{1}{2}r^2$	$\frac{1}{2}r(1-r)$	$\frac{1}{2}$	$-4D^2\delta$
11	$\frac{1}{4}r^2$	$\frac{1}{2}r(1-r)$	$\frac{1}{4}(1-r)^2$	$\frac{1}{4}$	$-4D\alpha + 4D^2\delta$
Σ	$\frac{1}{4}$	$\frac{1}{2}$	$\frac{1}{4}$	1	
g	$\alpha + \frac{1}{4}\delta$	$-\frac{1}{4}\delta$	$-\alpha + \frac{1}{4}\delta$		

Backcrosses

The backcrosses $BC_{A0} = P_{A0} \times F_1$ and $BC_{A1} = P_{a1} \times F_1$ are obvious choices to extend the information obtained in the F_2 cross, and again the linked marker provides additional information for the description of inheritance of the quantitative trait (Box 28). Without the marker locus, comparison of the phenotypic means of the F_2, F_1, BC_A, and BC_a crosses and the parental lines P_A and P_a may be used to estimate the additive and dominance effects relative to the F_2 population (Figure 7.8). This analysis between crosses may of course be used to strengthen the within-cross analysis of the linkage between marker and QTL. So much for the theory. In practice the analysis is more complicated because the environment must be controlled for so that the comparison among populations is not overwhelmed by environmental effects.

7.4.2 Inference

Mapping of the loci that determine the traits behind the variation of a quantitative character cannot rely on a hypothesis of only one QTL. The practical use of linkage disequilibrium mapping applies multiple linked markers in order to pinpoint the location of QTLs distributed in the genome. The implementation of the theory for diallel crosses is therefore geared towards multifactorial characters, allowing multiple QTLs and multiple markers. In addition, the crosses provide simultaneous information on the linkage relationships among the markers.

The statistical localization of QTLs exhibits an important bias: genes with variants of large effect have a large probability of detection, whereas genes with little effect go unnoticed. The principal goal of disclosing the genetic architecture of variation in natural populations therefore cannot be reached using these techniques, but they are effective in finding genes of major effects, should such exist. In quantitative genetics these are referred to as *major genes*. The more practical goal of finding major determinants of advantage in breeding thus seems

---

Box 28: Diallel backcross

We analyze the cross $BC_{A0} = P_{A0} \times F_1$ and assume the genotypic values $G_{AA} = \alpha + \frac{1}{4}\delta$ and $G_{Aa} = -\frac{1}{4}\delta$ ($m = \frac{1}{2}\alpha$) in order to make the analysis comparable to that of the F_2 cross (Table 7.6). The genotypic deviations are then $g_{AA} = \frac{1}{2}\alpha + \frac{1}{4}\delta$ and $g_{Aa} = -g_{AA}$, and the associated genotypic effects at the marker locus are

	AA	Aa	Σ	g
00	$\frac{1}{2}(1-r)$	$\frac{1}{2}r$	$\frac{1}{2}$	$2D\alpha + D\delta$
01	$\frac{1}{2}r$	$\frac{1}{2}(1-r)$	$\frac{1}{2}$	$-(2D\alpha + D\delta)$

where $D = \frac{1}{4}(1 - 2r)$ is again the linkage disequilibrium among gametes produced by the F_1 individuals. A manipulation similar to the one applied in the analysis of the F_2 cross produces

$$2\alpha + \delta = \frac{g_{00} - g_{01}}{2D} \,.$$

The F_2 cross and a backcross thus seem to provide sufficient information to estimate the recombination fraction between the QTL and the marker locus even in the presence of dominance at the QTL locus.

---

more attainable (see Chapter 8). As in any statistical draw of major effects out of a distribution of effects, we should expect that the located major genes upon further analysis rarely show the magnitude of effect that the first observation would seem to indicate.

Eye color is among the best known inherited characters in humans. The basic inheritance of blue and brown is Mendelian, with brown the dominant and blue the recessive trait. In passports, however, many more classes of eye pigmentation are listed (at least blue, grey, green, yellow, and brown), and many people have checkmarks in several of the classes (my passport has many checkmarks). In addition, various shades from close to colorless to heavily pigmented are common for most colors. Finally, two blue-eyed parents may indeed beget brown-eyed children without grounds for moral concern, so simple Mendelian inheritance is not the whole story.

The colors originate from various sources. Brown stems from melanin in melanocyte cells, and brown eyes are thus formed in much the same way as dark skin color. An iris with little melanin appears blue because light is reflected and refracted from the structures in the tissue in much the same way as a black summer starling appears violet in the sun. Eye color is therefore a complex character, and it should be no surprise that its variation involves many genes, for instance genes coding for melanosome proteins, associated signal proteins, and proteins involved in transport and uptake of melanin and its precursors (for a recent review, see Sturm and Frudakis 2004).

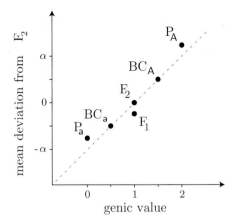

Figure 7.8: Analysis of the lines P_A and P_a and their filial- and backcrosses. The population means are shown relative to that of the F_2 population. The genic value is found from an additive model where allele A has effect 1 and a 0.

Zhu et al. (2004) subjected eye color variation in 502 twin families to a QTL analysis with a 5–10cM spacing between markers. One salient QTL was found on chromosome 15 close to the OCA2 locus, which codes for a melanosome protein. The character used in the investigation was eye color classified into three traits, and the curve shows the marker associations to this character along chromosome 15 as given by a likelihood score. The variation associated with the OCA2 locus accounted for about $\frac{3}{4}$ of the genotypic variance observed in this data.

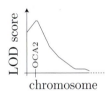

Many diseases show familial aggregation, and a typical example is breast cancer. Discrete genetics was added to the etiology of the disease by Williams and Anderson (1984), who in an analysis of 200 Danish pedigrees segregating breast cancer showed that the inheritance was consistent with an autosomal dominant allele that gave susceptibility to the disease. The frequency of the allele was estimated at about 0.01, and carriers of the allele had a much greater risk of developing cancer. Furthermore, the hereditary cancers showed early onset, and almost all the very early cancers occurred in genetically predisposed individuals, as shown in the figure. The frequency of the putative liability allele was estimated to be a little more than 0.1 among older women with breast cancer. Williams and Anderson (1984) found that the cancers attributed to the liability allele were more severe than the sporadically occurring cancers, and that most breast cancers in males were hereditary.

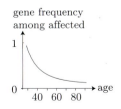

Using marker genes Hall et al. (1990) located a gene in the q sector of

Table 7.7: Association of breast cancer with markers

family	associated markers	breast cancer early	breast cancer late	ovarian cancer	sporadic cases
1	e	20	4	22	7
2	b+d	14	8	2	0
3	e	8	2	1	0
4	d	7	3	1	0
5	g	3	1	0	0
6	a+g	4	1	0	0
7	a	5	2	0	1
8	a	4	0	0	1

Simplified after Miki et al. (1994).

chromosome 17 that has a dominant allele segregating with breast cancer in families. Soon after Miki et al. (1994) cloned and sequenced the gene BRCA1. At the same time Wooster et al. (1994) located a gene BRCA2 on chromosome 13 that also has dominant breast cancer–associated alleles.

Miki et al. (1994) located BRCA1 within section 17q using seven markers in that section (shown in the linkage map to the right). Eight families were investigated and association between the disease and the markers was found. In each family most breast cancer victims carried one or two characteristic markers as shown in Table 7.7. The exceptional cases are classified as sporadic in the table. The breast cancer cases carrying the family marker are separated into early onset cases (taken as those diagnosed in women under the age of 50 years) and late onset cases. Thus the data clearly show the characteristic high frequency of early onset of hereditary breast cancers.

Using molecular genetic techniques Miki et al. designated the causal BRCA1 gene (shown by the bracket including marker d). Its gene code revealed a so-called zinc-finger protein that interacts with DNA. In five of the families the BRCA1 allele associated with the cancer cases showed mutations likely to cause malfunction. For instance, when Miki et al. looked within the translated sequence they found one with a deletion of 11 base pairs, one with an insertion of one base pair (frameshift), and one with a stop codon.

BRCA1 cancer liability alleles show pleiotropic effects, including susceptibility to ovarian, peritoneal, and prostate cancers. In the investigation by Miki et al. (1994) observations of ovarian cancer were included and shown to associate with the family markers (Table 7.7).[8] Breast cancer is a truly polygenic trait, and since the discovery of BRCA1 and BRCA2 a score of genes conferring

---

[8]The ovarian cancers in families 3 and 4 were diagnosed in individuals with breast cancer, and so was one of those in family 2. The sporadic cancers include both breast and ovarian cancers in individuals not carrying the family-specific marker.

susceptibility have been described (see reference 114480 in OMIM 2006).[9] The effects of the BRCA1 and BRCA2 liability alleles are nevertheless sufficiently large to be noted as dominant Mendelian factors in the analysis by Williams and Anderson (1984). Thus, they are major genes.

Exercises

Exercise 7.2.1 Show that $V_G = V_A + V_D$.

Exercise 7.2.2 At an additive two-allele locus in a population with random mating, show that the additive variance, and therefore the heritability, is maximal for equal gene frequencies.

Exercise 7.2.3 Show that $C_{\mathcal{HS}} = \frac{1}{4}V_A$.

---

[9]The OMIM database rests on the work of Victor A. McKusick (2006).

Part II

Variation and Selection

The phenomena addressed in Chapters 2, 4, 5, and 6 are fundamental to population genetics because they describe the effects of the basic genetic processes of Mendelian segregation, mutation, and recombination on the distribution of genetic variation in populations. They describe and develop the theoretical basis of the neutral theory of molecular evolution. This theory stems from the observation of high levels of molecular genetic variation within and between populations and species. The recognition of systematic regularity in molecular differentiation among species had a special influence on the formulation of the neutral theory. These evolutionary regularities are known as molecular clocks, and were originally described as a constant rate of amino acid substitution through the evolution of proteins (Zuckerkandl and Pauling 1962). Kimura (1968, 1969*b*) and King and Jukes (1969) showed that this constancy could be seen as a sign of evolution by substitution of selectively neutral variants, and the neutral theory of molecular evolution was born. Kimura, the main champion of the theory, explicates and discusses it in a book and a review article from 1983. This fundamental theory has been and remains very influential in modern evolutionary genetics. It is sometimes lambasted as heretical in relation to Darwin's theory of evolution, usually by authors who fail to acknowledge the full impact of indirect Mendelian inheritance. It is today an integral part of the neo-Darwinian synthetic theory of evolution. The role it plays in evolutionary genetics is similar to that played by the law of gene frequency conservation in population genetics.

Tests of hypotheses of neutrality abound in the literature. Are these really investigations probing Kimura's theory of neutral molecular evolution? In some respects they are, but in others they are not. The neutral theory has a status similar to Mendelian inheritance in that it describes observations of molecular genetic variation when only segregation, mutation, recombination, and population structure influence the presence and distribution of variants. In this way the neutral theory forms the basis for observations of phenomena in nature. Nobody would refute Mendel's first law if their fruit flies failed to show the proper segregation in an experiment. Rather, the conclusion would be that something disturbed the experiment, for instance different probabilities of survival of the different types. In the same way, failure to describe population data or data on variation among species by applying the neutral theory suggests that something, natural selection for instance, has influenced the variation through time, but the inference is specific to the variation observed. The neutral theory as a general contribution to insights into molecular evolution is tied to fundamental properties of genetic transmission. As a theory reflecting observations in natural populations, it is refuted if it fails generally and repeatedly, that is, if additional phenomena are regularly required to reflect deductions from empirical investigations. Such failure seems quite unlikely because evolutionarily important physiological functions have to be demonstrated for a majority of molecular variants.

In this part we will see that such convincing data are exceedingly difficult to produce. The main difference between inference based on Mendelian segregation and the neutral theory is that the latter addresses such complicated biological scenarios that it is often difficult to formulate the correct expectations for a particular set of data. The main difficulty in this respect is that the theory is not naive; it acknowledges Darwinian evolution and the widespread occurrence of deleterious mutations (Lewontin 1974). This reservation may seem a rather trivial clause, but deleterious alleles do in fact cause disturbance throughout the genome, disturbance that may mimic selection on the variation that we observe (a point elaborated in Chapter 9).

In Darwin's theory, phenotypic variation that influences survival and reproduction will be subject to natural selection. Individuals that have phenotypes well adapted to the environment in which they live on average leave more descendants. This works as an evolutionary force when parents produce offspring with phenotypes resembling their own more than those of the population at large. Indirect inheritance tends to corrupt this argument because genes and not phenotypes are transmitted from parents to offspring. Genetic linkage in particular disturbs the simple Darwinian argument on the evolution of phenotypes (Chapter 9).

Molecular population genetics is concerned with the description of genetic variation in a population and the ensuing evolution of the genetic material through time. The interest in natural selection therefore focuses on outcomes of selection on genetic variation—on induced genotypic selection. The problem in molecular population genetics is therefore that induced selection may affect the evolution of genetic variants that are truly neutral in the sense that they have no discernable effect on the survival and reproduction of individuals in the population. The variants may merely be associated with alleles at another locus with phenotypic effects that influence the "well-being" of individuals. Such associations among alleles, which are not functionally related, may usually be assumed to be of haphazard origin, and the influence of selection on neutral variants is therefore merely a stochastic disturbance.

Nevertheless, random associations among genes are very helpful in the analysis of genomic sequences because associations become increasingly likely the closer genes are situated on the linkage map. Phenotypically characterized genes, for instance diseases-causing genes, can thus be related to linked molecular markers by the observation of population associations. This is an important method for localizing the genomic position of such genes.

Chapter 8

Effects of Selection

In Darwin's evolutionary theory natural selection seems formulated as exerting pressure on heritable variation. However, natural selection should rather be considered a consequence of variation in a character that influences the survival and reproduction of the individual. The variation in vertebra count of *Zoarces viviparus* is heritable, with a heritability of about 0.8 (Figure 7.1 on page 182). If the number of vertebrae affects, say, the survival of the individual, the distribution of vertebrae among parents of the next generation will deviate from that among their peers at birth and evolution will result. Selection of the parents of the next generation among fish with a high number of vertebrae will result in an offspring population with a high mean number of vertebrae.

The basis for natural selection on the number of vertebrae may be hard to imagine, but the character conveys individual variation in body form; individuals with a high number of vertebrae have a tendency to appear long and slender, whereas individuals with a low number look more chubby (Figure 8.1). Thus, the character vertebra count may reflect a morphological variation that would otherwise be hard to quantify. In addition, a character like degree of slenderness is expected to be highly influenced by the condition of the fish. Much of

Figure 8.1: Sketches of two individuals of *Zoarces viviparus*, 26cm and 25cm long, respectively. The top individual had 121 vertebrae, and that at the bottom 105. They differ also in number of pigment spots and number of hard rays. From Øresund and Roskilde Fjord, Sealand, Denmark. From Schmidt (1918).

216

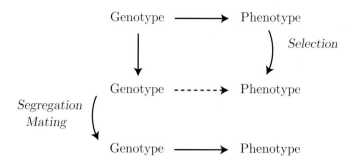

Figure 8.2: A simplified view of the neo-Darwinian theory of evolution.

the variation in slenderness would therefore not be heritable, and the character would then exhibit a high environmental variance, making it more difficult to study. The character for which the trait variation is the primary cause for natural selection is thus inherently difficult to distinguish in the fog of more or less correlated characters (Gould and Lewontin 1979).

The neo-Darwinian description of quantitative inheritance decouples natural selection from the record of this action. For instance, if the various phenotypes have different probabilities of survival to the age of reproduction, then the phenotypic distribution in the population changes through life. The phenotypic variation is heritable, so various genotypes give rise to phenotypes that may on average be different. The change in phenotypic distribution may thereby induce a change in the genotypic distribution in the population, and this induced change will determine the phenotypic distribution in the offspring population (Figure 8.2). The population genetic theory of selection studies this *induced genotypic selection* on genetic variation, and in this chapter we consider selection on the genotypic variation in one locus. A common observation in breeding applications of selection on quantitative characters is that the variance of a character does not change much over time. The biometric description of quantitative inheritance provides a good approximation for the effects of selection during a limited number of generations when constancy in the variance components is assumed during the course of the experiment (Figure 8.3). With polygenic inheritance mediated by variation at a number of loci in Robbins proportions, the effects of selection at any one of the loci is sufficiently attenuated to assume little change in gene frequency, and therefore to expect an approximately constant contribution of the locus to the genotypic variance. Indirect inheritance thus makes the constant-variance assumption reasonable. However, in evolutionary theory we have no general justification for making this assumption, although it may still provide reasonable approximations for some processes. The full treatment of evolution of quantitative characters requires multilocus theory. We will consider some aspects of this theory in Chapter 9, but in general it is very complicated, and a comprehensive treatment has yet to be achieved. Approximate results far better than the biometric description have been made, however

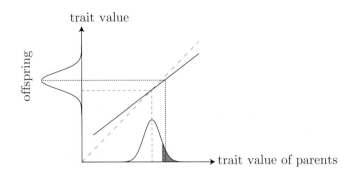

Figure 8.3: The effect of selection on a quantitative character. The parents with a trait value in the shaded part of the distribution are kept for further breeding (see Section 8.3).

(Bürger 2000).

Variation in the probability of survival from conception to maturity is the simplest form of selection, in particular when we may assume that probability of survival to maturity of an individual depends only on its phenotype and not on the phenotypic composition of the population. An example where this assumption seems sensible is the color variation in the peppered moth (*Biston betularia*) that influences each individual's risk of falling victim to bird predation (see Kettlewell 1973). The moth is common, and in pristine forest a light-colored *typica* morph is found, while a variant black *carbonaria* morph occurs in polluted forests. The moths are predated upon when perching individually on tree trunks in the daytime. The *typica* morph is cryptic on the lichen-covered trunks in pristine forest. Pollution kills these epiphytes[1] and the dark trunk emerges, making the *carbonaria* morph cryptic. The mortality of *typica* is distinctly higher than that of *carbonaria* in polluted areas, whereas the opposite is true in pristine habitats.

The morphss of *B. betularia* are expressions of simple Mendelian traits. The *carbonaria* trait corresponds to a dominant allele, while a *typica* individual is a recessive homozygote. *Carbonaria* was first observed in the middle of the nineteenth century near Manchester, England. The variant became the predominant morphs in this population by the end of the century, and only about 2 percent of the moths were of the original *typica* morphs. The increase of the *carbonaria* continued well into the twentieth century in industrial areas of England and Scotland. The diminished pollution in the last quarter of the century caused widespread decline of the *carbonaria* morph (Cook 2003). Such melanic polymorphisms have been described in many insects, and many seem to be adaptations to life in polluted areas (Kettlewell 1973).

Similar polymorphisms are found in the garden snail (*Cepaea nemoralis*), and these are again influenced by bird predation. In shady habitats snails are dark, while lighter shells are found in lighter habitats (Cain and Sheppard 1954).

---

[1]Lichens are very sensitive to atmospheric pollution.

C. nemoralis varies in color as well as in bandedness—caused by variation at several genes. Local populations are polymorphic for these traits, while keeping their superficial appearance (reviewed by Jones et al. 1977), that is, in dark or shaded habitats snails are usually dark or heavily banded, and in light habitats we find light-colored and thin-banded shells. Birds preying on the snails generate a search image of the prey, resulting in a tendency to prefer the more common types in the habitat. This behavior generates so-called frequency-dependent selection on the morphs; this particular form of frequency dependent selection is called apostatic selection. Frequency-dependent selection is in general selection where the fitness of the individual depends on the composition of the population. Selection may in addition depend on population density (see Christiansen 2004). Cryptic coloration may not be the only cause of selection on body color variation in *B. betularia* and *C. nemoralis*. For instance, solar heating and other climatic variables may have different impacts on the individual (Cowie and Jones 1983, True 2003).

Much of the theory in Chapters 2–6 is formulated in terms of two-allele variation, but in those settings the theory is equally applicable to multiallelic variation. The assumption of neutrality renders the alleles equivalent (Section 2.2.1 on page 30). When the focus is on the effects of selection, all alleles that contribute to the variation subject to selection have to be monitored. We treated the industrial melanism in *Biston betularia* as being due to the segregation of two alleles determining the *typica* and *carbonaria* morphs. This provides a good description of the initial increase in the frequency of melanic individuals, but somewhere in the process a third morph, *insularia*, appeared. This is determined by an allelic gene that is dominant to the *typica* allele and recessive to the *carbonaria* allele. The *insularia* morphs has a distinctly higher fitness than the *typica* morphs in polluted areas, but seems to be slightly inferior to the *carbonaria* morphs. We would thus expect the *insularia* allele to increase at a slower pace than the *carbonaria* allele in a mainly *typica* population. The frequency of melanic individuals is expected to increase more slowly in a population where both melanic morphss are present than in one where only the *carbonaria* morphs is an alternative to *typica*. Separating the genes into a *typica* allele and a melanic allele is therefore insufficient to describe the evolution in polluted populations. Shell color in *Cepaea nemoralis* is also determined by multiple alleles. They are, in decreasing order of dominance, C^B, C^{DP}, C^{PP}, C^{FP}, C^{DY}, and C^{PY}, and the phenotypes are various shades of brown, pink, and yellow.[2] The genetic determination of bandedness and band appearance involves several loci with two or more alleles.

8.1 Selection Components

Viability selection is but one simple mode of action in natural selection. Variation among genotypes in their probability of survival to maturity (viability or

---

[2] Brown, dark pink, pale pink, faint pink, dark yellow, and pale yellow.

8.2   Viability Selection

Differential survival of genotypes is attributed to viability selection. We assume that no sex difference exists, for instance by presupposing that each individual may act both as male and female, a hermaphroditic or monoecious organism. For two alleles the survival to adulthood of the zygotes of genotypes AA, Aa, and aa are described by the probabilities w_{AA}, w_{Aa}, and w_{aa}. To simplify the arguments we assume that reproduction occurs by random mating.

The genotypic frequencies among zygotes are in Hardy–Weinberg proportions, p^2, $2pq$, and q^2, and after selection the genotypic frequencies are

$$\frac{p^2 w_{AA}}{W}, \quad \frac{2pq w_{Aa}}{W}, \quad \text{and} \quad \frac{q^2 w_{aa}}{W},$$

where

$$W = p^2 w_{AA} + 2pq w_{Aa} + q^2 w_{aa}$$

is the probability of survival in the population, also called the *average fitness*.

The gene frequency among sexually mature adults, and therefore that expected among the zygotes of the next generation, is then

$$p' = \frac{p^2 w_{AA} + pq w_{Aa}}{W}.$$

The probability of survival of allele A is $W_A = p w_{AA} + q w_{Aa}$, and the recurrence equation may be expressed in terms of this so that

$$p' = \frac{p W_A}{W}. \tag{8.1}$$

This is strictly an equation describing the transmission of alleles; the average fitness is $W = p W_A + q W_a$, where $W_a = p w_{Aa} + q w_{aa}$ is the probability of survival of allele a. W_A and W_a are also known as the average fitnesses of the alleles.

If allele A is absent from the population ($p = 0$) it stays absent ($p' = 0$) as we assume an isolated population and ignore the effect of mutation. The population is therefore at equilibrium when $p = 0$ and when $p = 1$. These two states are called the trivial equilibria, or in a more descriptive term, the *monomorphic equilibria*.

For multiple alleles $A_1, A_2, \ldots, A_k$, the description of the change in gene frequencies is in essence the same as for two alleles except for more elaborate bookkeeping of the population state. The frequency of allele A_i is p_i, $i = 1, 2, \ldots, k$, and the population state is described by the vector $\boldsymbol{p} = (p_1, p_2, \ldots, p_k)$. For two alleles we only have to monitor one gene frequency, say, p, the other being $q = 1 - p$, whereas $k - 1$ gene frequencies are needed to describe the state of a k allele polymorphism, and as $p_1 + p_2 + \cdots + p_k = 1$ we can specify $p_k = 1 - p_1 - p_2 - \cdots - p_{k-1}$. The number of genotypes is $\frac{1}{2} k(k+1)$, so we need that number of survival probabilities. The viability of genotype $A_i A_j$ is w_{ij}, and we assume that the genotypes $A_i A_j$ and $A_j A_i$ have the same probability of survival, and therefore $w_{ij} = w_{ji}$. This description of selection is conveniently summarized in the matrix

$$
\mathbf{W} = \left\{
\begin{array}{cccc}
w_{11} & w_{12} & \cdots & w_{1k} \\
w_{21} & w_{22} & \cdots & w_{2k} \\
\vdots & \vdots & \ddots & \vdots \\
w_{k1} & w_{k2} & \cdots & w_{kk}
\end{array}
\right\},
$$

called the *fitness matrix*, and as $w_{ij} = w_{ji}$ the matrix is symmetric.

The assumption of reproduction by random mating produces the genotype frequencies among newly formed zygotes in Hardy–Weinberg proportions, where genotype $\mathsf{A}_i\mathsf{A}_j$ occurs in the frequency $2p_ip_j$ when $i \neq j$ and the homozygote $\mathsf{A}_i\mathsf{A}_i$ in the frequency p_i^2. The *average fitness of allele* A_i is then

$$
W_i = \sum_{j=1}^{k} w_{ij}p_j, \quad i = 1, 2, \ldots, k,
$$

because with Hardy–Weinberg proportions allele A_i occurs in the genotype $\mathsf{A}_i\mathsf{A}_j$ with the probability p_j. The recurrence equation then becomes

$$
p_i' = \frac{p_iW_i}{W}, \quad i = 1, 2, \ldots, k, \tag{8.2}
$$

where

$$
W = \sum_{i=1}^{k} W_ip_i = \sum_{i=1}^{k}\sum_{j=1}^{k} w_{ij}p_ip_j
$$

is the average fitness. An allele is therefore expected to increase in frequency when it is favored by selection in the sense that its average fitness exceeds the average fitness of the population. Again, the absence of an allele, say, A_i, is a permanent state if we disregard mutation and immigration, and the k allele system then simplifies to a $k - 1$ allele system. The k allele model therefore exhibits k monomorphic equilibria.

In a generation where the fitness W_i of allele A_i is larger than the average fitness W in the population, its frequency is expected to increase ($p_i' > p_i$). In the same way, when $W_i < W$, then $p_i' < p_i$ and the frequency of allele A_i decreases. For two alleles, the principle simplifies in that the frequency of allele A increases ($p' > p$) when $W_{\mathsf{A}} > W_{\mathsf{a}}$ and decreases ($p' < p$) when $W_{\mathsf{A}} < W_{\mathsf{a}}$. The description of the effects of selection on a two-allele polymorphism was established in the 1920s by R.A. Fisher (1922) and J.B.S. Haldane (1932). The dynamic results for the two-allele model in a large population exhibit three qualitatively different scenarios, which are easily presented in terms of relations among the genotypic fitnesses. We therefore describe the dynamical effects of viability selection assuming a very large population size, that is, we contemplate a deterministic description of selection.

The first case is *directional selection*. For multiple alleles, directional selection for allele A_i occurs when $W_i > W$ for all gene frequency configurations $\boldsymbol{p}$

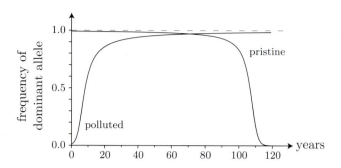

Figure 8.4: The change in gene frequency of a dominant allele. The curve marked "polluted" assumes $w_{AA} = 2w_{aa}$ and an initial gene frequency of 0.01. The curve marked "pristine" assumes $w_{AA} = \frac{1}{2}w_{aa}$ and an initial gene frequency of 0.99.

with $p_i > 0$, and directional selection against the allele occurs when $W_i < W$ for all $\boldsymbol{p}$ with $p_i > 0$. For two alleles, directional selection always prevails when $W_A > W_a$ for all gene frequencies $0 < p < 1$, or when $W_A < W_a$ for all polymorphic gene frequencies. Directional selection against allele a occurs when $w_{AA} \geq w_{Aa} > w_{aa}$ and then the frequency of allele A increases to fixation in a population where both alleles are present. On a gene frequency axis the dynamics may be shown as below, where • shows a stable equilibrium and ∘ an unstable one. A stable equilibrium is a state to which the population returns following

any small perturbation of the genetic composition.[3] An unstable equilibrium is a state where some small perturbations lead to lasting changes in the genetic composition of the population. The stable equilibrium for directional selection is *globally stable*, as the population returns to this equilibrium from any state except, of course, when starting from the monomorphic equilibrium where allele A is absent. Directional selection against allele A occurs when $w_{AA} \leq w_{Aa} < w_{aa}$, and allele a will eventually be fixed in the population ($p = 0$ is a globally stable equilibrium). After establishing these two-allele results, Haldane (1924) analyzed the rapid change in the population frequency of the *carbonaria* morph in the population near Manchester and showed that it corresponds to a fitness ratio of $w_{AA}/w_{aa} \approx 2$ with $w_{AA} = w_{Aa}$ (Figure 8.4).[4] While the *carbonaria* morph replaces the *typica* morph the population is in a period of transient polymorphism.

Directional selection for allele A_i in the multiple-allele model results in its fix-

---

[3]The mutation–balance equilibrium in equation (4.2) on page 78 is thus a stable equilibrium in a large population.

[4]Further numerical examples of the dynamics of directional selection may be obtained on the web (Holsinger 2006).

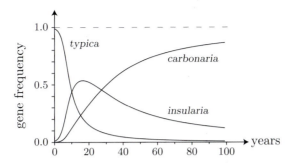

Figure 8.5: Example of changes in the frequency of the three alleles at the *carbonaria* locus in *Biston betularia*. Viabilities of *carbonaria* and *typica* as in Figure 8.4. The viability of *insularia* is 90 percent of that of *carbonaria*. The initial frequency of the *carbonaria* allele is 0.001 and that of the *insularia* allele 0.01.

ation in the population, and $p_i = 1$ is thus a globally stable equilibrium because it is reached from any state where allele A_i is present ($p_i > 0$). On the other hand, directional selection against allele A_i results in its loss from the population. In that case, further analyses of the fate of the population are conveniently made in the $k - 1$ allele model with the alleles $A_1, A_2, \ldots, A_{i-1}, A_{i+1}, \ldots, A_k$. This description of the consequences of directional selection encompasses the two-allele results in that the loss of an allele requires that we continue with the one-allele model. The melanism in *Biston betularia* is an example of variation determined by one locus with three alleles. Elaborating on the example in Figure 8.4, Figure 8.5 includes the *insularia* allele assuming a viability of the *insularia* morph of 0.9 relative to that of the *carbonaria* morph. The scenario of the figure is that the *insularia* allele is introduced first and has reached a frequency of 0.01 at the time when the *carbonaria* allele appears at a frequency of 0.001. After initial domination of *insularia* among the melanic morphs, the directional selection favoring the *carbonaria* is evident.

When selection in the two-allele model is not directional, the heterozygote viability is outside the interval formed by the homozygote viabilities. Then the allele fitnesses becomes equal, $W_A = W_a$, for the gene frequency

$$\hat{p} = \frac{w_{Aa} - w_{aa}}{2w_{Aa} - w_{AA} - w_{aa}} . \tag{8.3}$$

No change in gene frequency occurs when $W_A = W_a$, so this gene frequency describes an equilibrium in the population. The equilibrium is unique and is characteristic of the remaining two cases of Fisher's and Haldane's analyses of the two-allele model

Inspection of equation (8.2) shows that equilibria in the k allele model either have $\hat{p}_i = 0$ or $\hat{W} = \hat{W}_i$ for every $i = 1, 2, \ldots, k$, where $\hat{W}$ and $\hat{W}_i$ signify the average fitness and the average allele fitness at the equilibrium. To look for k allele equilibria we assume $\hat{p}_i \neq 0$ for $i = 1, 2, \ldots, k$ and ask whether we can

---

Box 29: *Multiallelic equilibria

The gene frequencies at an equilibrium where all k alleles are present in the population may be found as the solution to the equations $\hat{W} = \hat{W}_i$, $i = 1, 2, \ldots, k$. These may be written as

$$
\left\{
\begin{array}{cccc}
w_{11} & w_{12} & \cdots & w_{1k} \\
w_{21} & w_{22} & \cdots & w_{2k} \\
\vdots & \vdots & \ddots & \vdots \\
w_{k1} & w_{k2} & \cdots & w_{kk}
\end{array}
\right\}
\left\{
\begin{array}{c}
\hat{p}_1 \\
\hat{p}_2 \\
\vdots \\
\hat{p}_k
\end{array}
\right\}
=
\left\{
\begin{array}{c}
\hat{W} \\
\hat{W} \\
\vdots \\
\hat{W}
\end{array}
\right\}.
$$

If we consider $\hat{W}$ a constant, this may be viewed as a system of linear equations, and we then want to solve

$$
\left\{
\begin{array}{cccc}
w_{11} & w_{12} & \cdots & w_{1k} \\
w_{21} & w_{22} & \cdots & w_{2k} \\
\vdots & \vdots & \ddots & \vdots \\
w_{k1} & w_{k2} & \cdots & w_{kk}
\end{array}
\right\}
\left\{
\begin{array}{c}
x_1 \\
x_2 \\
\vdots \\
x_k
\end{array}
\right\}
=
\left\{
\begin{array}{c}
c \\
c \\
\vdots \\
c
\end{array}
\right\}.
$$

Such a system of equations has exactly one solution, say $\hat{x}_1, \hat{x}_2, \ldots \hat{x}_k$ (except for degenerate cases, where it may have zero or infinitely many). The equilibrium solution is found by normalizing the solution to frequencies that sum to one, that is,

$$
\hat{p}_i = \frac{\hat{x}_i}{\sum_{j=1}^{k} \hat{x}_j}, \quad i = 1, 2, \ldots, k.
$$

This is a solution to the gene frequency equations, but it is only interesting when $\hat{p}_i \geq 0$ for all $i = 1, 2, \ldots, k$. In conclusion, at most one equilibrium exists that segregates all k alleles in the population (except for degenerate cases). Similarly, for a given subset of the alleles at most one equilibrium exists that segregates these alleles (again with the above exception).

---

find gene frequencies that satisfy all the equations $\hat{W} = \hat{W}_i$, $i = 1, 2, \ldots, k$. The solution of these equations is outlined in Box 29; the basic result is that at most one equilibrium exists that segregates all k alleles in the population. But alas, we have to add the clause "except for degenerate cases" where curves of equilibria exist—a possibility that may arise in particular models that we shall ignore here. The reason is that the solution is obtained by solving a system of k linear equations with k variables, and such a system generally provides a unique solution. However, the unique solution $\hat{\boldsymbol{p}}$ in Box 29 is only interesting when $\hat{p}_i > 0$ for all $i = 1, 2, \ldots, k$.

Directional selection against an allele excludes the existence of an equilibrium where all k alleles are present in the population. As a result the population goes

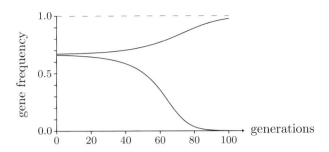

Figure 8.6: The expected change in the frequency of allele a when $w_{AA} = 1.2w_{Aa}$ and $w_{aa} = 1.1w_{Aa}$ for initial gene frequencies of 0.66 and 0.67.

on losing alleles as long as some allele is subject to directional selection against it. Finally, unless an allele is favored by directional selection, we are left with a subset of the k alleles, none of which is exposed to directional selection, and for this subset a gene frequency equilibrium exists (Kingman 1961). Uniqueness of the equilibrium holds for any subset of the k alleles in that, for a given subset of alleles, at most one equilibrium exists at which the population segregates all of these alleles. An equilibrium with two or more alleles present in the population is a *polymorphic equilibrium*.

For two alleles the equilibrium (8.3) exists when the heterozygote viability is either larger or smaller than both homozygote viabilities. The case where the heterozygote fitness is smaller than those of both homozygotes is called *underdominant selection*, that is, $w_{AA} > w_{Aa}$ and $w_{aa} > w_{Aa}$. When allele a is very common and allele A is rare, then $W_a \approx w_{aa}$, $W_A \approx w_{Aa}$, and allele A decreases in frequency. On the other hand, allele A increases in frequency when it is sufficiently common. The direction of change in gene frequency changes at $\hat{p}$, and therefore $W_A < W_a$ when $p < \hat{p}$ and $W_A > W_a$ when $p > \hat{p}$. The equilibrium $\hat{p}$ is thus unstable (a small perturbation away from the equilibrium frequency grows, Figure 8.6), and the monomorphic equilibria are stable (illustrated below assuming a symmetric model where $w_{AA} = w_{aa}$ and therefore $\hat{p} = \frac{1}{2}$).[5] Any real

population therefore cannot stay at the polymorphic equilibrium for underdominant selection, because random genetic drift will cause the gene frequency to deviate from the equilibrium gene frequency even in a very large population.

In Section 10.2.1 (on page 327) we consider the fitness effects of structural karyotypic mutations. Many of these provide good examples of underdominant selection. An example is the selection on the three karyotypes in a translocation polymorphism. The normal homozygote and the translocation homozygote

---

[5]Numerical examples of the dynamics of under- and overdominant selection also may be obtained on the web (Holsinger 2006).

both have normal fertility, whereas the translocation heterozygote has a lowered fertility. This simple kind of selection caused by variation in fertility may be crudely modeled as a lowered viability of the translocation heterozygote. The prediction from the results for underdominant selection is that a translocation polymorphism is unstable and the population will end up with one or the other karyotype.

The last of the three cases described by Fisher and Haldane is *overdominant selection*, where the heterozygote fitness is higher than that of both homozygotes, that is, $w_{AA} < w_{Aa}$ and $w_{aa} < w_{Aa}$. Both alleles increase in frequency when they are sufficiently rare because the common genotype has a lower fitness than the heterozygote that carries the bulk of the minority alleles. In a large population both alleles are therefore protected from loss, a state called *protected polymorphism* (Prout 1968). In accordance with this, an allele is said to be protected if it increases when rare at all equilibria where it is initially absent. The monomorphic equilibria are unstable and the equilibrium $\hat{p}$ is globally stable, in

that the equilibrium is reached from every polymorphic state (again, $w_{AA} = w_{aa}$ is assumed in the drawing).

The classical example of overdominant selection is found in human populations exposed to the malaria parasite *Plasmodium falciparium*. Sickle cell anemia is a recessive disease caused by abnormal hemoglobin due to the allele Hb^S at the locus coding for the β-chains of the hemoglobin molecule. The allele Hb^S produces a polypeptide different from the normal one produced by allele Hb^A at one position in the amino acid sequence (Box 30). The aberrant polypeptide causes lethal anemia in the homozygote, and the heterozygote $Hb^A Hb^S$ may under stress suffer a mild anemia with little consequence for its survival or reproduction. This should cause directional selection against the Hb^S allele, but the disease allele also conveys partial resistance against the malaria parasite. Allison (1964) investigated populations in West Africa where malaria causes mortality and showed that the $Hb^A Hb^S$ heterozygote has a higher viability than the homozygote $Hb^A Hb^A$. Each homozygote thus suffers excess mortality due to a disease, one that has genetic causes, and one that is characteristic of the local environment. The result is overdominance in viability, giving high frequencies of the lethal allele Hb^S (Figure 8.7). Variation in malaria susceptibility due to variation in hemoglobin β is actually multiallelic (Box 30). In addition, synonymous variation in the DNA sequence is known (Table 4.3 on page 78). The nonsynonymous variation in Table 4.3 is the sicklecell polymorphism (of the sequences shown, 12 haplotypes are sampled in Gambia and the last 3 in England (Fullerton et al. 2000)).

Fisher's and Haldane's theories of viability selection have provided a well-developed intuition to generations of population geneticists for the dynamics of one-locus variation in a randomly mating population. The theory for viability selection on variation due to the segregation of multiple alleles at one locus is

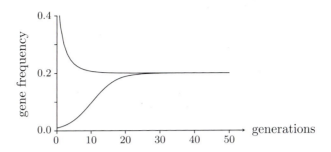

Figure 8.7: The expected change in the frequency of the Hb^S allele when $w_{\mathsf{aa}} = \frac{3}{4} w_{\mathsf{Aa}}$ ($w_{\mathsf{AA}} = 0$) for an initial gene frequency of 0.01 and one of 0.5.

equally well understood at a high level of detail, but the necessity of multi-dimensional formulation of the dynamics impedes its elementary exposition. The stability of an equilibrium is, for instance, determined by the properties of the fitness matrix (Kingman 1961). Excellent descriptions of the theory of multiallelic systems can be found in many textbooks of population genetic theory, for instance Ewens (1979, 2004).

In spite of these technical complications, the qualitative aspects of multiallelic dynamics are simple. When no interior equilibrium exists, selection will cause at least one allele to be lost from the population. Otherwise one equilibrium exists with all alleles in positive frequencies. This equilibrium is either globally stable or unstable, just like the interior equilibrium (8.3) in the two-allele model. When it is unstable at least one allele is lost, and two or more locally stable equilibria exist. The simple relations among the genotypic fitnesses in the two-allele model, however, do not have useful generalizations to conditions that determine the three principal dynamical possibilities for multiple alleles. Example 8.2.2 below shows that simple overdominance with four alleles may produce an unstable four-allele equilibrium with two stable equilibria segregating two alleles each. Simple overdominance is therefore insufficient to maintain all alleles in the population. The condition for stability of a completely polymorphic equilibrium is a series of conditions involving the principal determinants of the fitness matrix (Kingman 1961), and these have yet to be formulated as intuitively appealing fitness relations among genotypes. On the other hand, the existence and stability of a polymorphic equilibrium require some heterozygotes to be superior in fitness.

The evolutionary dynamics of all one-locus models with viability variation among genotypes in a randomly mating population are nevertheless very simple. The average viability of individuals in the population increases through time (Fisher 1930b, Wright 1931, Kingman 1961). We will only illustrate this principle for the two-allele model. The average fitness may be studied as a function of the gene frequency p. For $p = 0$ we have $W = w_{\mathsf{aa}}$, and for $p = 1$ it is

---

Box 30: Hemoglobin β variants and malaria

The Hb^S allele differs from the normal allele Hb^A of the gene coding for the β-chains of the hemoglobin molecule at amino acid position 6 (see reference 603903 in OMIM 2006):

$$Hb^A \quad Val—His—Leu—Thr—Pro—Glu—Glu—$$

$$Hb^S \quad Val—His—Leu—Thr—Pro—Val—Glu—$$

(also shown in Table 4.3). It is common in malaria-ridden areas of Africa. Allele Hb^C, common in West Africa, also differs from Hb^A at position 6:

$$Hb^C \quad Val—His—Leu—Thr—Pro—Lys—Glu—$$

Hb^D occurs in Africa and Southeast Asia. Like Hb^C, it was initially identified by protein electrophoresis and later characterized by amino acid change. It covers two proteins, however. One, Hb^D Punjab, has a replacement of Glu with Gln at position β 121, and the other, Hb^D Ibadan, has a replacement of Thr with Lys at position β 16. Hb^E, common in Southeast Asia, codes for a replacement of Glu with Lys at position β 26.

Other genetic changes involving hemoglobin are known to convey resistance, for instance regulatory variants like those causing thalassemias (see, e.g., Cavalli-Sforza et al. 1994).

$W = w_{AA}$. The change in W as a function of p is described by

$$\frac{dW}{dp} = 2pw_{AA} + 2(q-p)w_{Aa} - 2qw_{aa} = 2\left(W_A - W_a\right).$$

Thus, the average fitness is an increasing function of p when $W_A > W_a$ and decreasing when $W_A < W_a$, and the change in gene frequency is therefore in the direction where the average fitness increases.[6] The average fitness is at a minimum or a maximum at the equilibrium (8.3), and the equilibrium is stable when the average fitness is at a maximum and unstable when the average fitness is at a minimum. This result may be written into the recurrence equation in the gene frequencies in a formula due to Wright (1931):

$$p' - p = \frac{pq}{2W} \frac{dW}{dp}. \tag{8.4}$$

This equation is referred to as Wright's equation.

Example 8.2.1 Consider a locus with four alleles, A_1, A_2, A_3, A_4, where the genotypic fitness of the homozygote $A_i A_i$ is $w_{ii} = w$ and that of the heterozygote $A_i A_j$ is $w_{ij} = 1$ for i and $j = 1, 2, 3, 4$, and $i \neq j$ (Table 8.4). The model thus

---

[6] * This does not prove that the average fitness increases between generations. For overdominance we also have to show that when $p < \hat{p}$ then $p' < \hat{p}$, and vice versa.

Table 8.4: Example of one-locus four-allele fitnesses

	A_1	A_2	A_3	A_4
A_1	w	1	1	1
A_2	1	w	1	1
A_3	1	1	w	1
A_4	1	1	1	w

only distinguishes homozygotes and heterozygotes. The average fitness of allele A_i and the mean fitness in the population are then

$$W_i = 1 - (1 - w)p_i \quad \text{for } i = 1, 2, 3, 4,$$
$$W = 1 - (1 - w)(p_1^2 + p_2^2 + p_3^2 + p_4^2).$$

At equilibrium the four alleles have equal average fitness and we can write

$$1 - (1 - w)\hat{p}_1 = \hat{W},$$
$$1 - (1 - w)\hat{p}_2 = \hat{W},$$
$$1 - (1 - w)\hat{p}_3 = \hat{W},$$
$$1 - (1 - w)\hat{p}_4 = \hat{W},$$

because at equilibrium the four equal allele fitnesses must be equal to the average fitness in the population. The equilibrium with all alleles present therefore has $\hat{p}_1 = \hat{p}_2 = \hat{p}_3 = \hat{p}_4$, because the four alleles are indistinguishable in their fitness effects. Thus, $\hat{p}_i = \frac{1}{4}$, $i = 1, 2, 3, 4$ (see Figure 8.8, left).

The mean fitness has a local maximum at the equilibrium $\hat{p}_i = \frac{1}{4}$ when $w < 1$ and a minimum when $w > 1$. Thus, the equilibrium is stable with overdominance and unstable with underdominance, and with underdominance the four vertices ($p_i = 1$ for $i = 1, 2, 3,$ or 4) are stable equilibria.

Analysis of stability of the monomorphic equilibrium $\hat{p}_4 = 1$ proceeds by assuming that the gene frequencies p_1, p_2, and p_3 are small. The common genotype is A_4A_4, and the alleles A_1, A_2, and A_3 are so rare that we may ignore the possibility that they meet in gamete union. The only extra genotypes are then the rare types A_iA_4, $i = 1, 2, 3$. The change in gene frequency of the A_1 allele is then approximately given by the recurrence equation

$$p_1' \approx \frac{w_{14}}{w_{44}} p_1, \tag{8.5}$$

in that the average fitness is close to w_{44}, the fitness of common type. Formally the approximation is obtained by ignoring all terms in the recurrence equation that are at least of the order of $p_i p_j$, $i, j = 1, 2, 3$. With the current fitnesses (Table 8.4) we get

$$\frac{w_{14}}{w_{44}} = \frac{1}{w}.$$

This relative fitness is less than one when $w > 1$ and greater than one when $w < 1$. Allele A_1 will thus increase in frequency when $w < 1$ (the rare A_1A_4 type

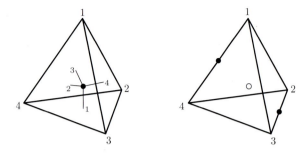

Figure 8.8: Examples of equilibria for four alleles. Left: Example 8.2.1. Right: Example 8.2.2.

has a higher fitness than the common A_4A_4 type); the equilibrium is therefore unstable. The same is true for alleles A_2 and A_3. The equilibrium is unstable to the introduction of any of the absent alleles. For $w > 1$ all of the alternative alleles decrease in frequency if introduced, and the equilibrium is therefore locally stable.

Due to the pronounced symmetry of the fitness matrix, the result holds for all four monomorphic equilibria at the vertices of the simplex. $\square$

Example 8.2.2 Consider a locus with four alleles where the genotypic fitnesses are as in Table 8.5. All homozygotes have the fitness w^2. The heterozygotes

Table 8.5: Another example of one-locus four-allele fitnesses

	A_1	A_2	A_3	A_4
A_1	w^2	w	w	1
A_2	w	w^2	1	w
A_3	w	1	w^2	w
A_4	1	w	w	w^2

A_1A_4 and A_2A_3 have fitness 1, with the remaining four having fitness w. The average fitnesses of the four alleles are

$$
\begin{aligned}
W_1 &= w^2 p_1 + w p_2 + w p_3 + p_4, \\
W_2 &= w p_1 + w^2 p_2 + p_3 + w p_4, \\
W_3 &= w p_1 + p_2 + w^2 p_3 + w p_4, \\
W_4 &= p_1 + w p_2 + w p_3 + w^2 p_4.
\end{aligned}
$$

These four fitnesses are equal when $p_i = \frac{1}{4}$, $i = 1, 2, 3, 4$, and only one equilibrium with all alleles present exists. The equilibrium with all alleles present is thus $\hat{p}_i = \frac{1}{4}$, $i = 1, 2, 3, 4$.

The mean fitness in the population is

$$W = 2(p_1p_4 + p_2p_3) + 2w(p_1p_2 + p_1p_3 + p_2p_4 + p_3p_4) + w^2(p_1^2 + p_2^2 + p_3^2 + p_4^2),$$

and this never has a local maximum at the equilibrium $\hat{p}_i = \frac{1}{4}$ where the value is $\hat{W} = \frac{1}{4}(1+w)^2$. Thus, the equilibrium is always unstable, even though for $w < 1$ the fitnesses of the homozygotes are lower than the fitnesses of all heterozygotes, and in this sense fitness is overdominant.

For $w < 1$ two locally stable equilibria exist and are stable, namely $\hat{p}_1 = \hat{p}_4 = \frac{1}{2}$ and $\hat{p}_2 = \hat{p}_3 = 0$, and $\hat{p}_1 = \hat{p}_4 = 0$ and $\hat{p}_2 = \hat{p}_3 = \frac{1}{2}$ (see Figure 8.8, right). These two-allele equilibria both have $\hat{W} = \frac{1}{2}(1 + w^2)$, and this is higher than the mean fitness at the four-allele equilibrium with all alleles present because $\frac{1}{4}(1 + w)^2 < \frac{1}{2}(1 + w^2)$ for all $w \neq 1$.

The four two-allele equilibria $\hat{p}_1 = \hat{p}_2 = \frac{1}{2}$ ($\hat{p}_3 = \hat{p}_4 = 0$), $\hat{p}_1 = \hat{p}_3 = \frac{1}{2}$ ($\hat{p}_2 = \hat{p}_4 = 0$), $\hat{p}_3 = \hat{p}_4 = \frac{1}{2}$ ($\hat{p}_1 = \hat{p}_2 = 0$), and $\hat{p}_2 = \hat{p}_4 = \frac{1}{2}$ ($\hat{p}_1 = \hat{p}_3 = 0$) always exist. They are stable as two-allele equilibria when $w < 1$, but their average fitness is $\frac{1}{2}w(1 + w)$, which is then less than $\frac{1}{4}(1 + w)^2$, and they are therefore not stable in the four-allele context. For $w > 1$ they are unstable two-allele underdominant equilibria.

Analysis of stability of the vertex $\hat{p}_4 = 1$ proceeds as in the previous example. The change in gene frequency of the A_1 allele is again approximately given by equation (8.5), which with the current fitnesses is governed by

$$\frac{w_{14}}{w_{44}} = \frac{1}{w^2}.$$

This is less than one when $w > 1$ and greater than one when $w < 1$. The current fitness matrix is less symmetric, and we must also consider the recurrence equations of the remaining two alleles. These are governed by the relative fitnesses

$$\frac{w_{i4}}{w_{44}} = \frac{1}{w}, \quad i = 2, 3,$$

and this is again less than one when $w > 1$ and greater than one when $w < 1$. Similar analyses pertain to the equilibria $\hat{p}_1 = 1$, $\hat{p}_2 = 1$, and $\hat{p}_3 = 1$. At any of the monomorphic equilibria any of the rare alleles thus increase in frequency when introduced if $w < 1$, and the equilibria are unstable. For $w > 1$ any of the rare alleles decrease in frequency when introduced at any of the monomorphic equilibria. With underdominance, $w > 1$, the four monomorphic equilibria are therefore locally stable. $\square$

8.2.1   Viability selection in *Zoarces*

The only component of selection left to analyze in the population sample including mother–offspring combinations in the example on page 219 is differential viability. *Zoarces viviparus*, however, matures at age two years and may live for at least a decade. Thus, we cannot assume nonoverlapping generations. The ages of adult individuals caught in the period 1971–74 in the sample shown in Tables 8.1 and 8.3 were determined from otoliths, and the genotypic effects of differential survival may then be seen as changes through time in a birth cohort of adults. In addition, this allows the stability of the population to be assessed by comparing the genetic composition across cohorts (Christiansen and Frydenberg 1976). No such heterogeneities were observed, and the composition of the sample is therefore fairly represented by the adult females and males in Table 8.3. Comparison reveals no sex difference. Viability selection may therefore only occur between unborn offspring and the first adult age classes observed, and it causes a difference between the expected Hardy–Weinberg proportions and the genotypic frequencies in adults.[7] Analysis of this difference confirmed the indication in Table 8.1 of a difference between mothers and offspring. The proper comparison, however, is between the Hardy–Weinberg proportions, corresponding to the total observed gene frequencies in the sample, and the total observed genotypic frequencies in adults of the sample. This comparison allows us to estimate the relative viabilities of the genotypes. The estimation procedure assumes stability in the population, and therefore the resulting relative fitnesses correspond to an equilibrium given by the total gene frequencies of 0.36 and 0.64. Relative to 1 for the heterozygotes, the viabilities become 1.07 and 1.04 for the homozygotes. We thus observe underdominance, which is at odds with a stable population (for further discussions, see Christiansen et al. 1977).

8.3   Selection on a Quantitative Character

In evolutionary theory selection is defined in relation to phenotypes, morphological characters in particular, so let us return to our simple model of quantitative genetics in Chapter 7. The quantitative character shows normal-distributed trait values, and we assume that the variation is determined additively by the genotypes at a large number of independent loci. Without selection the distribution among offspring is exactly like that among the parents (Figure 7.3 on page 187), but when viability depends on phenotype, the phenotypic distribution among individuals changes (we will assume discrete generations and that selection acts as differential viability before offspring mature and become parents). If the phenotypic distribution without selection is Gaussian, selection has to be quite special to produce a normal distribution among parents. Experience shows, however, that the population of offspring of selected parents often can be reasonably assumed to exhibit phenotypic variation that follows the normal distribution. As in Chapter 7 we assume the environmental variance to be the same in all generations.

The experience comes from selection experiments on neutral or nearly neutral characters. Thus the experimenter manipulates a population like that in Figure 7.3 on page 187—undisturbed, it patiently reproduces the same phenotypic distribution among mature adults every generation. Now suppose that the experimenter selects as parents the fraction of mature adults that have a pheno-

---

[7]Viability selection does not in general produce a deviation from Hardy–Weinberg proportions in adults, see Exercise (8.2.1). We accepted that the population is stable, however.

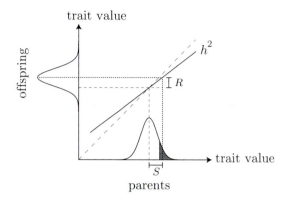

Figure 8.9: The effect of truncation selection on a quantitative character.

typic value above a threshold T, that is, $P \geq T$, and assume them to mate at random. This form of selection is called *truncation selection*. The phenotypic effect of selection among parents may then be described by the change in mean value, that is, $S = \mathrm{E}(P_\mathcal{P}|P_\mathcal{P} \geq T) - \mathrm{E}\,P_\mathcal{P}$, referred to as the *selection differential*. The *selection response* seen among offspring is also a change in phenotypic mean value, and that is $R = \mathrm{E}(P_\mathcal{O}|P_\mathcal{P} \geq T) - \mathrm{E}\,P_\mathcal{P}$. This is a reasonable measure because the mean in the offspring population is expected to be the same as that among parents had selection not occurred. The expected response is 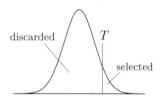 readily calculated from the midparent–offspring regression because both parents of any offspring are drawn from the selected population. Figure 8.9 shows the calculation, and because the regression slope is $\beta_{\mathcal{MO}} = h^2$ (see page 193), we get the basic equation for response to selection as

$$R = h^2 S. \tag{8.6}$$

This equation holds for any kind of selection—the assumption of truncation selection just renders visualization easier.

To describe the full parent–offspring transition with selection we only need to add the assumption that the phenotypic variance is unchanged between generations, selection or not. This assumption is not sensible in all situations with any kind of selection. For instance, for a character determined by one locus with additive allele effects, extreme truncation selection may result in an offspring population where most individuals are of the same kind of homozygote. The difference in variance between the two generations is then V_G. Experience is, however, that the assumption works satisfactorily.

The response equation may be used to predict the results of selection on the basis of heritability estimates obtained from kin correlations. The experiments may then be used to test the prediction in equation (8.6), and thereby check the

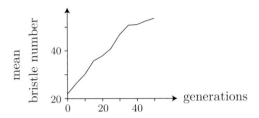

Figure 8.10: Response to truncation selection with constant intensity for high number of abdominal bristles in a laboratory population of *Drosophila melanogaster*. Data from Jones et al. (1968).

ability of the simple quantitative transmission models to describe the population effect of selection. The selection differential S and the response R are observable in selection experiments, and from this we can estimate the *realized heritability*

$$h^2 = \frac{R}{S} \tag{8.7}$$

and compare it to the heritability estimated from family analyses.

Breeding efforts usually rely on the cumulative effects of selection over many generations, and the real interest is therefore whether the description of transmission before the breeding program provides a good indicator of its ultimate success. The definition of selection in terms of a threshold is of no use when addressing this question because the response to selection with a fixed threshold would select a higher and higher proportion of mature individuals to be parents. We need a fixed threshold relative to the distribution, that is, we should chose T such that

$$i = \frac{S}{\sqrt{V_P}}$$

is constant, and this is accomplished by selecting the same proportion to be parents. In terms of this measure, called the *intensity of selection* measure, the response equation reads

$$R = h^2 i \sqrt{V_P}.$$

For fixed intensity i the response to selection is predicted to be the same every generation, still keeping in mind the assumption that both the additive and phenotypic variance are constants throughout the experiment. The cumulated response is then predicted to show a linear increase. Figure 8.10 shows the results of such an experiment, where the response in bristle number is linear within the first 35 generations, with a change of 0.8 bristles per generation (this experiment will be further discussed in Section 9.3.1).

8.3.1   Genetic effects of selection

The genetic changes due to selection cause changes in the genotypic variance in addition to the changes in the genotypic value. We study this in a one-locus

---

Box 31: Viability with truncation selection

Assume normal-distributed environmental deviations and that the truncation point is at $E(P)+T$. The viability by truncation selection of a genotype with deviation g is then

$$w(g) = \frac{1}{\sqrt{2\pi V_E}} \int_T^\infty \exp\left(-\frac{(\xi - g)^2}{2V_E}\right) d\xi.$$

We may approximate the difference in fitness of a genotype with deviation g by a Taylor expansion around $g = 0$, which is the position of the phenotypic mean value. We thus need

$$\left.\frac{\partial w(g)}{\partial g}\right|_{g=0} = \frac{1}{V_E\sqrt{2\pi V_E}} \int_T^\infty \xi \exp\left(-\frac{\xi^2}{2V_E}\right) d\xi = \frac{S(0)w(0)}{V_E},$$

where $S(g)$ is the mean phenotypic deviation of selected individuals with genotypic deviation g. The approximation of the genotypic fitness is then

$$w(g) \approx w(0) + g\frac{S(0)}{V_E}w(0) = w(0)\left(1 + g\frac{i}{\sqrt{V_P}}\right)$$

and the relative fitness is therefore given by equation (8.8).

---

model where the alleles A and a influence a character under truncation selection. The phenotype given the genotype is assumed to be Gaussian distributed, with the variance equal among genotypes. We refer to this variance as the environmental variance. To quantify the effect of selection on the allelic variation, we need to find the induced genotypic fitnesses. The viability of a genotype is given by the fraction of individuals with a phenotypic value above the threshold T, and that is given by the area under the normal-distribution curve in the tail from T and out to the right. The viabilities of genotypes whose val-

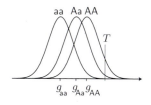

ues differ by a small amount are calculated in Box 31. The fitnesses of the three genotypes relative to a hypothetical genotype with value equal to zero (the population mean) are approximately given by

$$w_{\mathsf{AA}} \approx 1 + \frac{g_{\mathsf{AA}}i}{\sqrt{V_E}}, \quad w_{\mathsf{Aa}} \approx 1 + \frac{g_{\mathsf{Aa}}i}{\sqrt{V_E}}, \quad \text{and} \quad w_{\mathsf{aa}} \approx 1 + \frac{g_{\mathsf{aa}}i}{\sqrt{V_E}} \tag{8.8}$$

when the genotypic deviations are small. We will use these, and in the following assume small allele effects in our calculations. The mean fitnesses of the alleles are therefore approximately given by

$$W_A \;=\; 1 + \frac{i}{\sqrt{V_E}}\left(pg_{AA} + qg_{Aa}\right) \;=\; 1 + \frac{i}{\sqrt{V_E}}\,\alpha_A,$$

$$W_a \;=\; 1 + \frac{i}{\sqrt{V_E}}\left(pg_{Aa} + qg_{aa}\right) \;=\; 1 + \frac{i}{\sqrt{V_E}}\,\alpha_a,$$

and the population mean fitness is $W = 1$ (of course). The change in gene frequency in a generation is therefore determined by α, which is the additive effect of the two alleles causing the variation under selection ($\alpha = \alpha_A/q = -\alpha_a/p$, see Table 7.3 on page 196). By using Wright's equation (8.4) on page 230 we then obtain the change in gene frequency as a function of the additive effects of the alleles as

$$p' - p = \frac{i}{\sqrt{V_E}}\,pq\alpha. \tag{8.9}$$

If we view fitness as a quantitative character, a quite straightforward calculation then produces the additive effect of fitness as

$$\alpha_w = \frac{i}{\sqrt{V_E}}\,\alpha,$$

and the additive variance in fitness of truncation selection on a character is then

$$V_{TA} = \frac{i^2}{V_E}\,V_A \tag{8.10}$$

for a locus with a small contribution to the variation in the character. Equation (8.10) is thus valid for many loci in the additive model.

After this excursion into the population genetics of truncation selection, we return to the reason we made these calculations, namely the question of how truncation selection influences the genotypic variance components, the additive variance in particular. The key parameter is the additive effect α. To study the effect of selection on variance we must focus on the genotypic mean values, and these may be specified in terms of an additive contribution of each allele, ε_A and ε_a, and a dominance parameter, say, η:

$$G_{AA} = 2\varepsilon_A + \eta, \quad G_{Aa} = \varepsilon_A + \varepsilon_a - \eta, \quad \text{and} \quad G_{aa} = 2\varepsilon_a + \eta. \tag{8.11}$$

Any genotypic values for a one-locus two-allele model may be written in this way. In these terms the additive values are

$$\alpha_A \;=\; q(\varepsilon_A - \varepsilon_a) + 2q(p - q)\eta, \tag{8.12}$$

$$\alpha_a \;=\; -p(\varepsilon_A - \varepsilon_a) - 2p(p - q)\eta, \tag{8.13}$$

which merge into the additive effect $\alpha = \varepsilon + 2(p - q)\eta$ where $\varepsilon = \varepsilon_A - \varepsilon_a$.

The additive effect is constant for the additive alleles model ($\eta = 0$), and we get directional selection because the change in gene frequency ($p' - p$) is proportional to pq (Figure 8.11, top left)—a proportionality also present in the

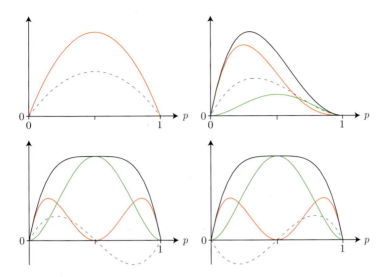

Figure 8.11: Change in the genotypic variance in four different regimes. Top left: the additive alleles model; top right: dominance of allele A; bottom left: symmetric overdominance; bottom right: symmetric underdominance. Black: genotypic variance; red: additive variance; green: dominance variance. Variances are scaled to show the maximal genotypic variance as equal in the four drawings. The dashed curves indicate $p' - p$ on an inflated scale. Compare to Figure 7.6 on page 200.

genotypic variance $\big(V_G = V_A$; see equation (7.18) on page 196$\big)$. The course of evolution due to truncation selection is thus that the genetic variance after a maximum at equal gene frequencies vanishes as the advantageous allele fixes. In addition, the rate of change is maximal when the additive variance is maximal.

In general, evolution may be a lot more complex. For instance, if $\varepsilon_A = \varepsilon_a > 0$ and $\eta < 0$ we get symmetric overdominance, where the population settles at a stable equilibrium with $V_A = 0$ and V_G maximal and equal to V_D (Figure 8.11, bottom left). In this symmetric case with $\eta > 0$ we get underdominance with an unstable equilibrium (Figure 8.11, bottom right). The variation in the variance components is very similar for over- and underdominance, but again the additive variance follows the change in gene frequencies per generation. This illustrates the second part of Fisher's fundamental theorem of natural selection, namely that the rate of change in gene frequency equals the additive variance in fitness (Fisher 1930b).[8] The simple symmetric variation in the additive variance in Figure 8.11, top left, is characteristic for the additive alleles model. General directional selection exhibits additive and dominance variances with maximal sizes for $p \neq \frac{1}{2}$. Figure 8.11, top right, shows the variance components for abso-

---

[8]Fisher showed this in a continuous time model with weak selection. For the discrete generation counterpart, see Ewens (1979).

lute dominance, which is obtained with either $\eta = -\frac{1}{2}(\alpha_A - \alpha_a)$ (A dominant) or $\eta = \frac{1}{2}(\alpha_A - \alpha_a)$ (A recessive).

The conclusion from these simple arguments is clear. For a quantitative character determined by additive two-allele loci, selection will in the long run change the population mean at the expense of the hereditary variation in the population. This may seem to be a qualification of the Mendelian solution to Darwin's variation problem, but it is not. Darwin was aware of the fact that changing the appearance of the population by natural selection was paid for by a loss in hereditary variation of the character. Darwin's problem was that his conception of heredity carried a large cost in hereditary variation. He nevertheless formulated the hypothesis that variation was replenished by variations of small effects unrelated to the process of natural selection—that is, by what we today describe as random mutations in the genome.

Virtually every phenotypic character shows heritable trait variation, so the question remains of how this variation is maintained. A balance between random genetic drift and mutation is the most prudent explanation. This explanation is unacceptable to many evolutionary biologists, and it is clearly objectionable as a universal explanation. The hypothesis has nevertheless been entertained by evolutionary biologists for a long time—in essence already by Johannes Schmidt (1917, 1918) in his interpretation of morphological variation in populations of *Zoarces viviparus*. Additive variance in fitness can only persist if the resulting selection is counteracted by mutation. We will return to this possibility in Section 8.4, ignoring it for now, and only note that otherwise, the only possibility for genotypic variation is dominance variance. The observed variation of a heritable quantitative character may, however, have a significant additive component even if it is under selection and the associated loci are maintained at a stable equilibrium. Selection may not have a simple relation to the variation in the observed character, and a character influenced by additive loci segregating additive alleles may well show genotypic variation (necessarily additive) with the allelic variation maintained by overdominant selection. The simplest example of this is so-called optimizing or stabilizing selection, where a certain phenotypic value has the highest fitness, which declines as the phenotypic value deviates more and more from the optimum value. If at a two-allele locus the genotypic value of the heterozygote is closest to the phenotypic optimum, then overdominance in genotypic fitnesses result.

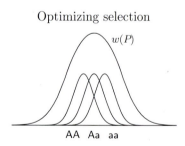

Optimizing selection

$w(P)$

AA Aa aa

Scharloo et al. (1977) compiled components of the phenotypic variance for five characters in *Drosophila melanogaster*. The most striking difference among these is the ratio between the additive and dominance variances. One of the characters "Ci^D expression" is quite special, both in its etiology and its genetic constitution. The dominant mutant allele Ci^D causes shortening of the fourth wing vein. It is situated on the very small fourth chromosome, and through a crossing procedure its chromosome was introduced into a wild-caught population without disturbing the rest of the genome. Thus, the expression of the mutant Ci^D was measured in a genome that had not previously been exposed to it. The genetic constitution of the character Ci^D expression may be described exclusively by the additive variance. The mutant Ci^D is deleterious, and selection for low expression should therefore be expected (given, of course, that the deleterious effect is correlated with the level of its morphological expression). The existence of a sizeable additive component and a small dominance component is therefore as expected, but it is noteworthy that it is extremely small.

The characters "abdominal bristle count" and "thorax length" are obvious and easily quantified characters that have been used in many experimental studies. They have a reasonably high heritability (part of their usefulness) and they typically have low dominance variance. The last two characters considered by Scharloo et al., "ovariole number" and "egg production," are obviously related to individual fitness. As such they are expected to show a low additive variance, and indeed a comparatively high proportion of the genotypic variance is dominance variance. Here the surprise is in the high level of additive contribution to measures closely related to the primary measure of fitness: procreation. Ovariole number and egg production in the laboratory may not be as closely related to the long-term success of a genotype as one would immediately expect. The total energy allocated to eggs may reduce the survival of the female, causing a lower number of eggs actually produced, and a high egg production may be achieved by making the eggs smaller, causing low survival of the offspring. Equilibration of a character at $V_G = V_D$ happens only in theory or in a situation where the character we measure is fitness, and that only happens in dreams.

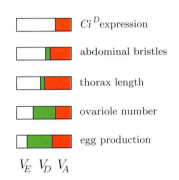

Apart from the trend in the genetic variance components, the analysis by Scharloo et al. (1977) shows that the characters related to individual fitness have a lower environmental variance. The environmental variance is difficult to compare among characters and experiments. Typical variations in the environment experienced by individual flies are likely to vary among experiments, and the sensitivity of genotypes to such variations may also vary with the character. Nevertheless, the trend seems to reflect a general phenomenon (Houle 1992).

8.3.2 Fitness maximization

The average fitness is at a local maximum when the population is at a stable equilibrium. This fitness maximization principle is an important part of Fisher's fundamental theorem of natural selection (Fisher 1930b). It states that "the rate of increase in fitness of any organism at any time is equal to its genetic variance in fitness at that time." Fitness is viewed as a quantitative character, and the genetic variance in question is the additive variance V_A that describes the part of the variation in fitness that gives rise to correlations in fitness between parents and their offspring (equations (7.9) and (7.10) on page 193). A corollary to Fisher's fundamental theorem is that the additive variance at equilibrium is zero.

Fisher's fundamental theorem holds for any number of alleles, but for simplicity we focus on the case of just two alleles. The genotypic values are the viabilities, and the population mean is the mean fitness (Table 8.6). The additive

Table 8.6: Variation in fitness

Genotype	AA	Aa	aa
frequency	p^2	$2pq$	q^2
fitness	w_{AA}	w_{Aa}	w_{aa}
additive value	$2W_A$	$W_A + W_a$	$2W_a$

effects of the two alleles are simply $W_A - W$ and $W_a - W$, so that $\alpha_w = W_A - W_a$. The additive variance (Fisher's genetic variance) in fitness is then

$$V_A = 2pq \left(W_A - W_a \right)^2$$

(equation (7.18) on page 196). At equilibrium, $W_A = W_a$ and therefore $V_A = 0$. Fisher's genetic variance thus vanishes at equilibrium, but the genotypic variance need not do so (Figure 8.11). At a polymorphic equilibrium the genotypic variance is positive and equal to the dominance variance.

Viewing fitness as a quantitative character, we can consider locus A as a SNP marker as in Section 7.4 (page 202), that is, describe the genotypic effects as in Table 8.7. The effect of the locus on fitness may then be described by the co-

Table 8.7: QTL of fitness

Genotype	AA	Aa	aa
frequency	p^2	$2pq$	q^2
Γ	0	1	2
fitness	w_{AA}	w_{Aa}	w_{aa}

variance between fitness and genotype, $\mathrm{Cov}(w, \Gamma) = -2pq\alpha_w$. Price (1972) saw that this covariance describes the effect of viability selection, in that Wright's equation (8.4) on page 230 may be written as

$$p' - p = \frac{-\mathrm{Cov}(w, \Gamma)}{2W}, \tag{8.14}$$

providing a more exact expression of the sketch in Figure 8.2 (page 217) of the neo-Darwinian view of evolution by natural selection. Equation (8.14) is referred to as Price's formula. Equation (8.9) on page 238 describing the effect of truncation selection is written in the form of Price's formula.

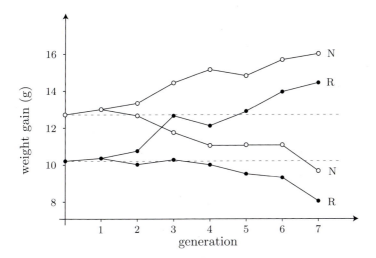

Figure 8.12: Response to selection for growth rate in the mouse (*Mus musculus*). The results of selection for high and low weight gain are shown in two environments. Normal diet: ○, reduced protein diet: ●. The response is shown as deviations from the control lines (dashed lines). Data from Nielsen and Andersen (1987); after Christiansen and Feldman (1986).

8.3.3   An experiment on growth rate in *Mus*

Genotype–environment interaction is the subject of a study by Nielsen and Andersen (1987) in a laboratory population of the house mouse, *Mus musculus*. Mice were selected for weight gain after weaning from three to nine weeks of age. The experiment applied selection in two environments defined by diet. In one environment, designated N, the mice were given feed with a normal protein content. In the other, designated R, the mice were fed a reduced protein diet. Four kinds of selection lines and two unselected controls were established, each with three replicate lines. Lines selected for high and low growth were kept in each environment, and so were unselected control lines. Measurement and selection of weight gain posed a challenge, as the youngsters resided in litters under maternal care. Maternal effects and the influence of common environment of litters were potential problems, both in the practical execution of the experiment and in the interpretation of results. The solution is selection within litters.[9] After suitable standardization, one female and one male were selected from each litter to be parents of the next generation. In the high-selection lines the female and male with the highest growth were chosen, and in the low-selection lines those with the lowest weight gain were kept. In the control lines a male and a female were chosen at random.

The eighteen lines were established in generation 0 and selection was begun

---

[9]The heritability of the within-litter deviation has properties very similar to those of the population-based heritability (see Falconer and Mackay 1996).

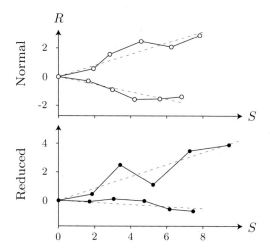

Figure 8.13: Regression of response on selection differential for growth rate in the mouse (*Mus musculus*). The abscissa shows the cumulative selection differential, the ordinate cumulative response, and the dashed line the regression. Normal diet: ∘, reduced protein diet: •. After Nielsen and Andersen (1987).

in generation 1. Each line consisted of eight mating pairs, and matings of least relationship[10] were made. Figure 8.12 shows the response to selection during the first six generations of selection. The deviation of the phenotypic averages in the selection lines[11] are shown as deviations from the corresponding controls.

Selection for high and low growth was successful in both environments, and the responses seem comparable. The response to high selection, however, was larger than that to low selection. How does this pattern look compared to the prediction from the response equation (8.6)? The answer is that the figure shows response, but a comparison to the response equation would require the selection differential as shown in Figure 8.13. From these comparisons the realized heritability (equation (8.7)) may be estimated from the regression lines. The divergence between the high and low lines is 0.33 ± 0.05 in the "normal" environment and 0.26 ± 0.08 in the "reduced" environment.

The environment clearly influences the amount of weight gained by young mice (Figure 8.12). The average growth in the control lines was 12.5g and 10.5g on normal and reduced diet, respectively. The phenotypic response to selection seems very comparable, but is the response durable in the sense that the changes in weight gain can be transferred to other environments?

To investigate this question, individuals in generation 7 were separated into two lots, one of which was raised in the environment of the line, the other in the alternative environment. The results are sketched in the drawing to the right.

---

[10]Among the possible mating patterns of the eight females and eight males selected, the one with the lowest average consanguinity coefficient was chosen.

[11]The line averages averaged across replicates.

Responses are shown as deviations from the dashed line, corresponding to the level of the unselected control lines in the rearing environment. The observed growth rate is shown by ○ for the lines selected on a normal diet, and by ● for those selected on the reduced diet—the rearing environments are designated N and R. The transferred mice show a clear regression towards the control level. The lines selected on the normal diet show a tiny response in the reduced environment compared to the response of the lines selected in that environment. The only transferable response is that of the high selection line in the reduced environment.

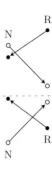

The genetic divergence of the high and low lines brought about by selection in the normal environment thus does not emerge as phenotypic differences in the environment of reduced-protein diet—a clear example of genotype–environment interaction. The low line selected in the reduced environment shows genotype–environment interaction of the same magnitude. The only exception is the high line in the reduced environment, which shows a general improvement.

8.4 Mutation and Selection

The equilibrium $\hat{p}_1 = \hat{p}_2 = \frac{1}{2}$ in the four-allele model of Example 8.2.2 on page 232 is stable for $w < 1$ as long as the alleles A_3 and A_4 are absent from the population. Mutation has an interesting qualitative effect on such equilibria because mutational production of the absent alleles will perturb the population away from the equilibrium, and selection will cause the new allele to increase in frequency. On the other hand, the equilibrium $\hat{p}_1 = \hat{p}_4 = \frac{1}{2}$ with $w < 1$ is stable in the four-allele context, and selection will push the gene frequencies back towards the stable equilibrium. The result of recurrent mutation is thus a balance between the production of new alleles by mutation and their elimination by selection, a *mutation–selection balance* equilibrium.

For two-allele variation and directional selection, a stable monomorphic equilibrium results without mutation. With mutation the deleterious allele is pro-

duced, so the monomorphic state is no longer an equilibrium. The result in a large population is a mutation–selection balance equilibrium close to the mono-

morphic state. Mutation–selection balance therefore describes the occurrence of genetic diseases at low frequencies in human populations. The incidence of dominant alleles causing a lethal condition is, of course, a direct reflection of the mutation rate because carriers of the allele die before they can pass it on to offspring. For less serious dominant conditions, for instance achondroplastic dwarfism (see reference 100800 in OMIM 2006), the condition is seen as truly hereditary. In a study comprising 94,000 newborns, Mørch (1941) observed 10 newborn achondroplastic dwarfs, three of whom had an achondroplastic parent. Even though transmission occurs, most affected children were sporadic cases, most likely due to new mutations and suggesting a mutation rate of 4×10^{-5} mutations per generation. The observation of 3 transmitted cases should then provide information on the selection balancing the rate of production of new cases by mutation.

The incidence of recessive lethal conditions is comparable to that of dominant diseases, which is in the range of 10^{-4} to 10^{-6} (or lower). The occurrence of an afflicted child is therefore highly unlikely to be due to a novel mutation. Mutations trickle into the population and stay there without consequence until two mutant genes happen to meet in an individual. An example of a lethal recessive disease is phenylketonuria.[12] It occurs in about one in 10,000 newborns in Denmark, and presupposing random mating the gene frequency is 0.01. Only about one in 100 of the deleterious alleles is therefore manifested in an individual with the disease. The waiting time until a mutant shows its effect is geometrically distributed. The expected waiting time until elimination from the population by the appearance in an afflicted individual was 100 generations before the condition could be treated.

Selection on such dysfunctional alleles is likely to be directional, unless of course, they have other effects, as in sickle cell anemia, where the Hb^S allele has a so-called pleiotropic effect on susceptibility to malaria. We thus model mutation–selection balance by one-locus two-allele directional selection where the probability of survival of genotypes AA, Aa, and aa are $w_{AA} = 1 \geq w_{Aa} > w_{aa}$ relative to that of the normal homozygote AA. Mutation modifies the recurrence equation for the gene frequency compared to that describing the action of selection. In higher animals, however, survival and the processes of segregation, mating, and reproduction are usually not influenced by mutation, and we shall assume that. The copies of the genes transmitted to the gametes are thus the same whether or not mutation has taken place, and the frequency p^* of copies of parental genes of allele type A among breeding adults is therefore still given by the recurrence equation (8.2) on page 222, so

$$ p^* = \frac{pW_A}{W}, \quad \text{where} \quad W_A = p + w_{Aa}q, \text{ and } W = p^2 + 2pqw_{Aa} + q^2 w_{aa}. $$

The genes in the gametes produced may nevertheless have mutated to the alternative allele type, and the frequency of allele A among gametes is therefore given by the recurrence equation (4.1) on page 78:

$$ p' = (1 - \mu_A)p^* + \mu_a q^*. $$

When the mutation rates μ_A and μ_a are sufficiently small compared to the effects of selection, the result is a globally stable mutation–selection balance

---

[12]A typical inborn error of metabolism (see reference 261600 in OMIM 2006), it causes severe mental retardation and used to be lethal. Symptoms are now alleviated by dieting.

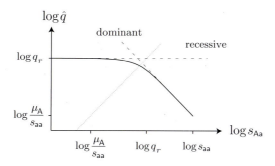

Figure 8.14: The equilibrium gene frequency $\hat{q}$ for $\mu_{\mathsf{a}} = 0$. The approximations given by equations (8.15) and (8.16) are indicated by the dashed lines. Axes are logarithmic. The dotted diagonal line is the line of identity.

equilibrium, which is perturbed only slightly away from the monomorphic equilibrium. For directional selection and small mutation rates we can therefore ignore the mutation of the rare allele, and we shall assume that $\mu_{\mathsf{a}} = 0$ in our analysis of the interaction between directional selection and mutation.

The equilibrium gene frequency of allele a is close to zero, and by ignoring higher-order terms in $\hat{q}$ and the mutation rates, we get an approximation of the equilibrium gene frequency as

$$\hat{q} \approx q_D = \frac{\mu_{\mathsf{A}}}{1 - w_{\mathsf{Aa}}} \quad \text{for} \quad w_{\mathsf{Aa}} \ll 1 \qquad (8.15)$$

with $\hat{W} \approx 1 - 2\mu_{\mathsf{A}}$. A genetic disease with $w_{\mathsf{Aa}} \ll 1$ is called a dominant disease because the trait of interest is that of the heterozygote. The recessive case where $w_{\mathsf{Aa}} = 1$ is simpler—the exact equilibrium gene frequency is

$$\hat{q} = q_r = \sqrt{\frac{\mu_{\mathsf{A}}}{1 - w_{\mathsf{aa}}}} \qquad (8.16)$$

with $\hat{W} = 1 - \mu_{\mathsf{A}}$. These results on the one-locus mutation–selection equilibrium are due to Haldane (1927, 1937). The validity of the approximations is illustrated in Figure 8.14 in terms of the selection coefficients $s_{\mathsf{Aa}} = 1 - w_{\mathsf{Aa}}$ and $s_{\mathsf{aa}} = 1 - w_{\mathsf{aa}}$. For the recessive equilibrium (8.16) to give a good description, we need s_{Aa} to be well below q_r, and for the dominance equilibrium (8.15) to provide a good description, s_{Aa} should be well above q_r. The dominance approximation in equation (8.15) may thus safely be used when $w_{\mathsf{Aa}} < 1 - \sqrt{\mu_{\mathsf{A}}}$.

Mutation–selection balance equilibria exist for any kind of selection that pushes towards the elimination of variant genes. For two-allele underdominant selection, disturbance of the dynamics due to small mutation rates $\mu_{\mathsf{A}} > 0$ and $\mu_{\mathsf{a}} > 0$ gives rise to two locally stable mutation–selection balance equilibria close to the monomorphic equilibria. Mutation–selection balance equilibria for multiple alleles have properties similar to those for two alleles (Clarke 1998).

For weak directional selection, where both selection coefficients are very small, the equilibrium frequency of allele a can be appreciable. If they are of the order of the mutation rate, the only equilibrium may be fixation of the deleterious allele ($\hat{q} = 1$). In such situations we thus cannot maintain the assumptions of unidirectional mutation and large population size. In general, mutation may introduce a major disturbance of the dynamics predicted by a model allowing weak selection only. Studies of the interaction of mutation and weak selection are an element in discussions of the maintenance of genetic variation in quantitative characters (see Bürger 2000). Interesting results evidently emerge only if the effect of random genetic drift is included.

8.4.1   Mutation, selection, and variation

Mutation is the source of genetic variability. The fixation of a mutation in a local population may allow it to persist in defiance of environmental changes—melanism in *Biston betularia* is a good example. Deleterious mutations are just the unavoidable cost of errors, and their presence is as fundamental to the population genetic description of variation as neutral mutations. The distinction between favorable, neutral, and deleterious mutations is obviously not objective—to say the least. Melanic morphs in *B. betularia* would be considered deleterious aberrations in a pristine habitat in exactly the same way that the majority of the mutants isolated by Kassen and Bataillon (2006) show an inferior growth rate in an environment without the antibiotic (Figure 4.1 on page 76). The background of deleterious alleles maintained at mutation–selection–balance equilibria is bound to cause disturbance in studies of particular genetic variants, and we will return to a more thorough discussion of this problem in Chapter 9.

The description of variation in Chapter 4 refers only to neutral mutations. Inference based on, say, the infinite alleles model therefore assume no selection. Deviations from predicted distributions may suggest that the variation causes selection, given that demographic effects may be excluded. This suggestion is not interesting unless we can argue that it is not caused by deleterious alleles in mutation–selection balance. The rightmost of Ewens' distribution examples (Figure 4.8 on page 98) could be viewed as a classical example of a large sample showing a monomorphic functional allele with a tail of rare deleterious alleles. Watterson (1978) discussed this problem in connection with his developments of the practical use of Ewens' sampling formula for inference on "interesting" selection.

8.5   Stability of an Equilibrium

The stability of an equilibrium is determined from an analysis of how selection, mating, and segregation modify the gene frequencies when the population is close to the equilibrium. The biological interpretation of the stability properties depends on the nature of the equilibrium, however. A population at an equilibrium where all alleles are present will regularly be perturbed away from the equilibrium state. Random genetic drift, for instance, will produce such perturbations. If the equilibrium is unstable we thus expect the population to move away from the equilibrium. If the equilibrium is stable we expect the population to remain close to the equilibrium, given that the population size is large enough to allow us to ignore random genetic drift.

A population at an equilibrium where some of the possible alleles are absent will behave differently. In *Biston betularia*, for instance, a monomorphic *typ-ica* population in an unpolluted area is stable to the introduction of *carbonaria* or *insularia* alleles, but in the event of heavy pollution the instability of the population state is not revealed until *carbonaria* or *insularia* alleles are introduced into the population. If we consider only the alleles that are present in the population, then the stability properties are the same as described above. This is called the *marginal stability* of the equilibrium, and it is the kind of stability that may be analyzed given the available variation in the population under study. But this is not enough to conclude that the equilibrium is stable. We must analyze the effect of introducing alleles not currently present in the population; we have to try the *external stability* of the equilibrium. This analysis has to be founded on variation, which may only be present in the population at very low frequencies.

The absent alleles may be present in other populations of the same species, and we may conclude that such alleles are likely to immigrate regularly into our population. The absence of the alleles is therefore an indication of the stability of the population state. If this is true, then we would predict that the "absent" alleles would be present at low frequencies due to a balance between immigration and selection that resembles a mutation–selection balance (we return to this phenomenon in Chapter 10).

At an externally stable equilibrium, mutation is expected to cause a mutation–selection balance where the "absent" alleles are present in supposedly low frequencies, depending on the mutation rate and the strength of selection against the alleles. From equation (8.1) (on page 222 and more generally equation (8.2)) the equilibrium with allele a fixed is externally unstable if $W_A > \hat{W}$ when allele A is sufficiently rare. If an absent allele has to be introduced by rare mutations, then the problem becomes one of a quite different nature. Mutation generates the variant allele that may establish a marginally stable equilibrium as externally unstable.

The waiting time until such a mutation occurs is expected to be long, of the order of the reciprocal mutation rate. It may especially be long compared to the time it takes for the mutant to become established in the population. The resulting process is called *long-term evolution*, and its dynamics is in many aspects different from the short-term processes studied in population genetics (Eshel 1991, 1996, Diekmann, Christiansen & Law 1996). Mutation among the alleles present in a protected polymorphism has a very small effect on the stable genetic equilibrium, unless, of course, selection is extremely weak. The role of mutation in long-term evolution is therefore to present new variants, but apart from this, mutation will have little influence on the gene frequency dynamics.

A mutant originates as a change in one particular gene, and a gamete carrying the mutant will therefore fuse with another that harbors an allele already in the population. Thus, we can suppose that the mutant materializes as one heterozygous individual in the population. In a stable population this individual can expect two offspring (if we consider neutral variation), and the probability that the mutant is lost before it is transmitted is then about $\frac{1}{4}$, simply due to Mendelian segregation. This argument is valid even in a very large population, and immediate loss is a likely event even for an advantageous allele. For such an allele the initial stochastic effect is not overcome until it multiplies to a number of descendants large enough to average out the inherent stochastic effects in segregation, survival, and breeding.

In a large randomly mating population the rare heterozygotes are effectively independent individuals all reproducing by mating the common homozygotes. The initial dynamics may thus be described by a so-called branching process (Haldane 1927, Fisher 1930a). The proliferation of the mutant, say, A, is described by its distribution of offspring. In Haldane's and Fisher's analyses the property we need is the fitness of allele A relative to that of the old allele a. For a rare mutant in a randomly mating population this is given by the fitness of the mutant heterozygote relative to that of the common genotype aa, that is,

$$\frac{W_A}{\hat{W}} \approx \frac{w_{Aa}}{w_{aa}}.$$

We actually need the selection coefficient of the rare mutant relative to the common allele, and this is given by

$$s = \frac{W_A}{\hat{W}} - 1 \approx \frac{w_{Aa}}{w_{aa}} - 1,$$

so the viability of the rare allele is $1 + s$ relative to that of the common allele. Haldane and Fisher conclude that the probability that an advantageous mutant survives in the population is about twice the selection coefficient of the mutant heterozygote relative to the common genotype in the population, given that the selection coefficient is not too large. With allele a fixed it is thus about $2s$.

This probability of survival is expected to be quite small in the majority of evolutionary circumstances. The occurrence of a proper mutation at an externally unstable equilibrium therefore does not guarantee that the population moves away from the equilibrium. An advantageous mutant may have to occur several times before it succeeds.

For weak selection (small s) the dynamics of a new mutant may be further illuminated by calculating the probability of ultimate fixation in the Wright–Fisher model (Box 32). The balance between selection and random genetic drift is, as expected, described by the parameter $4Ns$, and the results are shown in Figure 8.15. With no selection, $s = 0$, the probability of fixation is $\frac{1}{2N}$ because of gene frequency conservation. For $4Ns$ small it does not deviate much from that value. Mildly deleterious mutations in particular have only a probability of fixation slightly smaller than neutral mutations. Slightly advantageous

---

Box 32: Survival of a mutant

Selection may be introduced into the Wright–Fisher model by modifying the gene frequency by equation (8.1) (page 222) when forming the gamete pool. Fisher (1930b) and Wright (1931) analyzed this for the simple model where $w_{AA} = 1 + 2s$, $w_{Aa} = 1 + s$, and $w_{aa} = 1$ (the additive alleles model). The resulting probability of fixation in a monomorphic **aa** population is

$$\text{Prob}(\text{A fixes}) = \frac{2s}{1 - \exp(-4Ns)}.$$

In a large population this is approximately equal to $2s$, the value obtained by Haldane (1927) and Fisher (1930a). If $s = 0$ (neutrality) the probability of fixation of a mutant is $\frac{1}{2N}$. Relative to this value the probability of fixation becomes

$$\frac{4Ns}{1 - \exp(-4Ns)}.$$

The effect of selection can thus be expressed in terms of the parameter $4Ns$.

---

and slightly deleterious mutants under directional selection will therefore show substitution rates comparable to those of neutral mutations (Ohta 1973). For $4Ns \gg 1$ the approximation found by Haldane and Fisher is indeed excellent (Figure 8.15).

The interplay of random genetic drift and overdominant selection is rather complicated. The effect of random genetic drift without selection may be quantified by the rate of fixation of alleles when the distribution of the gene frequency of allele **A** has reached the distribution shown in Figure 2.13 on page 46. This rate is $\frac{1}{2N}$, that is, the fraction $\frac{1}{2N}$ of the polymorphic populations in Wright's island model will become monomorphic every generation. Robertson (1962) calculated the factor by which this rate was lowered by overdominant selection, and he called it the *retardation factor* (Figure 8.16). This factor depends on the strength of selection and on the equilibrium gene frequency. For a given gene frequency we may measure the strength of selection by $\frac{1}{2}(s_{AA} + s_{aa})$, which is given in terms of the selection coefficients relative to the heterozygote, namely

$$s_{AA} = 1 - \frac{w_{AA}}{w_{Aa}} \quad \text{and} \quad s_{aa} = 1 - \frac{w_{aa}}{w_{Aa}},$$

so the genotypic fitnesses are $1 - s_{AA}$, 1, and $1 - s_{aa}$. The surprise is that the retardation factor depends qualitatively on the equilibrium gene frequency. When the stable equilibrium maintains both alleles at high frequencies, then selection strongly lowers the rate at which polymorphism is lost. On the other hand, if one of the alleles is expected to occur in a low frequency at equilibrium, then selection accelerates the loss of the rarer of the alleles by random genetic drift (Figure 8.16). The gene frequency where the retarding effect of selection vanishes depends on the strength of selection.

The results show overdominant selection to be effective in maintaining poly-

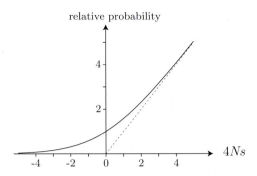

Figure 8.15: The relative probability of fixation of a new mutant in the Wright–Fisher model with additive selection. The dashed curve illustrates the large population approximation.

morphism when the effects of the two alleles are similar, but when the fitnesses of the two homozygotes are very different, then the effects of overdominant selection resemble those of directional selection.

The interaction of selection and random genetic drift may in general be evaluated using the average fitness of the alleles, in particular the selection coefficients of the alleles relative to the average fitness in the population. For two alleles these are given by

$$s_\mathsf{A} = 1 - \frac{W_\mathsf{A}}{W} \quad \text{and} \quad s_\mathsf{a} = 1 - \frac{W_\mathsf{a}}{W}.$$

Their average is zero and they may be summarized in the parameter s, such that $s_\mathsf{A} = sq$, $s_\mathsf{a} = -sp$, and $p' - p = spq$ (see equation (8.4) on page 230). Thus, in characteristic time the expected change in gene frequency per generation is $\mu(p) = 2Nspq$ (see Box 7 on page 44), and $2Ns$ therefore determines the effect of selection in a Wright–Fisher model (in accordance with the result in Box 32). In general, a rule of thumb is that random genetic drift prevails when $2N|s| \ll 1$, and that selection assumes a dominant role when $2N|s| \gg 1$. This principle should, however, be used with caution. For directional selection in particular, Figure 8.15 reveals a bias in fixation probabilities compared to random genetic drift even for small values of $2N|s|$.

The fitness of a rare recessive allele differs only slightly from that of the corresponding dominant allele (Exercise 8.4.1), and random genetic drift is therefore likely to offer a significant contribution to its evolutionary dynamics. The frequency of a deleterious recessive is thus expected to wander quite far from the mutation–selection balance equilibrium even in quite large populations (Wright 1937, Kimura 1955b). Figure 8.17 shows their results in a Wright–Fisher model— for a lethal recessive with a mutation rate of $\mu_\mathsf{A} = 10^{-5}$ per generation and an equilibrium gene frequency of 0.0032.

When 20 mutations are expected per generation, the distribution of the gene frequency resembles that for mutation–mutation balance (see Figure 4.2

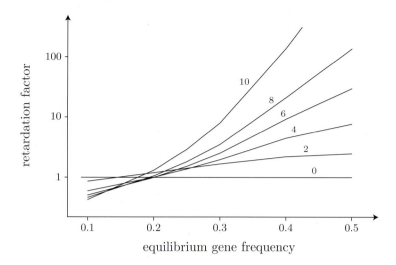

Figure 8.16: The influence of overdominance in selection on random genetic drift as described by the retardation factor in the Wright–Fisher model. The numbers on the curves are $N(s_{\mathsf{AA}} + s_{\mathsf{aa}})$. After calculations in Miller (1962).

on page 81), that is, the population stays close to the equilibrium and random genetic drift only causes minor excursions in a small neighborhood. For a small population where the waiting time between mutations is expected to be 20 generations, the gene frequency is close to the equilibrium with low probability, and with high probability it is much lower, showing that the deleterious allele is often absent. This is balanced by rare occurrences of a very high gene frequency. The expected frequency of the recessive, however, decreases as the population size becomes smaller, and as $2N\mu_{\mathsf{A}}$ becomes small it is approximately proportional to $\sqrt{N}$. The reason is that the level of inbreeding due to the limited population size increases quite rapidly as N decreases, and inbreeding causes a higher frequency of homozygotes, which in turn makes selection against the recessive allele more effective on the average.

The dynamics close to the mutation–selection balance equilibrium is slow. In Exercise 8.2.3 we calculated the time for selection to cause the frequency of a recessive lethal allele to be halved. With reference to Figure 8.17 a population with a frequency of $2\hat{q}$ of allele a is expected to reach the equilibrium frequency after $(2\hat{q})^{-1} = 158$ generations. Ignoring the effects of mutation and random genetic drift, this would require about 3000 years before the equilibrium frequency is restored in a human population. On the other hand, a population fixed for allele A is expected to reach $\hat{q} = 0.0032$ after more than 300 generations if we ignore the effects of selection and random genetic drift. This expected accumulation of mutations is of course independent of the population size. Combining the effects of mutation and selection (still disregarding random genetic drift) makes movement of the gene frequency still slower. Using the results in Exercise 8.2.3,

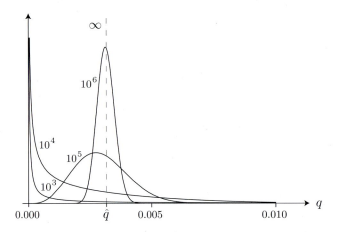

Figure 8.17: The distribution of the gene frequency of a lethal recessive with a mutation rate of 10^{-5} per generation for various population sizes, which are given at the curves as the number N of diploid individuals in the population. After Wright (1937).

we get

$$q' - q \approx \mu_A(1 - q) - \frac{q^2}{1 + q} \approx \mu_A(1 - q) - q^2(1 - q) \approx \mu_A - q^2$$

for q small. Thus, starting at $q = \frac{1}{2}\hat{q}$ the change in gene frequency per generation is about $\frac{3}{4}10^{-5}$, and starting at $q = (1 + \frac{1}{2})\hat{q}$ gives a change of about $\frac{1}{2}10^{-5}$ per generation. A population with an effective population size of 100,000 can therefore easily be trapped, for a very long time, with a 50 percent deviation of the gene frequency from the equilibrium at a locus with a lethal mutation rate of 10^{-5}. Variation among isolated local populations with an effective size of 10,000 can easily cover a range where the gene frequency varies by a factor 10. The gene frequency may show considerable variation even among large populations that have recently expanded—like human populations, for example. Indeed, a population expanding at a fairly slow rate is likely to conserve the large deviation in gene frequency expected in small historical populations (Figure 8.17). Expanded populations of indigenous people are expected to show major gene frequency differences that transgress the differences expected from the distributions in Figure 8.17. For instance, phenylketonuria is very common in Northern Europe, a frequency of about 2×10^{-4}, whereas among Ashkenazi Jews the frequency is less than about 5×10^{-6} (see reference 261600 in OMIM 2006). Cystic fibrosis[13] is common among Europeans and Ashkenazi Jews, frequency 3×10^{-4}, and rarer in Jews from Iran and Iraq, frequency 3×10^{-5}.

---

[13] A lethal recessive disease of the lungs (see reference 219700 in OMIM 2006).

8.6   The Coalescent and Selection

The *carbonaria* allele in *Biston betularia* was probably introduced as a unique mutation of the *typica* allele about 150 years ago, so we expect the sequences of *carbonaria* genes to be very homogeneous. The *insularia* genes are expected to be homogeneous in the same way. Genes of the *typica* allele, on the other hand, have been in the population for many generations. We therefore expect sequence variation among them. We have no reason to believe that this variation impacts the survival and reproduction of the moths, and for dynamical considerations of melanic evolution we may assume genes of the *typica* allele to be homogeneous.

Molecular variation at the carbonaria locus in an isolated monomorphic *typica* population is thus expected to adhere to the coalescent. Analysis of molecular variation in the *typica* allele in a polluted population showing melanic polymorphism is expected to reveal deviations from the coalescent. The coalescent tree should reveal deviations corresponding to a population that has experienced a period of decline, if it is indeed isolated. Otherwise, additional deviations caused by recurrent immigration might show. A similar analysis of the *carbonaria* allele has a very shallow coalescence: the genes are expected to coalesce within about 150 years. Formally it should show signs of a rapidly increasing population ($\hat{\theta}_T > \hat{\theta}_W$, see page 103), albeit sufficient variation is not expected to be present to draw this conclusion based on data. These expectations are of course formulated assuming that no recombination occurred between *carbonaria* and *typica* genes in the period since the *carbonaria* allele emerged.

In a sample over an area where both alleles are present, the general structure of the genealogy must be fairly simple: most of the *carbonaria* genes will coalesce before any noticeable amount of *typica* genes have found common ancestors, and all *carbonaria* genes will have coalesced before any of them coalesce with a *typica* gene. Thus, the probability of coalescence between two genes when tracing their parental copies depends on their allelic types, which means that the probability of coalescence in a sample depends on the composition of the sample. A description of a sample genealogy and variation subject to selection is contained in the so-called ancestral selection graph (Krone and Neuhauser 1997, Neuhauser and Krone 1997).[14]

The *carbonaria–typica* coalescence is fairly simple, but in general the process of coalescence in the presence of selection is very complicated. Just think of coalescence and overdominant selection on two alleles. If we can ignore recombination we may, as a first approximation, see the genes in the two alleles as genes in two isolated populations. For a young polymorphism we can use the same procedure as for the *carbonaria* genes, namely assume that the two alleles split by a unique mutation. After one of the alleles in the sample has coalesced to a common ancestor we can wait for the defining mutation to occur.

Overdominant selection on two alleles produces a stable equilibrium, and it may be long-lived, provided that the equilibrium gene frequencies are not

---

[14]Illustrations of the buildup of this graph are available on the web (Christensen et al. 2006).

extreme. The assumptions of no recombination and unique mutation may then become unreasonable, but we keep them in order to simplify further discussion of how a coalescent with selection may look.

Table 8.8 shows an extension of the simple overdominant model in Table 8.4 on page 231. All homozygotes have fitness w relative to the fitnesses of het-

Table 8.8: A simple one-locus k allele fitness matrix

	A_1	A_2	$\cdots$	A_k
A_1	w	1	$\cdots$	1
A_2	1	w	$\cdots$	1
$\vdots$	$\vdots$	$\vdots$	$\ddots$	$\vdots$
A_k	1	1	$\cdots$	w

erozygotes, which are all equal. Takahata (1990) analyzed the coalescence of this model and showed that it could be described as an allele coalescence indistinguishable from Kingman's coalescence (Takahata formulated his result in terms of Moran's model described in Section 2.7 on page 50), except that the coalescence time is long. This result resembles Wakeley's (1999) approximation to the coalescence in a widespread population presented in Section 5.2.2 (on page 124), in that significant deviations from the assumptions of the coalescent may be present without engendering qualitative deviations. The only deviations are quantitative because the coalescence time is long, and variation is therefore expected to be comparatively high. This coalescent of alleles can be absorbed in the usual gene coalescent experienced for neutral variation—only the mutation rate has to be considerably lower.

Takahata (1990) was inspired by the trans-specific polymorphism among MHC alleles. For instance, some MHC alleles in humans are more closely related to alleles in apes than to homologous human alleles. In addition, the histocompatibility types seem to occur in fairly uniform frequencies compared to the geographical variation in allele frequencies in enzyme polymorphisms disclosed by electrophoresis (Nadeau et al. 1988). Overdominant selection of MHC phenotypes seems in accordance with this observation, and Hughes and Nei (1988) presented support for this hypothesis through analysis of the sequences of MHC alleles.

The most spectacular class of trans-specific polymorphisms is found in self-incompatibility alleles in plants. This system of alleles prevents self-fertilization. The simplest is the gametophytic self-incompatibility system present in, for instance, the Solanaceae.[15] Richman et al. (1996) analyzed the allelic diversity in *Solanum carolinense* and *Physalis crassifolia* and included published sequences from alleles in other nightshades (*Solanaceae*) in the genealogy shown in Figure 8.18.

---

[15]Potato, tobacco, tomato, tomatillo, and the ornamental flowers of the genus *Petunia*.

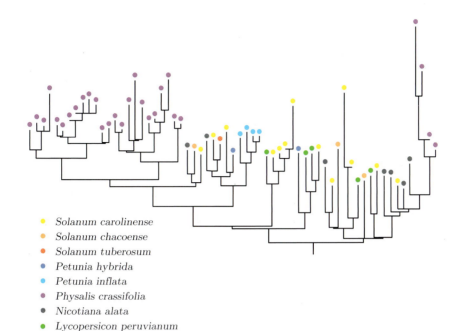

Figure 8.18: Genealogy of self-incompatibility alleles among Solanaceae. Redrawn after Richman et al. 1996.

The gametophytic self-incompatibility system is determined by one multiallelic locus[16] that is active both in the sporophyte and the gametophyte. Pollen carrying the allele S_ℓ can grow on a stigma of genotype $S_i S_j$ when $\ell \neq i$ and $\ell \neq j$. Pollen thus cannot fertilize ovules on plants of a genotype that includes the allele carried by the pollen. All plants will therefore be heterozygotes at the self-incompatibility locus, and the probability that allele S_ℓ can fertilize a random plant in the population equals $1 - p_\ell$ where p_ℓ is the frequency of the allele. A new allele S_ℓ introduced as rare can fertilize virtually every plant in the population. Pollen carrying a rare allele will fertilize more ovules than a common allele, and a rare allele will thus always increase in frequency if we assume that the genotypes of sporophytes and gametophytes at the locus only differ in the incompatibility relations. Selection therefore protects self-incompatibility alleles from loss, and populations are expected to segregate many alleles. Thirty to fifty alleles or even more are commonly observed. Equilibrium is reached when all alleles have the same probability of fertilization, that is, when all alleles have the same frequency, $\hat{p}_1 = \hat{p}_2 = \cdots = \hat{p}_k = \frac{1}{k}$. This equilibrium is stable. Selection is frequency dependent, and close to equilibrium, selection is approximately as in Table 8.8 with w close to zero.

The allele frequencies at this equilibrium become small when many alleles are

---

[16]The alleles are sequences consisting of at least two ORFs, one active in the stigma, the other in the anther.

present, and loss of alleles due to random genetic drift is therefore accelerated (Figure 8.16 on page 253). The probability of fertilization for a rare allele is about 1, which is only slightly higher than $1 - \frac{1}{k}$ expected for the common alleles. The probability of a new allele becoming established in the population is therefore very small, on the order of $\frac{1}{k}$. The extant number of alleles may therefore be understood as a balance between loss of alleles by random genetic drift and gain of alleles by mutation (Wright 1939, Vekemans and Slatkin 1994). This loss and gain process may be described by an allele coalescent, and it turns out to be very similar to Takahata's (1990) coalescent. In particular, it is indistinguishable from a coalescent of neutral variation except for the very long coalescent time.

In the gametophytic self-incompatibility system the alleles are symmetric, as each has exactly the same relation to the rest, just like in the model in Table 8.8 on page 256. In sporophytic self-incompatibility systems the probability that a pollen can fertilize a given ovule is determined by the genotype of the plant (the sporophyte) that produced the pollen and that of the plant that produced the ovule. Alleles in most of these systems do not exhibit symmetry because of dominance. Although the systems are more complex, their basic properties are the same: rare alleles are favored, selection becomes weaker with more alleles, and equilibrium frequencies become low with many alleles (Schierup et al. 1997). Nevertheless, the allele genealogies are expected to show little deviation from those of the gametophytic self-incompatibility system (Schierup et al. 1998).

These examples establish that selection does not leave a recognizable signature in the coalescent of a sample of genes. A sample of genes that have all been subject to strong directional selection is expected to show a shallow coalescent as long as the substitution event is recent compared to the coalescent time of a sample of genes not subjected to selection. A sample of genes taken during the substitution event may show a genealogy that deviates from the coalescent. A sample of sequences from a population polymorphic for alleles subjected to simple symmetric overdominance is expected to show a deep coalescent if the allelic polymorphism emerged further back than the coalescence time of a sample of genes not subjected to selection. If more than one gene is sampled from each allele, the gene genealogy of the total sample may deviate from the coalescent due to accumulation of neutral variation within the self-incompatibility alleles. The kind of deviations from the coalescent that may be observed in these two examples can, however, be mimicked by the genealogy of a sample from a geographically structured population (Chapter 5).

Allele symmetry thus allows the genealogy of a sample of allele sequences to be described by Kingman's coalescent in the face of selection, much in the same way that isotropic migration allowed the genealogy of a sample of neutral genes to be described by the coalescent in the face of deviations from panmixia (page 133). What about selection in a widespread subdivided population, then? Slade and Wakeley (2005) introduced selection into Wright's island model and showed that the sample genealogy is described by the ancestral selection graph with population size as given in equation (5.13) on page 125—that is to say, after the initial scattering phase. The effects of selection on the coalescent are thus expected to be similar to those discussed above.

Influence of selection is hard to uncover in extant variation, be it from genotypic frequencies or from gene genealogies. Phylogenetic analysis of nucleotide sequences may, however, present some anomalies that are very suggestive of the historical effects of selection. The key is heterogeneities in substitution rates coinciding with heterogeneities in the function of the sequence. For instance, Hughes and Nei (1988) found more sequence variation that produced amino acid variation than synonymous variation in the active site (the antigen recognition site) of MHC proteins. Usually synonymous variation is more abundant than nonsynonymous variation within the translated regions of functional genes. This suggests selection due to various correlations between the constitution of the antigen recognition site and the ability of the individual to cope with immunological challenges.

Their argument is not based just on the observation of an island of much amino acid variation within the coding sequence of DNA. It mobilizes a functional argument: the island coincides with the active site. The best illustration of this is that the opposite argument has been used earlier. The protein hormone insulin is made from a heavily modified sequence of 86 amino acids. First a peptide in one end is removed (yellow), then the remaining sequence folds

and joins stretches in the two red sections by covalent bonds (sulphur bridges between cysteine amino acids). Finally, the loop (orange) is removed. The sequence thus has four segments of which only the two red ones code for the mature hormone. Comparing the human and the rat gene revealed about five times as many nucleotide substitutions per site in the yellowish sections as in the red ones. The synonymous substitution rate was about equal in the insulin peptides and the loop, whereas the nonsynonymous substitution rate in the loop was about five times that in hormone coding stretches (see Li 1997). The interpretation of this observation is that most amino acid changes in the hormone are deleterious, whereas the constraints on the amino acid composition of the loop peptide are less stringent (Kimura 1983a). Thus, the high number of amino acid substitutions is taken as a sign of parts of the sequence where less specificity is needed.

8.6.1 Microcephalin alleles in humans

Microcephaly is a recessive genetic disease causing mental retardation. Affected individuals have small heads and small but otherwise normal-looking brains. The condition is caused by a mutation in one of several genes, and the gene microcephalin, MCPH1, on chromosome 8 is of interest here (see reference 607117 in OMIM 2006). Jackson et al. (2002) identified this gene and showed that it contains a stop codon in the 24th of 835 codons in afflicted individuals. They further argued that the gene was involved in fetal brain development.

To investigate the possible role of this gene in human evolution, Evans et al.

Table 8.9: Sequence variation in *Adh* genes in *Drosophila melanogaster*

S-consensus	T	C	C	C	C	C	T	C	C	·	C	T	A	G
S-Washington	·	T	T	·	A	A	C	·	·	**A**	·	·	·	·
S-Florida-1	·	T	T	·	A	A	C	·	·	**A**	·	·	·	·
S-Florida-2	A	·	·	·	·	·	·	·	·	**A**	·	·	·	·
S-Burundi	·	·	·	·	·	·	·	·	·	**A**	·	·	·	A
S-France	·	·	·	·	·	·	·	·	·	**A**	·	·	·	A
S-Japan	·	·	·	·	·	·	·	·	T	**A**	T	·	C	A
F-Washington	·	·	·	·	·	·	·	G	T	**C**	T	C	C	·
F-Florida	·	·	·	·	·	·	·	G	T	**C**	T	C	C	·
F-Bu	·	·	·	·	·	·	·	G	T	**C**	T	C	C	·
F-France	·	·	·	·	·	·	·	G	T	**C**	T	C	C	·
F-Japan	·	·	·	·	·	·	·	G	T	**C**	T	C	C	·

Data from Kreitman (1983).

(2004) determined the sequence variation in 27 normal individuals. They noted a high-frequency haplotype characterized by having the amino acid histidine in the protein at a position corresponding to codon 314 (Evans et al. 2005). This histidine allele, D say, occurs in a frequency of about 0.7 in human populations based on a sample of 178 genes representing the diversity of human populations. The D allele has no effect on brain size (Dobson-Stone et al. 2007).

Evans et al. (2006) analyzed 178 sequences of the chromosome region spanning six of the fourteen exons of the gene and found 86 different sequences. Little variation was seen among the 124 D allele sequences compared to that among the 54 non-D haplotypes. The average pairwise divergence within the non-D allele was about 10 times that found within the D allele, and the divergence between D and non-D alleles was about 30 times the D allele divergence. In addition, one third of the 178 determined haplotypes were identical and of allele type D, whereas the remaining 85 haplotypes were of low frequency no matter the allele type. The D allele thus shows the characteristics of an allele under recent directional selection.

8.6.2 *Adh* alleles in *Drosophila melanogaster*

Kreitman (1983) investigated sequences of both the electrophoretic alleles discussed in Section 4.3.1 on page 86. The full exon variation is shown in Table 8.9. The electrophoretic difference between the products of alleles Adh^F and Adh^S is determined by the site shown in bold type and omitted from the consensus sequence in the table. This site is at the second position in a codon AAG (lysine) in Adh^S and ACG (threonine) in Adh^F. All the remaining single-site differences are silent. Kreitman also describes sequence variation in the three introns, two transcribed but untranslated regions, and two flanking regions of the gene. Table 8.9 is therefore insufficient for a proper phylogenetic analysis of the sampled *Adh* genes.

However, a clear difference between the two alleles is apparent in the coding sequence. The genes of the Adh^F allele are homogeneous, in contrast to the genetic heterogeneity among the genes of the Adh^S allele described in Section 4.3.1. The coding sequence of Adh^S from Japan is rather close to that of the Adh^F genes, whereas the remaining five Adh^S genes share none of the six bases where Adh^F deviates from the Adh^S consensus sequence. The additive distance tree among the Adh^S genes may thus be extended to a parsimonious tree comprising all the observed sequences.

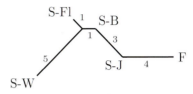

This tree is not additive, however. The difference G $\longleftrightarrow$ A between S-Fl and S-W, and S-B is also present between S-J and F.

The ubiquity of the Adh^F and Adh^S alleles in populations has attracted much interest, and many indications have been found that the polymorphism is maintained by natural selection. For instance, the frequency of the Adh^F allele increases from north to south in eastern Australia and from south to north in eastern North America (Berry and Kreitman 1993, Oakeshott et al. 1982). Each cline alone does not suggest selection (Section 5.2.4, in particular page 131), but two parallel independent poleward clines are significant. As all populations in these clines are polymorphic, the most immediate suggestion is that the two alleles are maintained by overdominant selection, with selection varying with the environment. Assuming overdominant selection, we can view the divergence pattern as an old Adh^S allele and the fairly recent Adh^F allele that has been kept in polymorphism for quite some time.

Further diversification of the sequenced genes is apparent in the noncoding regions (Kreitman 1983). Recombination seems to play a role when longer stretches of the sequence are investigated. For instance, F-Florida has the typical Adh^S sequence in intron 1, and S-Florida-2 is relatively similar to the typical Adh^F sequence in intron 1. The pattern of variation shown by the last of the varying sites in Table 8.9 may, of course, be ascribed to recombination as an alternative to the supposition of a double mutation at that site.

Exercises

Exercise 8.2.1 Assume the viabilities $w_{AA} = 1$, $w_{Aa} = w$, and $w_{aa} = w^2$. Assume random mating and find W, W_A, and W_a.

This selection model is often referred to as gametic selection. Is that reasonable?

Exercise 8.2.2 In the *Biston betularia* example (directional selection and dominance) write and simplify equation (8.1). Argue that $p' \approx 2p$ in polluted areas when the *carbonaria* allele is rare. Finally, find by numerical calculations the number of generations it takes for q to double in a pristine area when the *typica* allele is rare, say, $q = 0.001$.

Exercise 8.2.3 Consider a locus with two alleles where A is dominant and a is recessive and assume that $w_{aa} = 0$, that is, the recessive homozygote is lethal. Find the number of generations it takes for q (the frequency of a) to reach half the original value.

The calculations should, of course, assume Hardy–Weinberg proportions among zygotes. Is this assumption important?

Exercise 8.2.4 Show that $\hat{p}$ is the only solution to the equation $W_A = W_a$, and show the solution is a valid gene frequency when w_{Aa} is in the interval between w_{AA} and w_{aa}.

Exercise 8.2.5 Prove Wright's formula, equation (8.4).

Exercise 8.2.6 Could there be other reasons for the observed excess of homozygotes in *Zoarces viviparus*?

Exercise 8.3.1 Show that the genotypic values G_{AA}, G_{AA}, and G_{AA} uniquely determine ε_A, ε_a, and η in equations (8.11). Also calculate α_A and α_a to show equations (8.12) and (8.13) and find the additive variance. Finally, calculate the dominance variance, and show that it has maximum value for equal gene frequencies.

Exercise 8.3.2 Consider the viabilities (8.8) for a locus with additive allele effects. Show that the viabilities normalized to that of the heterozygote may be approximated as $w_{AA} = (1 + a)w_{Aa}$ and $w_{aa} = (1 - a)w_{Aa}$ when α is small. Find a.

Exercise 8.3.3 Assume the fitnesses $w_{AA} = w$, $w_{Aa} = 1$, and $w_{aa} = w$. Find W_A and W_a, and show that the equilibrium is $\hat{p} = \frac{1}{2}$. Find the additive, the dominance, and the genotypic variances.

Exercise 8.3.4 Show that the covariance between the additive values in Table 8.6 and fitness equals V_A.

Exercise 8.3.5 Show that
$$p' - p = \frac{pq\alpha_w}{W}$$
and relate this formula to our analysis of selection in Section 8.2.

Exercise 8.3.6 Show that $\text{Cov}(w, \Gamma) = -2pq(W_A - W_a)$, see Table 8.7.

Exercise 8.3.7 ∗ Show that Price's formula (8.14) holds in a population where the genotypic frequencies are not in Hardy–Weinberg proportions.

Exercise 8.4.1 Find the viability of the alleles A and a at the mutation–selection equilibrium when a is recessive and deleterious.

Exercise 8.4.2 In Mørch's (1941) investigation of achondroplastic dwarfism he found that dwarfs survived normally, but they had rather few children. 108 investigated adult dwarfs had 27, and their 457 normal sibs had 582 children.

How does this compare with the incidence of achondroplastic dwarfism among newborn and a mutation rate of 7 in 188,000?

Exercise 8.4.3 * Find the general expression for the gene frequency at the mutation–selection balance equilibrium for directional selection for allele A and $\mu_a = 0$.

Exercise 8.5.1 Why is the statement "This expected accumulation of mutations is of course independent of the population size" on page 253 all right?

Exercise 8.6.1 The "of course" in the statement "The pattern of variation shown by the last of the varying sites in Table 8.9 may, of course, be ascribed to recombination as an alternative to the supposition of a double mutation at that site" on page 261 obviously refers to Hudson and Kaplan's (1985) four-gamete test. Which pairs of sites fulfill the criterion of the four-gamete test?

Chapter 9

Genomic Effects of Selection

Genes at different loci associate due to linkage in the transmission from parents to offspring—they interact in the formation of offspring populations (Chapter 6). In addition, genes may interact in their function, and if the variation in the corresponding phenotype is subject to natural selection, the effects of selection and the evolution of the phenotype cannot be understood by considering single loci. To get a full understanding of the effects of selection, we need to study selection on multiple loci, if not on whole genomes.

Bateson, Saunders, and Punnett observed not only linkage in their experiments with sweet peas (*Lathyrus odoratus*) in 1905, but also interaction in the function of genes. They crossed two true-breeding lines with white flowers. The F_1 progeny had purple flowers, and of 651 F_2 plants, 382 had purple flowers and 269 were white. They also crossed two true-breeding lines of *Salvia horminum*, one with pink flowers and one with white flowers. The F_1 plants had purple flowers, and in F_2 they counted 255 plants with purple, 92 with pink, and 144 with white flowers. Neither of these experiments adheres to Mendelian expectations, but they saw that both can be explained by assuming that two genes determine flower color, and that they interact to determine the trait of the individual.

In both crosses a new trait emerges in the F_1 progeny. If the alleles A and a are assumed to be at one locus and B and b at the other, this phenomenon can be explained by introducing one dominant allele from each parental line. The parental lines are thus AA bb and aa BB, which gives the F_1 as Aa Bb. The genotypes A- B- yield the purple trait. The flower colors of aa B- and A- bb are those of the parental lines, that is, both are white in the *Lathyrus* cross, and, say, aa B- is white and A- bb pink in the *Salvia* cross. This model may be summarized as in Figure 9.1, of course assuming that aa bb is white. The experiment is thus explained on the basis of simple modifications of the Mendelian $9:3:3:1$ segregation in F_2. The first experiment produced a $9:7$ ratio ($7 = 3+3+1$), and the second a $9:3:4$ ratio—both in agreement with the observed segregation.

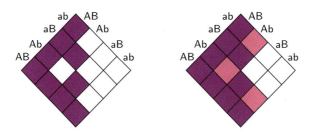

Figure 9.1: Bateson, Saunders, and Punnett's explanation of their observed F_2 segregation using Punnett's squares.

These deviant segregation ratios are caused by *epistasis*, a phenomenon due to interaction in the function of different genes. The development of genetics in the twentieth century is distinguished by the description and understanding of the phenomena of linkage and epistasis that were discovered by Bateson, Saunders, and Punnett in their experiments. The investigation of these two distinctly different phenomena, association among alleles of different genes when transmitted to offspring and interaction in their contribution to the phenotype, embraces the agenda of genetic research in the past century. The comprehension of both phenomena required characterization of the nature of the functional gene. The above segregations appear in crosses involving variation in two genes that produce enzymes in the metabolic pathway that produces purple pigment. Capital letters then represent functional genes and lowercase the allele comprising the nonfunctional genes. The gamete AB then ensures purple pigment, as does the presence of an A and a B. If enzyme A is placed before B in the pathway, the difference between the two kinds of segregation may just be whether the A product gives rise to color or not.

Selection on functionally related loci is of immediate interest. Some of the simplest interactions between genes are those between miRNA genes and their recognition sites within or close to the functional genes with which they interact. A more overt example is the color variation in *Cepaea nemoralis* that influences the risk of falling prey to birds. The cryptic aspects of the snail's appearance are influenced by shell color, banding pattern, and banding appearance, determined by more than ten loci. For genes that interact in their function, interaction in transmission is of course important, but in higher organisms the locus of a gene is usually unrelated to its function. There are exceptions, however. The major histocompatibility complex (MHC) in mammals is a cluster of closely linked genes involved in immune response (see OMIM 2006). Five of the genes related to crypsis in *C. nemoralis* are closely linked. The homeotic genes found in many multicellular animals are functionally related and occur in characteristic closely linked complexes. These *Hox* genes are phylogeneticly related, and each codes for a transcription factor that includes a characteristic 60-amino-acid DNA-binding domain, the homeobox. The *D. melanogaster Hox* genes are split in two fairly loosely linked clusters with three and five genes, respectively. These eight genes designate the identity of the body segments in the fly (Lewis 1978). Each regulates the expression of linked genes that control developments and interaction of tissues in the corresponding body segment (for a recent discussion, see Hueber et al. 2007). The differences in body plans of the four arthropod classes, insects, crustaceans, millipedes, and spiders, are reflections of different expression patterns of their *Hox* genes (Angelini and Kaufman 2005). Such a fully competent cluster is usually formed by closely linked genes. Mammals have four such clusters, the number of genes being 11, 10, and 9, with 9 in two of the clusters, and the full complement of mammalian genes may be sorted into 13 phylogenetic groups (Maconochie et al. 1996). The primitive vertebrate-like Amphioxus has a *Hox* cluster of 14 genes, which by comparison to the clusters of *D. melanogaster* is sorted into seven groups. These seven groups are all represented in the two long mammalian clusters, whereas the two 9-gene clusters contain only four and five of the groups, respectively (Lemons and McGinnis 2006). Other homeobox gene families exist, for instance the *Rhox* gene cluster on the X chromosome of the mouse (MacLean et al. 2005), involved in sperm maturation and other developmental processes related to reproduction.

The corresponding homeotic genes in plants are the *MADS-box* genes (see Theissen et al. 2000). They contain an evolutionarily conserved 57-amino-acid domain, the MADS-box, which is situated close to a kinase domain of similar size separated by a variable intervening sequence. Unlike *Hox* genes, functionally related *MADS-box* genes are unlinked and distributed throughout the genome. A MADS-box gene without the kinase domain is involved in the forming of skeletal muscle tissue in vertebrates.

The development from a newly formed zygote to a fully differentiated animal involves genes scattered all across the genome that define anterior–posterior and dorsal–ventral dimensions of the embryo (Nusslein-Volhard and Wieschaus 1980). The function of these genes is based on the polarity of the zygote that was defined by maternal genes during maturation of the egg cell. The figure shows the presence in the *D. melanogaster* blastoderm of the regulatory protein

 produced by the *dorsal* gene, shown as dark matter. It is present in the cytoplasm on the dorsal side and in the nuclei on the ventral side. This protein is initially evenly distributed. Its redistribution is mediated by the so-called Toll protein through a signal pathway involving other proteins—interaction of genes mediating the distribution of the dorsal protein that defines one of the axes of the embryo in its further development. The Toll signal is also involved in the innate immune response of cells to viral infection, and variation in the *Toll* gene may therefore show pleiotropic effects in that a mutant may show aberrant dorsal–ventral differentiation as well as abnormal susceptibility to pathogens.

The genetics of the embryonic development of animals has mainly been studied in the nematode *Caenorhabditis elegans*, *D. melanogaster*, and the house mouse (*Mus musculus*). They are evolutionarily distant, and the early de-

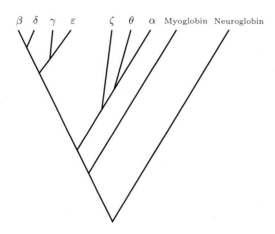

β δ γ ε ζ θ α Myoglobin Neuroglobin

Figure 9.2: Evolutionary relationship of human globins

velopment of their embryos looks rather different, but the genes and the way they interact are strikingly similar. A good example is the gene producing the *Drosophila* engrailed protein and the corresponding engrailed1 protein in the mouse. A nonfunctional gene in the mouse results in a severe defect of the brain, but insertion of a functional fly gene leads to a mouse with a normally developed brain (Hanks et al. 1998). The *Drosophila* protein thus rescues the genetically deficient mouse.

A particularly interesting case of pleiotropy is exhibited in humans by the protein syncytin coded by the *Env* gene of an old endogenous retrovirus of the family HERV-W. Endogenous retrovira are relicts of retroviral infections (see Bannert and Kurth 2006). They show high diversity and comprise about 8 percent of the human genome (see Villesen et al. 2004). These retrovira are related to MSRV, a virus associated with multiple sclerosis (their phylogeny defines the HERV-W family (Blond et al. 1999)), and syncytin is involved in the inflammation of nerves characteristic of multiple sclerosis (Antony et al. 2004). Expression of syncytin was noted in the placenta where the protein mediates the fusion of the embryo and the uterine wall in the formation of the placenta (Mi et al. 2000, Mallet et al. 2004).

The *Hox* and MADS-box genes are examples of *gene families*. Other examples of gene families that show intimate interactions abound. The human hemoglobin gene family consists of at least the genes coding for α, β, γ, δ, ε, ζ, and θ polypeptides (see OMIM 2006). The hemoglobins and neuroglobin diverged about 800 million years ago (Figure 9.2), and the split between the α and β related lineages occurred around 500 million years ago (Goodman and Moore 1973). Globin polypeptides interact in forming the tetrameric hemoglobin molecule, for instance α and β polypeptides in adults. Adult ($\alpha_2\beta_2$ and $\alpha_2\delta_2$) and fetal ($\alpha_2\gamma_2$) hemoglobin interact across the placenta. Two closely linked genes code for identical α proteins, and two closely linked genes code for γ proteins

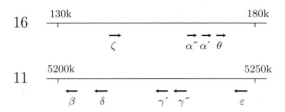

Figure 9.3: The placements of the hemoglobin genes on chromosomes 11 and 16.

that differ in one amino acid position. The α, ζ, and θ genes are found on chromosome 16, and the β, γ, δ, and ε genes reside on chromosome 11, both as closely linked clusters (Figure 9.3). Neuroglobin is involved in oxygen transport in the brain, and its locus is on chromosome 14. Myoglobin is found in muscles and is located on chromosome 22.

Gene families are based on evolutionary relationships, and they may or may not be closely linked. Closely linked clusters of functionally related genes are known from Mendelian analyses. They are referred to as *supergenes*. A classical example is the determinant of heterostyly in *Primula*[1] flowers. This chromosomal region determines not only the visible characters, the length of the style and the placement of the anthers, but also pollen size and pollen–style compatibility (Barrett 1992). It is hard to unravel just how these genes evolved from a recent common ancestor because their functions seem very different even though they are all involved in avoidance of self-fertilization. Mutants of MADS-box genes, however, have been shown to be able to change the phenotype quite drastically, and so the genes within the heterostyly supergene may well be a cluster of such genes (Christiansen 2000).

9.1   Multilocus Selection

The formulation of selection models for multiple loci is closely related to the formulation in Section 8.2 describing the effect of selection on multiple alleles at one locus, exept in terms of gamete types instead of alleles. The difference is that alleles are transmitted unchanged from parents to offspring, whereas gametes may be altered by recombination. The description of selection is thereby the same, whereas the description of segregation requires more work. The effect of viability selection on the change of the genotypic frequencies from newly formed zygotes to mature adult individuals is described entirely in terms of the gamete frequencies.

We will only cursorily discuss selection in multilocus systems in a large randomly mating population, and exemplify the emerging evolutionary phenomena by the two-locus two-allele model introduced in Section 6.2 on page 148. A more thorough discussion can be found in Ewens (1979, 2004), Bürger (2000),

---

[1]The two phenotypes in the heterostyly polymorphism are *pin*, with long style and short anthers, and *thrum*, with short style and long anthers.

and Christiansen (2000). Table 9.1 shows the viabilities of the ten genotypes in this genetic model (we assume $w_{ij} = w_{ji}$, $i, j = 1, 2, 3, 4$). The model sim-

Table 9.1: Two-locus genotypic viabilities

Maternal	Paternal gamete			
gamete	A_0B_0	A_0B_1	A_1B_0	A_1B_1
A_0B_0	w_{11}	w_{12}	w_{13}	w_{14}
A_0B_1	w_{21}	w_{22}	w_{23}	w_{24}
A_1B_0	w_{31}	w_{32}	w_{33}	w_{34}
A_1B_1	w_{41}	w_{42}	w_{43}	w_{44}

plifies if we assume that the two double heterozygotes have the same fitness, that is, $w_{14} = w_{23}$; we say that the fitnesses show *no position effect*. The fitnesses may then be expressed in terms of the combinations of one-locus genotypes (Table 9.2, where the subscripts refer to the two-locus gametes forming

Table 9.2: Two-locus viabilities without position effects

	B_0B_0	B_0B_1	B_1B_1
A_0A_0	w_{11}	w_{12}	w_{22}
A_0A_1	w_{13}	w_{14}	w_{24}
A_1A_1	w_{33}	w_{34}	w_{44}

the genotype). The forward equation (see Section 6.3.1) for the frequency of gamete A_0B_0 is then

$$W x_1' = w_{11}x_1^2 + w_{12}x_1x_2 + w_{13}x_1x_3 + (1-r)w_{14}x_1x_4 + rw_{23}x_2x_3.$$

W is again the average fitness in the population given by

$$W = \sum_{i=1}^{k} W_i x_i, \quad \text{where} \quad W_i = \sum_{j=1}^{k} w_{ij}x_j, \quad i = 1, 2, 3, 4,$$

are the average fitnesses of the gametes. Collecting the terms involving r provides the recurrence equations

$$
\begin{aligned}
W x_1' &= x_1 W_1 - rD w_{14}, \\
W x_2' &= x_2 W_2 + rD w_{14}, \\
W x_3' &= x_3 W_3 + rD w_{14}, \\
W x_4' &= x_4 W_4 - rD w_{14},
\end{aligned}
$$

where $D = x_1x_4 - x_2x_3$ is the linkage disequilibrium.

With no recombination ($r = 0$) the recurrence equations are the same as those for four alleles. This is as expected, because with no recombination the two-locus gametes are passed unchanged through meiosis and therefore behave like alleles, and the average fitness increases. For no selection, recombination breaks up associations and the gamete frequencies converge to Robbins proportions. The interesting phenomena in multilocus population genetics appear in the interplay of these two conflicting trends.

Directional selection is the simplest one-locus case. For two loci we may expect a similarly simple outcome. When $w_{11} < w_{13} < w_{33}$ the monomorphic A_0B_0 equilibrium ($\hat{x}_1 = 1$) is unstable to the introduction of allele A_1 in the form of gamete A_1B_0 and the A_1 allele is eventually fixed, at least if we assume no variation at the B locus. This equilibrium is unstable in the same way to the introduction of gamete A_0B_1 when $w_{11} < w_{12} < w_{22}$. If the realization of the instability requires a wait for successful mutations, then we can reasonably assume that one, say A_1, occurs first. Gamete A_1B_0 will thus increase to fixation, and a subsequent successful mutation at the B locus will cause the population to fix at $\hat{x}_4 = 1$ if $w_{33} < w_{34} < w_{44}$.

If $w_{11} > w_{13}$ and $w_{11} > w_{12}$, neither A_1B_0 nor A_0B_1 can increase in frequency when introduced in a low frequency into a monomorphic A_0B_0 population. The $\hat{x}_1 = 1$ equilibrium, however, may still be unstable to the introduction of the A_1B_1 gamete that occurs in the genotype A_1B_1/A_0B_0 when it is sufficiently rare. When $w_{11} < w_{14}$, this *cis* double heterozygote has a higher viability than the resident genotype, but only the fraction $\frac{1}{2}(1 - r)$ of its gametes will provide a source of double heterozygotes in the offspring population. The gametes A_0B_1 and A_1B_0 will also be produced, but as they occur in the genotype A_0B_1/A_0B_0 and A_1B_0/A_0B_0, they decrease in frequency due to selection and therefore remain rare. The frequency of A_1B_1 will thus change as

$$x_4' \approx \frac{w_{14}x_1x_4(1 - r)}{W} \approx (1 - r)\frac{w_{14}}{w_{11}}x_4.$$

In the first approximation we neglected the possibility of production of A_1B_1 by the *trans* double heterozygote, and that is a reasonable assumption because the genotype is formed by two rare gametes. The second approximation is similar in that $\frac{x_1}{W}$ only deviates from $\frac{1}{w_{11}}$ by terms of the order of the frequencies of the rare gametes. The rare complementary gamete A_1B_1 will therefore increase in frequency when

$$w_{11} < (1 - r)w_{14}. \tag{9.1}$$

The condition $w_{11} < w_{14}$ is thus only sufficient for absolute linkage. Recombination may thus keep an advantageous gamete from entering a population (Bodmer and Parsons 1962, Eshel and Feldman 1970). The principle of fitness maximization, established in Sections 8.2 and 8.3.2, is truly a one-locus result because with selection at two or more loci *the average fitness cannot be expected to increase* in general.

This result is particularly important for considerations on the evolution of communication. The characteristic sequence motif of an miRNA gene may only change in a way that allows it to conserve its function, that is, maintain sufficient affinity to the recognition motif in the target gene. Thus, a major change is expected to be deleterious unless it is paralleled by a corresponding change of the target. As such genes are usually loosely linked, evolution of their sequences is expected to occur by nudging small unimportant changes into their sequences one at a time. A high degree of evolutionary conservation is therefore expected in these genes.

Evolution of animal behavior is constrained in a similar way. Roelofs et al. (1987) studied a geographical polymorphism in the sex pheromones of the European corn borer *Ostrinia nubilalis* governed by three unlinked genes. The two pheromones produced by females are determined by a locus with two codominant alleles. Each homozygote produces a characteristic pheromone, and the heterozygote produces the two pheromones in about equal amounts. A second two-allele locus influences the receptors on the male antenna. The receptors of the two homozygotes each react strongly to one of the pheromones, and the heterozygote has both kinds of receptors. These two autosomal loci are unlinked. The third two-allele locus is W-linked and determines the reaction of the male to the receptor signal. The alleles are codominant, and each mediates acceptance of one of the two pheromones. Local populations are usually monomorphic with corresponding alleles at the three loci, but populations of mixed compositions are known.

9.1.1   Selection on one of the loci

Some aspects of the one-locus theory do survive, however. Selection on one of the loci is again described by equation (8.1) in that

$$p'_\mathsf{A} = \frac{p_\mathsf{A} W_{\mathsf{A}_0}}{W}, \quad \text{where } W_{\mathsf{A}_0} = p_\mathsf{A} w_{\mathsf{A}_0\mathsf{A}_0} + q_\mathsf{A} w_{\mathsf{A}_0\mathsf{A}_1}. \tag{9.2}$$

The only difference, albeit a major one, is that the genotypic viabilities at the A-locus depend on the gamete frequencies because

$$w_{\mathsf{A}_0\mathsf{A}_0} = \frac{w_{11}x_1^2 + 2w_{12}x_1x_2 + w_{22}x_2^2}{p_\mathsf{A}^2},$$

$$w_{\mathsf{A}_0\mathsf{A}_1} = \frac{w_{13}x_1x_3 + w_{14}(x_1x_4 + x_2x_3) + w_{24}x_2x_4}{p_\mathsf{A}q_\mathsf{A}},$$

$$w_{\mathsf{A}_1\mathsf{A}_1} = \frac{w_{33}x_3^2 + 2w_{34}x_3x_4 + w_{44}x_4^2}{q_\mathsf{a}^2}.$$

The one-locus fitnesses in equation (9.2) depend on the genetic background, and due to the viability differences of the two-locus genotypes, this genetic background also changes. The above one-locus genotypic fitnesses thus describe the change in gene frequencies during one generation only. We can calculate the fitnesses of the equation for the next generation when all changes in the gamete frequencies are known (Ewens 1976).

Equilibrium in the gene frequencies is reached when the gamete frequencies settle at an equilibrium. At such an equilibrium the gene frequencies are also at an equilibrium of equation (9.2), and the equilibrium is therefore given by

$$\hat{p}_A = \frac{\hat{w}_{A_1A_1} - \hat{w}_{A_0A_1}}{\hat{w}_{A_1A_1} + \hat{w}_{A_0A_0} - 2\hat{w}_{A_0A_1}}, \tag{9.3}$$

where the fitnesses are evaluated assuming gametic equilibrium. The gene frequency equilibrium is valid when $0 \le \hat{p}_A \le 1$, and it is unique if it exists given the equilibrium viabilities. However, equilibria may exist for different combinations of gene frequencies at the two loci. At an equilibrium where both loci are polymorphic, induced selection on any of the loci will exhibit either over- or underdominance (Ewens and Thomson 1977).

For two- or multilocus systems the immediate question is then: Will a polymorphic locus show overdominance at a stable equilibrium? The answer is that selection at the level of the genome is not that simple. The expectation may fail even for just two loci. Hastings (1982) gave an example: the fitnesses in Table 9.3 for $0.0095 < r < 0.01$ permit a stable equilibrium where both loci show underdominance in fitness. The underdominance in viability selec-

Table 9.3: Hastings' example

	B_0B_0	B_0B_1	B_1B_1
A_0A_0	0.97	0.98	1.04
A_0A_1	0.98	1	0.98
A_1A_1	1.04	0.98	0.99

tion, observed at an esterase polymorphism in a population of *Zoarces viviparus* (Section 8.2.1 on page 233), may thus after all be compatible with the observed genetic stability of the population.

9.1.2 Tight linkage

For closely linked loci we can view recombination as a disturbing factor much like mutation. This is particularly true for different sites within a gene with variation that causes selection. Assume that the population settles at a stable equilibrium, which in the two-locus sense is a *totally polymorphic equilibrium*, that is, an equilibrium where both loci are polymorphic. The occasional recombinant then corresponds to a mutation from one gamete to another. Thus, for sufficiently small r the population will settle at a selection–recombination balance equilibrium displaced a bit from the stable equilibrium in the four-allele model (Karlin and McGregor 1972), given, of course, that the equilibrium is totally polymorphic (such equilibria segregating two or three gametes may well exist for absolute linkage; see Example 9.1.2 below). The only circumstances where recombination will not have an effect is when the four-allele equilibrium happens to have gamete frequencies in Robbins proportions, but this is true only for very special models (Bodmer and Felsenstein 1967).

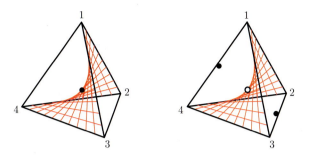

Figure 9.4: Equilibria for tight linkage (see Figure 6.4). Left: Example 9.1.1. Right: Example 9.1.2.

The pin and thrum supergenes in *Primula* (see page 268) are examples of a polymorphism where four two-allelic loci are in maximal linkage disequilibrium as only two of the 16 gamete types are present. In natural populations aberrant phenotypes occasionally appear that are due to recombination within the supergene. The sporophytic self-incompatibility alleles (see page 258) are another example of variation in a supergene. An incompatibility type is characterized by a pollen and a stigma trait. In *Brassica* these are determined by three closely linked genes, two expressed in stigma, one in pollen (Nasrallah 2000). The self-incompatibility alleles should therefore be regarded as haplotypes, but it is convenient to treat them like alleles because recombinants have not been found and are presumably rare.

Example 9.1.1 Consider two loci with two alleles subject to selection as given by the fitnesses in Table 9.4. The model only distinguishes double homozygotes

Table 9.4: Simple overdominance in two-locus viabilities

	B_0B_0	B_0B_1	B_1B_1
A_0A_0	w	1	w
A_0A_1	1	1	1
A_1A_1	w	1	w

from the rest of the genotypes. The equilibrium with all gametes present for absolute linkage is given by $\hat{x}_i = \frac{1}{4}$, $i = 1, 2, 3, 4$ (Example 8.2.1 on page 230). These gamete frequencies are in Robbins proportions (see Figure 9.4, left). Recombination thus does not disturb this equilibrium, and it always exists. It is stable for absolute linkage and $w < 1$, and it is in fact stable when $w < 1$ for any frequency of recombination, $0 \le r \le \frac{1}{2}$. □

Example 9.1.2 Consider two loci with two alleles subject to selection as given by the fitnesses in Table 9.5. The model distinguishes double heterozygotes

Table 9.5: Symmetric multiplicative two-locus viabilities

	B_0B_0	B_0B_1	B_1B_1
A_0A_0	w^2	w	w^2
A_0A_1	w	1	w
A_1A_1	w^2	w	w^2

with fitness 1, single heterozygotes with fitness w, and double homozygotes with fitness w^2. The model is multiplicative in that the contribution of a heterozygote at any of the loci is 1, and that of a homozygote is w.

The stable equilibria for $w < 1$ and $r = 0$ (Example 8.2.2 on page 232) are $\hat{x}_1 = \hat{x}_4 = \frac{1}{2}$ and $\hat{x}_2 = \hat{x}_3 = 0$, and $\hat{x}_1 = \hat{x}_4 = 0$ and $\hat{x}_2 = \hat{x}_3 = \frac{1}{2}$. Both loci are polymorphic at these equilibria and they both have maximal linkage disequilibrium because either the *cis* gametes or the *trans* gametes are present at equilibrium. At the equilibrium with the *cis* gametes a tiny bit of recombination introduces the *trans* gametes into the population, but because selection will favor the *cis* gametes, the *trans* gametes will remain at a low frequency in a recombination–selection balance equilibrium (see Figure 9.4, right). The linkage disequilibrium will therefore remain high for tight linkage (as in the *Primula* example). □

The equilibrium for no recombination is at a local maximum for the average fitness in the population. A population that settles at a selection–recombination equilibrium slightly away from this equilibrium cannot be at an extreme value of the average fitness. If the population is perturbed from the stable selection–recombination equilibrium towards the corresponding stable equilibrium for absolute linkage, then the average fitness may decrease every generation as it returns towards the selection–recombination equilibrium. In the selection model in Table 9.5 with $w < 1$, for instance, a population starting with $x_1 = x_4 = \frac{1}{2} - \epsilon$ and $x_2 = x_3 = \epsilon$ and ϵ very small will show a steady decrease in fitness every generation if recombination is allowed (Moran 1968), once again emphasizing that fitness cannot be expected to increase with multilocus selection.

The viabilities in Table 9.5 provide an example of the *multiplicative viability model* (Bodmer and Felsenstein 1967). It models a situation where survival due to genotype is independent at the two loci. The biological motivation for this may be that the variation in genotypic survival at the two loci is expressed at different life stages of the individual. Imagine, for instance, two loci in an insect where the A locus is acting during the larval stage and the B locus during the pupal stage. When both loci influence viability only during the stage where they are transcribed, then the actions of the loci are independent. The survival probabilities are thus given as products of the probabilities of surviving through each of the two stages. With overdominance in both stages, we have in general that an equilibrium stable for tight linkage exhibits high linkage disequilibrium— a so-called *high complementarity equilibrium*. Independence in survival is thus no guarantee of evolutionary independence.

The viabilities in Table 9.4 provide an example of the *multiplicative interaction model* (Christiansen 1988), a model with multiplicative interaction in the selection coefficients. These models always possess a totally polymorphic equilibrium that is stable for all values of the recombination frequency.

9.1.3   Loose linkage and weak selection

If selection is weak and recombination becomes dominant, then the gamete frequencies will rapidly converge to a state close to the (red) surface of Robbins proportions. Such a state is described as being in *quasi linkage equilibrium* (Kimura 1965, Nagylaki et al. 1999, Bürger 2000). As a first approximation we may thus assume Robbins proportions and then study the change in gene frequencies by equation (9.2). This equation now simplifies to

$$p'_\mathsf{A} \approx \frac{p_\mathsf{A}}{\tilde{W}} \left(p_\mathsf{A}\tilde{w}_{\mathsf{A}_0\mathsf{A}_0} + q_\mathsf{A}\tilde{w}_{\mathsf{A}_0\mathsf{A}_1}\right),$$

where $\tilde{W}$ is the average fitness in the population evaluated assuming Robbins proportions, that is, evaluated from the average one-locus genotypic fitnesses

$$
\begin{aligned}
\tilde{w}_{\mathsf{A}_0\mathsf{A}_0} &= w_{11}p_\mathsf{B}^2 + 2w_{12}p_\mathsf{B}q_\mathsf{B} + w_{22}q_\mathsf{B}^2, \\
\tilde{w}_{\mathsf{A}_0\mathsf{A}_1} &= w_{13}p_\mathsf{B}^2 + 2w_{14}p_\mathsf{B}q_\mathsf{B} + w_{24}q_\mathsf{B}^2, \\
\tilde{w}_{\mathsf{A}_1\mathsf{A}_1} &= w_{33}p_\mathsf{B}^2 + 2w_{34}p_\mathsf{B}q_\mathsf{B} + w_{44}q_\mathsf{B}^2.
\end{aligned}
$$

If the average fitnesses of the genotypes show directional selection, we get a simple directional response, and if this pattern persists as the gene frequencies evolve at the B locus, locus A fixes one allele.

If over- or underdominance prevails, the dynamics become more complicated and two-locus polymorphism may result if joint equilibria exist. Such an equilibrium is found from equation (9.3) by evaluating the fitnesses assuming Robbins proportions and equilibrium at the other locus. For given gene frequencies at the B locus, the equilibrium is unique, if it exists, but equilibria may exist for different combinations of gene frequencies at the two loci.

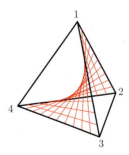

When an approximate gene frequency equilibrium is reached it is possible to calculate an approximation to the value of the linkage disequilibrium at equilibrium in the population:

$$\hat{D} \approx \frac{\hat{p}_\mathsf{A}\hat{q}_\mathsf{A}\hat{p}_\mathsf{B}\hat{q}_\mathsf{B}}{r} \left(\tilde{W}_1 - \tilde{W}_2 - \tilde{W}_3 + \tilde{W}_4\right),$$

where $\tilde{W}_i$, $i = 1, 2, 3, 4$ is the average fitness of gamete i in the population evaluated assuming Robbins proportions and gene frequency equilibrium. The approximation is good when r is large compared to the fitness differences. The equilibrium linkage disequilibrium is therefore proportional to a quantity that may be interpreted as an epistatic interaction in fitness at equilibrium.

A population in quasi linkage equilibrium has the property that *the average fitness increases* during its evolution (Kimura 1965). The reason for this characteristic of the weak selection model is that it exhibits a conglomerate of single-locus evolutionary behaviors. Before quasi linkage equilibrium is reached, there is no reason to believe that mean fitness should increase.

Example 9.1.3 For loose linkage and $w < 1$ the model in Table 9.5 has a globally stable equilibrium where $\hat{x}_i = \frac{1}{4}$. The average fitness at this equilibrium is $\hat{W} = \frac{1}{4}(1 + w)^2$, which is the maximum value at the linkage equilibrium surface. For absolute linkage, however, $\hat{x}_1 = \hat{x}_4 = \frac{1}{2}$ and $\hat{x}_2 = \hat{x}_3 = 0$ is a stable equilibrium with $\hat{W} = \frac{1}{2}(1 + w^2) > \frac{1}{4}(1 + w)^2$. Thus, the high complementarity equilibria for absolute linkage has an average fitness higher than that of the globally stable equilibrium for loose linkage. The average fitness of the globally stable equilibrium cannot be a global maximum because the average fitness at the point $x_1 = x_4 = \frac{1}{2}$ and $x_2 = x_3 = 0$ is higher.

The weak selection model is particularly attractive as a model describing selection on quantitative characters. Heritable variation in a quantitative character is usually described using a polygenic model in which the single genes are assumed to have a small influence on the trait variation. Selection on such a character induces selection on the genotypic variation, and this selection may be assumed weak if the gene loci are loosely linked. In such a population the response to selection results in increasing mean fitness. In case of truncation selection on an additive character, it results in change in the mean in the direction of selection. The results of the analysis of multilocus weak selection are thus in agreement with observations in quantitative genetics in general and animal husbandry in particular.

In evolutionary discussions of quantitative characters, optimizing selection has played a key role. Truncation selection or other kinds of directional selection continue to generate response as long as genetic variance is present, and such selection therefore stops only when variation is depleted. Optimizing selection could then be responsible for the extant quantitative variation, but the answer is not that simple. Wright (1935) showed that for weak selection equivalent to Gaussian optimizing selection at two-allelic loci, at most one locus can remain polymorphic. This form of selection therefore cannot maintain much variation. Inclusion of mutation and random genetic drift, in addition to allowing for more alleles per locus and different forms of optimizing selection, can allow for a bit more polymorphism. Recent discussions of such aspects of directional and optimizing selection may be found in Barton and Keightley (2002) and Bürger (2000).

The analysis of truncation selection in Section 8.3.1 on page 236 may be extended to two loci. Consider two loosely linked loci with additive effects on a quantitative character, each locus segregating two alleles with additive effects.

The genotypic fitnesses may then be represented approximately as in Table 9.6

Table 9.6: Two-locus additive viabilities

	BB	Bb	bb
AA	$1 + a + b$	$1 + a$	$1 + a - b$
Aa	$1 + b$	1	$1 - b$
aa	$1 - a + b$	$1 - a$	$1 - a - b$

for small additive effects at both loci (see Exercise 8.3.2 on page 262). The selection coefficients a and b are small, so we may assume quasi linkage equilibrium. Both loci are therefore under directional selection, leading to ultimate fixation of one allele. For simplicity we assume $a > 0$ and $b > 0$, thus implying ultimate fixation of gamete AB.

At some point in this process, assume the gene frequencies p_A and p_B of alleles A and B, and suppose the linkage disequilibrium is small and equal to D. In the next generation the linkage disequilibrium becomes

$$D' \approx -abp_A q_a p_B q_b - rD\big(a(p_A - q_a) + b(p_B - q_b)\big) + (1 - r)\frac{D}{W^2} \, ,$$

where the mean fitness is $W = 1 + a(p_A - q_a) + b(p_B - q_b)$. The contribution of selection to the linkage disequilibrium is thus given by the first term, which is always negative. The contribution of the second term is more complicated. When the favored alleles are both rare the sign of their contribution is the same as that of the parental linkage disequilibrium, whereas when both are common the sign is opposite. When $p_A > \frac{1}{2}$ and $p_B < \frac{1}{2}$ or vice versa, the sign is indeterminate.

For sufficiently close linkage the linkage disequilibrium quickly becomes negative (Felsenstein 1965), and numerical simulations show that it usually ends up that way. The assumption of weak selection ensures that the additive variance in the underlying quantitative character is proportional to that of the selection coefficients in Table 9.6. In this simple model the additive variance may be written as the variance in the selection coefficients due to locus A, plus that due to locus B, plus the covariance in selection coefficients due to the two loci. The covariance has the sign of the linkage disequilibrium, which is negative, and selection therefore results in a decrease of the additive variance (Felsenstein 1965). This result is also true when dominance is allowed, and Bulmer (1971) concluded in general that truncation selection will show a decrease in the realized heritability during the initial generations of directional selection.

9.1.4 Intermediate linkage

The dynamics are simple for the two extreme cases of absolute linkage and loose linkage with weak selection. In both cases the population settles at an equilibrium from any starting condition and no oscillatory or chaotic behaviors occur. This is not true when the effects of selection and recombination are comparable (Akin 1979, Hastings 1981).

Rather little is known about this situation in the general two- or multilocus model, and most of our knowledge is based on analyses of simple models. Most influential is the symmetric viability model suggested by Lewontin and Kojima (1960). The fitness of a genotype only depends on the loci where it is heterozygote (Table 9.7). The models in Tables 9.4 and 9.5 are special cases of the

Table 9.7: The two-locus symmetric viability model

	B_0B_0	B_0B_1	B_1B_1
A_0A_0	$1 - \alpha$	$1 - \gamma$	$1 - \alpha$
A_0A_1	$1 - \beta$	1	$1 - \beta$
A_1A_1	$1 - \alpha$	$1 - \gamma$	$1 - \alpha$

symmetric viability model. The symmetric model is special because it always exhibits an equilibrium with gene frequencies in linkage equilibrium, namely an equilibrium where $\hat{x}_i = \frac{1}{4}$, which is known as the *central equilibrium*.

Even this simple model shows rich behavior (Box 33). The model was constructed to have an equilibrium with gamete frequencies in Robbins proportions. The result, that Robbins proportions are attained for loose linkage, should therefore be viewed only as an indication that the state predicted in the weak selection approximation has been reached. Thus, for loose linkage the population will be at an equilibrium with a low amount of linkage disequilibrium. For close linkage the population will attain an equilibrium determined mainly by selection. If that equilibrium is totally polymorphic and located at the boundary of the gamete frequency domain, strong linkage disequilibrium results and it is a high complementarity equilibrium like those found in the model of Table 9.5. If it corresponds to a four-allele polymorphism, we expect lower levels of linkage disequilibrium (as in the model of Table 9.4).

These two cases are typical for nice overdominant symmetric models in which fitness increases with degree of heterozygosity and the difference in the fitnesses of the single heterozygotes is not too large. If these conditions are relaxed more equilibria may be stable and the dynamics depend on r in a complicated way (Box 34).

The multiplicative model in Table 9.5 (and in general for symmetric models with positive epistasis, see Box 34) exhibits stable equilibria for tight linkage, where linkage disequilibrium prevails. A similar result holds for multiple loci (Christiansen 1988). It is therefore tempting to compare the two- and multilocus results as corresponding in the same way as the one- and two-locus results correspond (Section 9.1.1 on page 271). Inferring a fitness matrix for two loci from multiple loci at a high-complementarity equilibrium due to symmetric viability will not give a fitness matrix like that in Table 9.7. The multilocus linkage disequilibrium produces a position effect in the two-locus fitness matrix. Nordborg et al. (1995) considered a two-locus model with symmetric viabilities

---

Box 33: *Symmetric equilibria

The model in Table 9.4 for $w < 1$ represents the dynamics of symmetric models with $\alpha > |\beta - \gamma|$ and negative epistasis, that is, $\alpha > \beta + \gamma$. In these models the central equilibrium is stable for all values of r, $0 \leq r \leq \frac{1}{2}$. The model in Table 9.5 for $w < 1$ represents the dynamics of symmetric models with $\alpha > |\beta - \gamma|$ and positive epistasis, that is, $\alpha < \beta + \gamma$. In these models the central equilibrium is unstable for tight linkage, for $0 \leq r \leq \frac{1}{4}(\beta + \gamma - \alpha)$, and then the equilibrium

$$\hat{x}_1 = \hat{x}_4 = \frac{1}{4}\left(1 + \sqrt{1 - \frac{4r}{\beta + \gamma - \alpha}}\right),$$

$$\hat{x}_2 = \hat{x}_3 = \frac{1}{4}\left(1 - \sqrt{1 - \frac{4r}{\beta + \gamma - \alpha}}\right),$$

and the one with the *cis* and *trans* gametes interchanged are stable. These are high complementarity equilibria. For $r = \frac{1}{4}(\beta + \gamma - \alpha)$ the high complementarity equilibria fuse with the central equilibrium and for $\frac{1}{4}(\beta + \gamma - \alpha) \leq r \leq \frac{1}{2}$ the central equilibrium is stable.

---

except for a difference in survival between the two double heterozygotes, and they inferred dynamics more complicated than those described in Box 34.

9.2   Selection Experiments in *Drosophila*

Population genetic effects of selection on Mendelian characters as well as on quantitative characters in experimental populations have been widely studied. The studied changes can be due either to "natural selection," where the changes in the frequencies of traits of a character are passively followed in an experimental environment, or to "artificial selection," where the experimenter actively selects some traits by discarding individuals with other traits. In both kinds of experiments, however, individuals and not traits, and genomes and not genes, are selected, so we cannot neglect concurrent changes in other characters or at other loci.

Among the classical experimental organisms in genetics, *Drosophila melanogaster* is the one most widely used in population genetic studies in the laboratory. It has been used in the study of both Mendelian and quantitative characters. The following two experiments illustrate this. Many economically interesting characters in livestock are quantitative, and although *D. melanogaster* is excellent for investigating the principal aspects of selection in animal breeding (e.g., Section 9.5.1), aspects related to a particular character are better illuminated in more closely related organisms like the house mouse, *Mus musculus*, which is a widely used model for mammals (Section 8.3.3).

Box 34: *Equilibria in the symmetric model

When the symmetric viability model deviates from nice overdominance while still allowing for stable totally polymorphic equilibria, the recombination frequency may have a strong and complicated influence on the dynamics of the model. For instance, with $\alpha = 0.05$ and $\beta = \gamma = 0.2$ the high complementarity equilibria are stable for $0 \leq r < 0.053$ and $0.083 < r < 0.088$ with the central point stable for $0.088 \leq r \leq \frac{1}{2}$. In addition, the four monomorphic

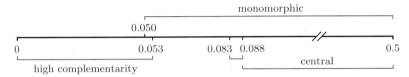

equilibria are stable for $0.05 \leq r \leq \frac{1}{2}$. For $0 \leq r \leq 0.05$ the two high complementarity equilibria are the only stable ones. For $0.05 < r < 0.053$ the high complementarity equilibria are stable concurrently with the four monomorphic

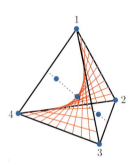

equilibria. For $0.053 \leq r \leq 0.083$ the monomorphic equilibria are the only stable ones. For $0.083 < r < 0.088$ the high complementarity equilibria are again stable simultaneously with the monomorphic equilibria. And for $0.088 \leq r \leq \frac{1}{2}$ the central point is stable simultaneously with the monomorphic equilibria.

The details of these transitions in the dynamics are not particularly interesting because they pertain to an exceedingly simple model. The message is the complexity of the changes in the dynamics as the recombination frequency varies. Totally polymorphic equilibria may be stable simultaneously with monomorphic equilibria, and monomorphic equilibria may be the only stable equilibria for some recombination frequencies even if totally polymorphic equilibria are stable for both tighter and looser linkage. For $\beta \neq \gamma$ the symmetric viability model allows yet another dynamic possibility, namely that stable totally polymorphic equilibria exist for tight linkage while only monomorphic equilibria are stable for loose linkage.

The original analysis by Lewontin and Kojima (1960) of the two-locus symmetric viability model uncovered the central equilibrium and the high complementarity equilibria. The demonstration of the possible gap in the stability of the high complementarity equilibria was made by Ewens (1968), and it is therefore known as Ewens' gap. The completion of the analysis is due to Karlin and Feldman (1970) and Nordborg et al. (1995). A review of the theory of the symmetric viability model is given by Christiansen (2000).

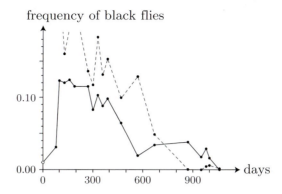

frequency of black flies

Figure 9.5: Observations in two experiments on ebony trait variation in *Drosophila melanogaster*. Solid line: initial $q = 0.1$. Dashed line: initial $q = 0.9$. Data from Frydenberg (1964).

9.2.1   An experiment using the *ebony* allele

Frydenberg (1964) studied selection in laboratory populations of *D. melanogaster* that varied in body color due to the presence of the recessive trait ebony, giving black flies. The *ebony* locus is on chromosome III, and the alleles are e and $+$. The source of the variation was two true-breeding strains, ebony and wild type, kept in the laboratory for many years. The populations were started from 100 flies from one of the strains and 25 flies that were F_1 of the two strains. Of the 250 genes in the initial generation, 25 were thus of the minority allele, giving it a gene frequency of 0.10. The flies were kept in a population cage with regular supplies of fresh food. Frydenberg estimated the generation time to be about 15 days with an effective population size on the order of 500. The frequency of the ebony trait was followed by collecting samples of eggs at various times in the evolution of the population. The eggs were reared under optimal conditions to maintain very low mortality, and the frequency of the ebony trait among the flies that developed from these eggs was taken as an estimate of the frequency in the population. Changes in composition of the population during the experiment are thus due to "natural selection."

Frydenberg made five experiments, three with a low ($q = 0.1$) and two with a high ($q = 0.9$) initial frequency of the e allele at the start of the experiment.[2] We discuss the results for two of these experiments. The first sample of a population initiated with $q = 0.1$ showed an ebony frequency of about 3 percent (solid line in Figure 9.5), and already in the second sample this had increased to about 12 percent. The first 100 days thus showed an increase in the frequency of the ebony flies from an expected value of 0.01 (estimated assuming random mating) to about 0.12. This is surprising because e is ex-

---

[2]In the *ebony* experiment the chromosome II recessive markers cn and bw were introduced into the stocks to allow the experimenter to detect contamination by foreign flies. Chromosome II was made identical in the two strains by a balancer technique.

pected to be a deleterious allele. The population, however, remained at the high frequency for 160 days, suggesting the action of overdominant selection on the three genotypes at the *ebony* locus (compare to Figure 8.7 on page 229). Had the experiment been terminated after these ten or so generations, at that time a reasonably long experiment, the conclusion for the dynamics would have been rather straightforward.

Frydenberg continued the experiment, however, and observed that the frequency of the ebony decreases a bit over the next months. From day 300 to 390 it stayed at a level of about 0.09 before decreasing again over a period of about 200 days to a level of around 0.03, which lasted about a year until the sample at day 1000. Finally, after 1060 days the ebony trait had disappeared from the population.

The second experiment, started at a gene frequency of 0.9, showed a similar evolution in the frequency of black flies. The ebony frequency decreased to just below 0.3 at day 80—a considerable drop from an expected initial frequency of 0.81. It oscillated around 0.2 for the next hundred days and decreased to about 0.15, where it stayed the following two hundred days (dashed line in Figure 9.5). From day 400 to 600 the ebony frequency stayed at a slightly lower level before a decline to very low frequencies during the final year of the experiment. The final three experiments showed similar results, not in detail but in the presence of seemingly indeterminate changes in the genetic composition of the population. All experiments ended in the loss of the ebony trait.

The observations in the first experiment clearly show that knowledge of the population frequency of ebony is insufficient to predict its future evolution. After 80 days 0.03 was observed, and after a rapid increase the frequency remained high for almost a year. The trait frequency subsequently reached 3 percent again and stayed there for a long period. A straightforward explanation of this indeterminism is that the alleles at the *ebony* locus are associated with particular alleles at other loci that do not produce a detectable phenotype, but are subject to selection. Such associations influence the fitness of the ebony and wild-type flies, but recombination changes the linkage disequilibrium between the *ebony* alleles and the alleles of the associated fitness effects. If we view fitness as a phenotype we may see the *ebony* allele as a marker and a QTL of major effect, and the associated fitness effects are due to other QTLs in the vicinity of the *ebony* locus.

Direct observation of gene frequencies is, of course, impossible with dominance, but estimation may be possible if we know the mating pattern in the population. In the face of ignorance the only reasonable choice is to make the simplifying assumption that mating occurs at random and that ebony and wild-type flies have the same expected number of offspring.[3] We thus assume that the genotypic frequencies among the eggs in the samples are in Hardy–Weinberg proportions and the frequency of ebony flies is expected to be the square of the gene frequency. The square root of the ebony frequency is then a crude esti-

---

[3]These assumptions that ignore some aspects of selection may introduce a serious bias in the gene frequency estimates if those components of selection are indeed acting (see Christiansen and Prout 1999), but this does not seriously impede the present arguments.

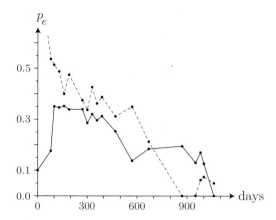

Figure 9.6: The estimated gene frequencies of the *ebony* allele in the two experiment in Figure 9.5 (solid line: initial $q = 0.1$; dashed line: initial $q = 0.9$). After Frydenberg (1964).

mate of the gene frequency (Figure 9.6). With this estimation procedure the pattern of variation in gene frequency corresponds closely to that in the phenotype frequency in Figure 9.5. We can nevertheless read the "equilibrium" gene frequency after day 100 as 0.35 in the first experiment, corresponding to the selection coefficients $s_e \approx 2s_+$, where $w_{++} = 1 - s_+$, $w_{+e} = 1$, and $w_{ee} = 1 - s_e$. In the second experiment the decrease in q is rather smooth over the first year, showing no clear "equilibrium" interruptions (Figure 9.6). The frequency drops rapidly to a gene frequency of around 0.5 and stays there for about 50 days, followed by a slide that seemed like a prolonged hovering around 0.4. The interim equilibria, should such exist, correspond to selection coefficient ratios between 1 and 2. The putative QTLs and their initial associations with the marker are the same as in the first experiment, but the erosion of linkage disequilibrium depends on the particular recombination events. As expected, these add significant stochasticity to the process, and the differences between the two experiments add to the credibility of our description.

The complex behavior of multilocus systems under selection was further illustrated in the last two years of the experiments. The differences between them increased. Although they both underwent a plunge in the gene frequency of the *ebony* allele, it occurred at different times and in different ways. Typically for an inherently stochastic process, however, the general patterns in the two experiments are fairly similar, albeit neither in the detail nor in the timing of events.

9.2.2 An experiment using *white* alleles

Barker (1977) made similar experiments with two mutant alleles at the sex-linked *white* locus in *D. melanogaster*. Two laboratory strains, *white* and *white-*

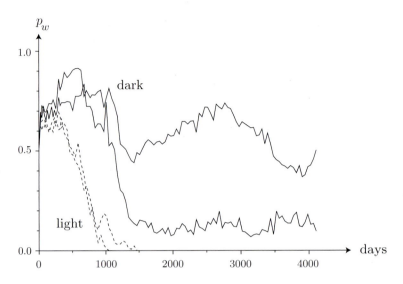

Figure 9.7: The gene frequency in the experiment using the *white* mutants in *Drosophila melanogaster* in two environments. After Barker (1977).

blood, carrying the alleles w and w^{bl}, respectively, were used. These strains are expected to be as different as the strains in the *ebony* experiment, and to diminish the genetic differences they were outcrossed to wild type population.[4] The genotypes ww, ww^{bl}, and $w^{bl}w^{bl}$ can be distinguished in the females because of intermediate dominance of the w^{bl} allele, and the gene frequencies in both males and females can therefore be estimated in samples from the populations. Barker's ambition was to achieve a deeper understanding of Frydenberg's results, and he added two dimensions to the experiments, namely population size (large versus small) and two environments (normal conditions with light and dark periods daily versus conditions with complete darkness). Population size had little influence on the evolution. Figure 9.7 shows the development of the four experiments with large population size.

The initial behavior resembles that in the *ebony* experiment, but as time passes, only the populations kept in light conditions retain the similarity, in that the w allele is lost from the population. The white-eyed flies are blind, and ww^{bl} and $w^{bl}w^{bl}$ individuals thus seem to have an advantage in the light environment. In the light environment blindness is a deleterious trait, and w is thus in this respect a sex-linked recessive deleterious allele, which is eventually lost from the populations. As selection diminishes the w frequency selection occurs almost exclusively in males because most w alleles in the females will be carried with the common w^{bl} allele. In males, however, the frequency of individuals with white eyes equals the gene frequency of w, and the elimination

---

[4]Contamination markers were not needed because the genotypes at the w locus are easily distinguished from wild type flies. The two strains were at any rate backcrossed to a wild type strain for three generations before the founding stocks were synthesized.

of the deleterious allele is very rapid as soon as it is decoupled from most of the alleles at linked loci with effects on fitness.

The experiments under the dark conditions take a very different course because the w and w^{bl} alleles are expected to be virtually equal in fitness. The morphological alleles are therefore merely neutral markers of the chromosomal region containing the w locus. After the initial balanced polymorphism the *white–white-blood* polymorphism becomes volatile, and the two populations show a very different course of evolution, most likely because of different histories of recombination events around the w locus.

The dark populations were maintained and observed for about five years after the loss of polymorphism in the light populations (Figure 9.7), and in the full 11 years of evolution we see another aspect that is consistent with our multilocus interpretation of the experimental results. At the beginning the populations quickly reach the first intermittent equilibria, but as time passes the definition of equilibria becomes more and more difficult because transition is slower and stability weaker. In the last six years of the experiment each of the populations, when viewed alone, seems to be subject to random genetic drift. The continued influence of the genetic background and its associated fitness effects on the *white* alleles emerge from a comparison of the two populations. The lower one seems to be at a stable equilibrium, while the upper one is much more volatile.

9.3 Hitchhiking

The focus of the experiments in Sections 9.2.1 and 9.2.2 is the observable genetic variation, namely the frequency of the *ebony* phenotypes and the *white* allele expressions. We therefore see the results of the effects discussed in Section 9.1.1. The observed evolution in the populations may be viewed as revealing the influence of the changes in the genetic background on the gene frequency changes at the observed locus. In the *white* experiment the dynamics in the dark environment seems to suggest that the two alleles are neutral or nearly so, and hence the changes we see reflect mainly the dynamics of the genetic background. The *white* alleles are genetic markers of changes in the population of genomes. These are *hitchhiking* phenomena, where a genomic element initially in linkage disequilibrium with loci subject to selection is taken for a ride on the genetic changes at the loci subject to selection (Maynard Smith and Haig 1974, Thomson 1977).

The ebony and the white traits correspond to of diseases in humans, and they are extremely rare in nature. The e allele is deleterious and recessive. In Section 8.4 we found that the fitness of a rare recessive allele is very close to that of the common dominant allele (Exercise 8.4.1 on page 262). Linkage of the *ebony* locus to genes on chromosome III may thus carry the deleterious recessive allele to high frequencies. In the experiments of Barker and Frydenberg the observed alleles were subject to variation in the quantitative and qualitative aspects of selection, and as this phenomenon is due to the genetic background associated with the alleles, it may be described as hitchhiking. Frydenberg (1963) described such effects as *associated fitness* and defined them as "due to

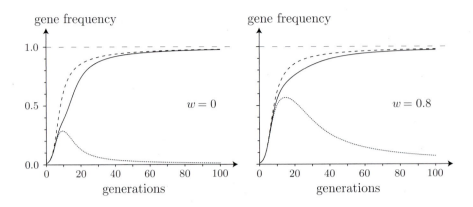

Figure 9.8: The effects of hitchhiking on a deleterious recessive allele due to increase in the frequency of the *carbonaria* allele in *Biston betularia*. The dotted curves show the frequency of the deleterious allele with viability w; the solid curves show that of the *carbonaria* allele. The dashed curve is a reference to Figure 8.4 on page 224 (corresponding to $w = 1$).

non-randomized linkage between the observed pair of allelic units and the entire remainder of genetic alternatives present in the same chromosome." He used the concept of associative overdominance to describe the initial period in the *ebony* experiment.

A new mutant is in complete linkage disequilibrium with all variable loci in the population because it is present only in one chromosome. Its dynamics are therefore highly influenced by associative fitness effects. The other side of the coin is that a new advantageous mutant increasing in frequency is expected to effect hitchhiking in its genomic neighborhood. The quicker it increases in frequency, the more widespread the effect because of the shorter time available for recombination to relieve the associations. As an example we may evaluate the influence of the increase in *Biston betularia* of the *carbonaria* allele on the frequency of an associated recessive deleterious allele. The scope of this example is quite general because the initial increase of any allele depends mainly on the viability of the heterozygote—emphasizing that the potential for a recessive allele to mediate hitchhiking is very small.

We use the selection model for the *carbonaria* locus as in Figure 8.4 on page 224, that is, $w_{AA} = 2w_{aa}$. The locus of the linked recessive deleterious allele b is 1cM away, and the homozygote has a viability w relative to the normal genotypes Bb and BB, that is, $w_{bb} = w\,w_{BB}$. We assume a multiplicative specification of two-locus genotypic fitnesses (see Example 9.1.2 on page 273). At the start of the iterations the frequency of gamete Ab was 0.01 and that of gamete aB 0.99. In the left-hand graph in Figure 9.8 the results are shown for allele b when assumed to be recessive lethal, and even then, hitchhiking on the A allele carries it to a frequency of almost 0.30. A more mildly acting allele

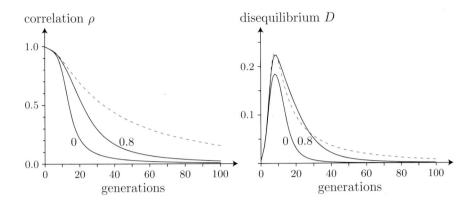

Figure 9.9: The association between the *carbonaria* allele and a recessive deleterious allele in the hitchhiking examples in Figure 9.8. The solid curves show
the association between the *carbonaria* allele and a deleterious recessive allele
(w as in Figure 9.8). The dashed curves show that for a neutral allele.

inducing a 20 percent mortality is carried to a frequency of above 0.5.

The lethal allele has virtually no effect early in the advance of the *carbonaria*
allele. It hitchhikes on the *carbonaria* allele almost as if it were selectively neutral. Later the lethal allele has a period where it retards the increase in frequency
of the *carbonaria* allele. Its frequency peaks in this period and then decreases
quite rapidly. The association between the loci, as measured by the gametic
correlation ρ, decreases slowly at first, followed by a period of rapid decrease
(Figure 9.9), showing that after the retardation effect ceases, the recessive allele
leaves of its own accord. The frequency of the milder acting recessive allele
peaks later. The retardation of the melanic type thus peaks later and persists
longer.

The pattern of variation in the linkage disequilibrium measure D shown in
Figure 9.9 underscores the difficulties of interpreting the D values mentioned
in Section 6.3 on page 152. Recombination erodes the association between the
loci, but selection produces a lot of linkage disequilibrium in the initial period.
However, the magnitude of linkage disequilibrium does not convey information
about association between the loci before its value is juxtaposed with the prevailing gene frequencies. These increase a lot in the beginning, thus extending
the range of possible D values. In a similar way, the D values decline faster
than the gametic correlation when the gene frequency of the deleterious allele
decreases.

A dominant deleterious allele, C say, may also hitchhike. Figure 9.10 shows
an example comparable to the ones in Figure 9.8. The drawing assumes the
relative fitness $v = 0.8$ of the heterozygote ($w_{Cc} = v\,w_{cc}$) and it is therefore
comparable to the recessive case with $w = 0.8$.[5] As expected, the fitness effect
shows immediately as a retardation of the increase of the *carbonaria* allele, but

---

[5]For $v < 0.5$ the AC gamete cannot increase in frequency.

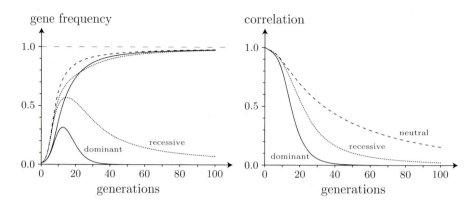

Figure 9.10: The effect of hitchhiking on a deleterious dominant allele compared to that of a recessive allele. In the left-hand drawing the solid curves show the increase of the frequency of the *carbonaria* allele and that of a linked dominant deleterious allele with $v = 0.8$. The dotted curves show the frequencies of the *carbonaria* allele and a recessive deleterious allele ($w = 0.8$). The dashed curve corresponds to the neutral case. The drawing to the right shows the development of the association between the *carbonaria* and the linked allele in the three cases.

after about 25 generations it overtakes the increase in the recessive case. The initial retardation is due to the strong association with the deleterious allele, but that is broken quite abruptly because of the immediate fitness difference between the gametes AC and Ac.[6]

Recombination must be viewed as a stochastic process in any real population. Retardation of selective advance of an advantageous allele is therefore expected to be episodic in much the same way as the shifts between intermittent equilibria in the *ebony* experiment. Temporal changes in the dynamics are due to the emergence of novel associations around the observed marker, and this requires the formation of a new advantageous constellation by recombination, followed by its establishment in the population. The episodic nature of the process thus becomes apparent when the waiting time till a new haplotype emerges is significant. Linkage disequilibrium degrades quickly for loci far from the selected locus, but as the associated piece becomes smaller, the waiting time until recombination becomes longer (see Section 6.3.5 on page 160). In addition, the waiting time is inversely proportional to the population size. The episodic nature of the associations in the *ebony* experiment is more pronounced than that in the comparable *white* experiment in the light environment. That is, when the width of associations becomes sufficiently narrow in order to foster intermittent equilibria in addition to the initial one, then selection on the variation at the

---

[6]The iteration in the figure assumes multiplicative fitness interaction between the loci as well as multiplicative interaction of the C alleles, so the fitness of CC is $v^2 = 0.64$. This simplifying assumption matters only in the fine details of the drawing.

marker locus dwarfs the associated fitness effects.

An additional source of hitchhiking emerges when the waiting time to the production and survival of a recombinant chromosome becomes significant, namely the uniqueness of the emerging recombinant. Its increase in the population will have an effect parallel to that of the increase of the *carbonaria* allele. The recombinant chromosome will therefore change associations among loci around the marker, but due to hitchhiking the location of the recombination event need not be a good predictor of new correlations among alleles around the marker. It even has the potential of extending the linkage disequilibrium bracket, but in the long run recombination is expected to progressively narrow the associated piece of chromosome unless, of course, other sources of linkage disequilibrium exist. The *white* experiment in the dark environment clearly shows progressively slower dynamics in the observed gene frequencies, but association effects are still significant after 3000 generations.

Genes are not isolated beans in a bag. They reside in a genome and have well-defined neighborhood structures on the chromosomes. We therefore cannot view the functional gene as a unit in evolutionary processes. It exists in an ever-changing relationship with its linked neighbors, with which it usually does not have physiological relationships. This loosely defines a *dynamical unit of selection*. We expect this unit to be larger than the typical functional gene in any natural population (Lewontin 1974) in agreement with the general pattern of linkage disequilibrium seen in humans (see Section 6.5 on page 177). Observed differences in survival and reproduction among the genotypes at a given locus therefore need not reflect the physiological variation caused by the functional differences among the alleles, should such exist. For substitutions of alleles at the locus to be deemed adaptive, the function of the new allele must reflect positively on the survival and reproduction of its carriers. In this sense the gene is the *functional unit of selection*. Its function may indeed depend on the variants present at other genes—its *genetic background*—and its influence on genotypic survival and reproduction therefore depends on the population variation in its genetic background. Nevertheless, selection brought about by its variation is the result of ensuing variation in its function (see Section 9.1.1). The residual effect of its dynamical unit of selection is stochastic, potentially giving rise to haphazard changes in its observed genetic effects on fitness.

9.3.1 An experiment on a bristle character in *Drosophila*

The dynamical unit of selection is in general determined by the population size in terms of the dynamical parameter $4Nr$ (Section 6.3.4 on page 159). Selection on the ambient variation in a population is thus expected to relate to this fundamental unit, and the genetic response to selection will be to favor subsets of loci that happen to be associated with average positive fitness effects. The result is random fixation of less favored alternatives among the alleles at some loci and the haphazard loss of alleles with desirable effects. The response to selection is therefore expected to be lower in smaller than in larger populations, both in terms of the rate of response and in the total response to continued

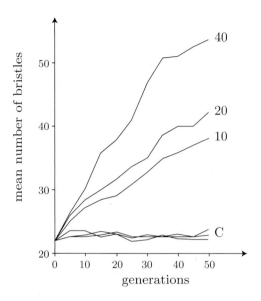

Figure 9.11: Response to selection for high number of abdominal bristles in a laboratory population of *Drosophila melanogaster*. Results for three selection lines (each with three replicates; means shown) are depicted with the number of parental pairs shown on the curves. Results for the three corresponding control lines (also means of three replicates) are marked C. Data from Frankham et al. (1968*a*); after Jones et al. (1968).

selection (Hill and Robertson 1966, Robertson 1970). The response obtained is therefore limited in relation to what we would have expected, had we known the genetic variation in the population.

These effects were demonstrated in short- and long-term response to selection on number of abdominal bristles in laboratory populations of *D. melanogaster* in experiments made by Frankham et al. (1968*a*) and Jones et al. (1968). They used flies from the population described in Figure 7.4 on page 188, but limited the bristle counts to one abdominal sternite, the fourth in males and the fifth in females. In the base population the mean and standard deviation of the counts were 17.7 ± 1.93 in males and 21.7 ± 2.02 in females. Sheridan et al. (1968) studied the transmission of traits of this character and estimated the heritability to be around 0.15. Three population sizes were used in that 10, 20, and 40 pairs of parents were selected in each generation. Five selection intensities were used. The parents of the next generation were chosen as 10, 20, 40, 80, or 100 percent

of the counted flies with the highest number of bristles. The number of flies counted thus varied from 10 to 400 in the various experimental lines. The lines where 100 percent of the flies were kept functioned as unselected controls in the experiment.

Figure 9.11 shows the results for the experiments in which 10 percent were selected (the response curve for 40 pairs is reproduced in Figure 8.10 on page 236). Each selection line shows the typical linear response for directional selection on a polygenic character, but the response per generation (the slope of the line) increases with the size of the selected population. The control lines were fairly constant throughout the experiment and no effect of population size was detected. The population size effect shown in Figure 9.11 reflects the influence when the full series of experiments was analyzed, and they thus provided beautiful empirical support of the effect predicted by Hill and Robertson (1966).

9.4   Inbreeding Depression

Inbred livestock individuals usually have lower values for production, survival, and fecundity than outbred individuals. This phenomenon, called *inbreeding depression*, was observed by humans early in the development of animal husbandry. Many cultures transferred this experience to humans to discourage or prohibit consanguineous marriages. Fecundity in *Drosophila melanogaster* shows inbreeding depression as illustrated in Figure 9.12 to the left (the egglaying rate of noninbred females is compared to that of females with an inbreeding coefficient of $\frac{1}{8}$ (data from Dahlgaard and Hoffmann 2000)). The figure shows another characteristic feature of inbreeding, namely that the phenotypic variance becomes larger—in this particular case more than twice as large.[7] The main interest in the present context is inbreeding depression in fitness, and to keep the discussion simple we shall assume that the effect is entirely on viability.

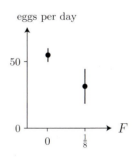

Figure 9.12: Fecundity in *D. melanogaster*

Effects of inbreeding are seen both at the genomic level and at the level of chromosomes or parts thereof. Dobzhansky et al. (1963) investigated inbreeding depression on chromosome II in a natural population of *Drosophila pseudoobscura*. They collected flies from wild populations and cloned their second chromosomes using classical techniques of *Drosophila* genetics (see Box 35). They subsequently combined these chromosomes to study their genotypic effects on fitness. The viability of wild type flies was measured in crosses, where cloned chromosomes (colored in Figure 9.13) were introduced with a balancer chromosome (black). The upper histogram shows the distribution of fitness among 208 crosses in which both parents carry the same wild-caught chromosome. A

---

[7]Indicated by a bar showing four times the standard deviation of the means in samples of 37 and 33 individuals, respectively.

number of crosses

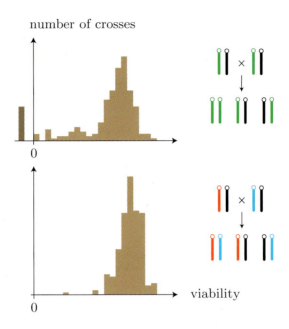

Figure 9.13: Relative viability of wild type *Drosophila pseudoobscura* in crosses with cloned wild-caught chromosomes. Top histogram: 208 crosses of flies with identical chromosome II balanced by the *Ba* chromosome. The number of crosses where the wild-caught chromosome carries a recessive lethal is shown by the dark column. Lower histogram: 210 crosses of flies with different chromosome II balanced by the *Ba* chromosome. Data from Dobzhansky et al. (1963).

fair proportion (dark column) of the crosses produce no wild type flies, so the cloned chromosomes carry recessive lethal genes in these crosses. The majority of the crosses, however, collect in a fairly normal distribution of high viability. Between these extremes an almost uniform distribution of crosses appear with subvital wild type homozygotes. The distribution of viabilities of heterozygotes for wild-caught chromosome II was estimated based on results from 210 crosses (lower histogram in Figure 9.13). Only one cross yielded obviously subvital heterozygotes, and the rest cluster in a fairly normal distribution of high viability.[8] The "normal" heterozygotes, however, have a clearly higher mean than the "normal" homozygotes shown above. Inbreeding depression on chromosome II is thus apparent as a distinctly lower viability of identical chromosomal homozygotes. The low mean is lower for two reasons: a minority of homozygotes clearly have a lower viability than nearly all heterozygotes, and a majority have fitnesses whose distribution largely overlaps that of the heterozygotes, but has a lower mean.

One possible source of inbreeding depression is mutation–selection balance

---

[8]The distribution is reminiscent of that in Figure 4.1 on page 76 showing the growth rates of bacteria carrying a new mutant.

Box 35: Cloning of *Drosophila* chromosomes

Dobzhansky et al. (1963) collected male flies and crossed them individually to females from the *Ba gl/Δ* stock. The genes *Ba* (*Bare*) and *Δ* (*Delta*) are morphologically dominant and recessive lethals, and *gl* (*glass*) is a morphological recessive allele. The *Ba gl* chromosome is a so-called balancer chromosome that suppresses recombination in chromosome II, and the stock may be maintained in a permanent heterozygous condition for chromosome II. The wild-caught males transmit their chromosomes without recombination, and harvesting a *Bare* male among the offspring from each cross provides a number of intact chromosome IIs independently collected in the natural population. Crossing each of these males to females from the balancer stock provides F$_2$ offspring that carry a clone of the wild-caught chromosome II in the father, and the *Bare* flies will transmit this chromosome unaltered. Crossing two *Bare* flies with identical chromosome II produces offspring which are expected to segregate in $\frac{1}{4}$ wild types, $\frac{1}{2}$ *Bare* heterozygotes, and $\frac{1}{4}$ *BaBa*, but this last class dies, and the segregation in the culture vial is then expected to be $\frac{1}{3}$ wild type to $\frac{2}{3}$ *Bare*. The observed segregation in the vial thus provides information on the probability of survival of the identical homozygote relative to that of the *Bare* heterozygote. Heterozygotes for chromosome II may be formed by crossing flies carrying different F$_2$ clones.

---

equilibria in the population. In a large randomly mating population we expect to observe Hardy–Weinberg proportions. A tiny proportion of consanguineous matings will not produce a noticeable deviation from these, though inbred individuals occur in the genotypic frequencies

$$\text{AA}: p^2 + fpq, \quad \text{Aa}: 2pq - 2fpq, \quad \text{aa}: q^2 + fpq,$$

where f is the inbreeding coefficient of the individual, and p and q are the gene frequencies in the population (see page 69). The average fitness of individuals with inbreeding coefficient f is then

$$\begin{aligned} W(f) &= (p^2 + fpq)w_{\text{AA}} + (2pq - 2fpq)w_{\text{Aa}} + (q^2 + fpq)w_{\text{aa}} \\ &= W(0) + fpq(w_{\text{AA}} - 2w_{\text{Aa}} + w_{\text{aa}}), \end{aligned} \tag{9.4}$$

where $W(0)$ is the average fitness in a randomly mating population (individuals with $f = 0$). Inbreeding depression in a predominantly randomly mating population occurs when $W(f) < W(0)$, which is when $w_{\text{AA}} - 2w_{\text{Aa}} + w_{\text{aa}} < 0$. The dominance effect of the two alleles on fitness $-w_{\text{AA}} + 2w_{\text{Aa}} - w_{\text{aa}}$ (see Table 7.3 on page 196) therefore determines the difference $W(f) - W(0)$, and inbreeding depression occurs at a locus with positive dominance in fitness. Positive dominance at one locus with two alleles is present when the fitness of the heterozygote Aa is above the line connecting the fitnesses of homozygotes AA and aa, as shown in the figure. For

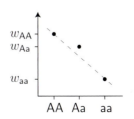

directional selection this means that the fitness of the heterozygote is closer to that of the advantageous homozygote than to that of the homozygote for the deleterious allele. As expected, recessive deleterious alleles always contribute to inbreeding depression.

Inbreeding depression affects an individual, and the single-locus effect of inbreeding is rarely interesting (unless, of course, we are interested in the effect of inbreeding on the prevalence of a genetic disease). We need to include the effect of inbreeding due to all loci with deleterious alleles. A crude approximation is that inbreeding depression results from mutation–selection balance equilibria when the dominance effects are on average positive. Inbreeding depression is also an effect of random genetic drift that emerges as a decrease in the average fitness (equation (7.19) on page 201)—of course given that the average dominance in fitness is positive. This effect is however diminished because selection against deleterious homozygotes becomes more effective as the population level of consanguinity increases (see Figure 8.17 on page 254).

In the 1960s Terumi Mukai investigated the fitness contributions of mutants in a series of mutation accumulation experiments in *Drosophila melanogaster*. The accumulation of mutations affecting viability was followed on chromosome II by a balancer technique similar to that used by Dobzhansky et al. (1963). In such experiments the original homozygote AA at a particular locus plays a special role and fitnesses are measured relative to that. These relative fitnesses are in classical population genetics given by the selection coefficient s of the homozygote aa, and the dominance h of allele A is expressed relative to that selection coefficient:

$$\frac{w_{\mathsf{aa}}}{w_{\mathsf{AA}}} = 1 - s \quad \text{and} \quad \frac{w_{\mathsf{Aa}}}{w_{\mathsf{AA}}} = 1 - hs.$$

Assuming that at most one mutant per locus occurs during the course of the experiment and neglecting lethal mutations, the observed average selection coefficient of new mutants was observed to be about $\bar{s} = 0.05$ and their dominance was about $\bar{h} = 0.4$ (Mukai et al. 1972). This corresponds to positive dominance in that $-w_{\mathsf{AA}} + 2w_{\mathsf{Aa}} - w_{\mathsf{aa}} = s(1 - 2h)w_{\mathsf{AA}}$, and these mutants will thus on average produce inbreeding depression. The inbreeding depression should, however, be calculated by the average dominance of the deleterious genes at equilibrium. Mukai et al. (1972) estimated this average at about $\bar{h} = 0.2$—still positive and smaller than the observed average, in accordance with the expectation that the equilibrium frequency of more recessive deleterious genes tends to be higher.

In the calculation of $W(f) - W(0)$ we did not assume that the deleterious allele occurs at a low frequency, so segregation of polymorphic but deleterious alleles under weak selection also produces inbreeding depression in case of positive dominance. The dynamics of such polymorphisms are complicated because the current frequency is determined by mutation, selection, and random genetic drift (see Bürger 2000). The presence of deleterious alleles in any population may therefore account for the universal occurrence of inbreeding depression in outbreeding organisms. Of course, more special phenomena may give rise to similar effects, for instance one-locus overdominance in fitness, and

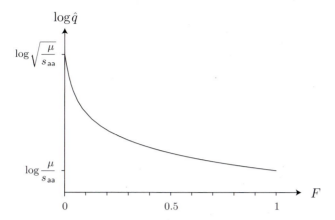

Figure 9.14: Equilibrium frequency $\hat{q}$ of a recessive deleterious allele in a population with general inbreeding coefficient F.

similar multiallelic or multilocus effects may produce inbreeding depression, but they are expected to be of minor importance (Lewontin 1974, Charlesworth and Charlesworth 1999).

9.4.1   Inbreeding in inbred populations

Many species, most notably plants, commonly allow consanguineous fertilizations as part of their mating system. In such species inbreeding exposes deleterious variation with positive dominance, and the deleterious alleles are thus subject to more intense selection. Their equilibrium frequencies should therefore be lower, as the influx of new mutants is not influenced by the mating system.

We may illustrate this by the dynamics of a recessive deleterious allele, which again has the simplest dynamics. The mutation–selection balance equilibrium $\hat{q}$ is the solution to the equation

$$s_{\mathsf{aa}}\big((1-F)\hat{q}^2 + F\hat{q}\big) = \mu,$$

where $s_{\mathsf{aa}} = 1 - w$ is the selection coefficient for the recessive phenotype relative to the normal phenotype ($w_{\mathsf{aa}} = w\,w_{\mathsf{AA}}$), and where F is the average inbreeding coefficient in the population. The expression for the equilibrium is a bit more complicated than that for $F = 0$ given in equation (8.16) on page 247. The change in equilibrium gene frequency as a function of F is shown in Figure 9.14. For $F = 1$ the effect of a deleterious mutation is expressed almost immediately, and the equilibrium value corresponds to that of a dominant allele in a randomly mating population (equation (8.15)). The equilibrium gene frequency of the deleterious allele is very sensitive to small deviations from random mating, and it decreases rapidly as a function of F, approximately as $\hat{q}(F) \approx \hat{q}(0) - \frac{1}{2}F$ for F small. For larger deviations we get the very robust approximation $\hat{q}(F) \approx \frac{1}{F}\hat{q}(1)$.

Inbreeding depression is again calculated from $W(F)$ in equation (9.4). A population, in which the general level of inbreeding is F, individuals with an inbreeding coefficient of f will show the inbreeding depression

$$W(f) - W(F) = -(f - F)\hat{p}(F)\hat{q}(F)(-w_{\mathsf{AA}} + 2w_{\mathsf{Aa}} - w_{\mathsf{aa}}).$$

For a rare recessive, the factor $\hat{p}(F)\hat{q}(F)$ decreases as $\hat{q}(F)$ decreases, which it does when F increases (Figure 9.14). The stronger effect of selection on deleterious homozygotes will lower the gene frequency in an inbred population, and the effect of a given increase in consanguinity will thus have but a smaller effect. The inbreeding coefficient of offspring from a consanguineous mating is $f(1+F)$, where the relationship is given by the consanguinity coefficient f (see Box 5). Such matings will thus only add $f - (1 - f)F$ to the inbreeding coefficient of offspring in the inbred population, compared to f in a randomly mating population. This again adds to the expectation that inbreeding depression is lower in an inbred population than in a comparable randomly mating population.

In Section 3.1.2 we discussed partially selfing plant populations and showed that the general level of inbreeding is $F = \frac{\alpha}{2-\alpha}$ in a population where the fraction α is produced by selfing and $1 - \alpha$ by random mating.

The average inbreeding coefficient of selfed offspring then becomes $\frac{1}{2-\alpha}$ and their average fitness is then

$$W_{\mathrm{self}} - W(F) = \frac{1 - \alpha}{2 - \alpha} \, \hat{p}(F)\hat{q}(F)(w_{\mathsf{AA}} - 2w_{\mathsf{Aa}} + w_{\mathsf{aa}}).$$

It is, however, more natural to relate W_{self} to the average fitness of outbred individuals, say, W_{non}, and for these we have $W(F) = (1 - \alpha)W_{\mathrm{non}} + \alpha W_{\mathrm{self}}$. Inserting this into the above, we get

$$W_{\mathrm{self}} - W_{\mathrm{non}} = -\frac{1}{2 - \alpha} \, \hat{p}(F)\hat{q}(F)(-w_{\mathsf{AA}} + 2w_{\mathsf{Aa}} - w_{\mathsf{aa}}).$$

With positive dominance the inbreeding depression will thus show as a lower fitness of individuals produced by selfing, and this difference persists as complete selfing is approached ($\alpha \to 1$).

The presence in plants of self-incompatibility alleles and other mechanisms of selfing avoidance may be seen as an adaptation to avoid the effects of inbreeding depression, and similar arguments have been used to explain incest taboos in animals (Bengtsson 1978). Alleles that promote a higher frequency of selfing are usually transmitted to a higher number of offspring because they have a higher transmission through pollen. The inbreeding depression among offspring counteracts this effect (for a recent review, see Charlesworth and Charlesworth 1998).

9.4.2 Hitchhiking and inbreeding depression

The rapid spread of the *carbonaria* allele in populations of *Biston betularia* in polluted habitats causes a segment of the original chromosome to be completely associated with the *carbonaria* allele. The *carbonaria* homozygote is therefore expected to suffer inbreeding depression and experience a fitness lower than that of the heterozygote that gives the same phenotype. The result is an episode of associative overdominance at the *carbonaria* locus. This effect is already suggested in the right-hand graph of Figure 9.8 on page 286. In fact, the slower than expected increase of the frequency of the *carbonaria* form by the end of the transient period was noted by Haldane (1924).

In this particular case, however, the expected rate of increase is hard to predict towards the end of the process. The decrease in the frequency of the recessive *typica* allele is very slow when it becomes rare (see Exercise 8.2.2 on page 261). In addition, one would expect the appearance of the *insularia* just before year 1900 to influence the pace of increase of melanic forms.

Inbreeding depression provides a general explanation for the results of the *Drosophila* experiments discussed in Sections 9.2.1 and 9.2.2 (pages 281 and 283). Laboratory stocks are usually highly inbred. The strains used as sources for the experiment were no exception, and despite elaborate crossing, a section of chromosome III containing the *ebony* locus is in complete association with the e and $+$ alleles originating from the original strains, and likewise, the w and w^{bl} alleles are in complete linkage disequilibrium with a surrounding block of chromosome I. Thus, we expect inbreeding depression in the homozygotes in both experiments with a resulting superiority of the heterozygotes, often referred to as *heterosis*. This heterotic effect produces the associative overdominance seen clearly in the beginning of the experiments.

This associative overdominance effect evidently depends on the average size of the piece of chromosome where homozygotes for the observed genetic variation are identical homozygotes. The average size of these pieces changes through the experiment. The size of the associated piece depends on recombination, but a recombinant chromosome should also be able to increase in frequency and therefore have a selective advantage. The dynamics of this process were analyzed by Christiansen (2000). The initial segment around allele e can only be broken down by recombination in the heterozygotes $e+$, and a recombinant chromosome will experience less inbreeding depression in the homozygote ee, but also less heterosis when present in genotype $e+$. The balance between these effects is hard to gauge, and the recombinant may as well be lost as assimilated into the population. However, in the *ebony* experiments the e allele occurs more often in $e+$ than in ee individuals, so the conservation of the heterotic effect may often win, retarding the breakdown of the block.

Once we get to the first overdominant equilibrium after 100 days (Figure 9.6 on page 283) calculations become a bit more straightforward. We shall for simplicity assume that only two chromosome pieces, marked by e and $+$, respectively, are present in the population. The two marker alleles then have the same average fitness ($W_e = W_+$). If the fitness effect of a piece broken off the block around allele e, say, relieves the inbreeding depression in genotype ee as much as the burden added to it by the genotype $e+$, then the recombinant is favored only if q, the gene frequency of allele e, is larger than p, that is, if $q > \frac{1}{2}$ (Christiansen 2000). In the experiments shown in Figure 9.6 on page 283 only recombinants of the $+$ chromosome are favored after the first 80 days. In the long run this effect will work to conserve inbreeding effects around the e allele relative to those around the + allele, but the blocks are still expected to grad-ually become smaller and selection therefore weaker. The final demise of the *e* allele reveals it as recessive and deleterious, and a residual block of inbreeding effects may add to the speed.

Inbreeding depression on the X chromosome is not expected to be as strong as it is on the autosomes. The effects of recessive deleterious genes are expressed in males. Their gene frequency is therefore much lower than the order $\sqrt{\mu}$ char-acteristic for autosomal recessive alleles and closer to the order of magnitude of μ characteristic of dominant deleterious alleles (equations (8.15) and (8.16) on page 247). Nevertheless, mildly deleterious genes are expected to accumu-late in laboratory stocks due to their extensive period of maintenance as small populations.

Inbreeding effects may also emerge in experiments with artificial selection on a character. In such experiments the population is usually carefully outcrossed, simply to ensure a reasonable response to selection. The population is finite, however, and the emerging linkage disequilibrium may produce inbreeding ef-fects (Robertson 1970). These may in turn give rise to phenomena like those seen in the *ebony* and *white* experiments and produce a balance between the artificial selection on the character and the fitness effects associated with its QTLs. This is an effective way of setting limits to selection, and a difference between the observed response to selection and that expected from observed covariance among relatives. Deleterious alleles hitchhike on the selected QTL alleles, and upon cessation of artificial selection, the obtained response will re-vert due to selection against these alleles. Such spontaneous regression of the character mean towards the state of the population before the experiment is commonly observed in experiments in *D. melanogaster* (for instance, Frankham et al. 1968*b*). Straightforward associative overdominance due to inbreeding ef-fects unrelated to the experiment may in itself stop the response to selection in small populations (Charlesworth 1991, Charlesworth et al. 1993*b*, Latter 1998). If the extent of linkage disequilibrium is sufficiently wide, then inbreeding ef-fects may be strong enough to overwhelm the effects of artificial selection, even while genetic analysis reveals plenty of hereditary variation. If arrest of the re-sponse to selection happens for this reason, then relaxation of artificial selection following regression may leave a population with a fairly constant phenotypic composition showing sluggish response to renewed artificial selection.

9.4.3   Inbreeding effects on a bristle character

Gilligan et al. (2005) studied the rate of loss of variation of bristle number in *D. melanogaster* due to inbreeding in small populations and compared it to the loss of genetic variation at electrophoretically defined loci. The laboratory pop-ulations were founded from a large population initiated from wild-caught flies and kept in the laboratory for several years. Their sizes varied from a few to 500 individuals, and variation was observed to decrease as a linear function of the expected identity coefficient of the population (calculated from equation (2.9) on page 31). The slopes of the lines were equal for the two characters and around $-\frac{3}{4}$. For neutral variation the equality of the slope is expected, but the expected value is -1, which is significantly steeper than the observed slope. The more sluggish response to inbreeding is probably due to associative over-dominance caused by inbreeding depression. Similar effects were observed by Charlesworth (1991) in a simulation study assuming small populations reproducing by partial selfing, and the observations in the *white* experiments point to similar expectations.

The genomic background of deleterious genes is expected to react to any disturbance of the population, thus causing hitchhiking affecting all other genetic variation present. *Any such disturbance of a population*, whether experimental or natural, *will thus cause responses in neutral variation that can only be interpreted as effects of selection.*

Kristensen et al. (2005) investigated the effect of inbreeding due to random genetic drift on the variation in bristle number in *Drosophila melanogaster* (see Section 9.3.1 on page 289). The experimental design may be described by referring to Wright's island model as applied in Section 7.3.2 (page 201). Two rates of random genetic drift were studied, corresponding to $N_e = 2$ and $N_e = 8$, and for each rate ten isolated populations were kept until the identity coefficient was expected to be $\frac{2}{3}$. The experiment was initiated by establishing 30 island populations with a large population size on each island. The founding population was established eight generations before the start of the experiment from wild-caught flies of diverse origin.

Ten of the populations were maintained as controls with a large population size for 23 generations. The 20 drifting populations were also maintained for 23 generations, initially with a large population size, but with a small size for the final generations to reach the target level of identity (18 generations for slow drift and 5 for fast drift). In generation 23, 104 pairs were collected from the offspring of each of the 30 island populations. These were mated, and the bristles on the parent male and two male offspring were counted, providing 104 father–son combinations to estimate components of the phenotypic variance in bristle number on each island.

The mean of the bristle counts was not found to differ in the three Wright's island populations, the control, the fast, and the slow inbreeding populations. The bristle character is known to have a very low dominance variance (see the figure on page 241), so this observation is in accordance with expectations (Section 7.3.2 page 201). The observed average phenotypic variance in the three Wright's island models is shown in the last row of Table 9.8. Random genetic drift decreases the variance, reaching the same value in the two small

Table 9.8: Variance components of a bristle character

N_e	large	8	2
V_A	0.75	0.29	0.18
V_R	1.03	1.13	1.25
V_P	1.78	1.42	1.43

populations. This is as expected, given the same identity coefficients (equal to $\frac{2}{3}$), and assuming the additive alleles model. The decrease in V_P should therefore reflect a decrease in V_A, but the observed additive variances for slow and fast drift are significantly different, and this difference is balanced by a difference in the residual variance V_R, which is interpreted as the environmental variance V_E of the character.

The variation in the environmental variance may reflect unwanted changes in the environment because of different population densities or increased sensitivity to the environment (Kristensen et al. 2005). That does not explain the simultaneous change in V_A, however. In the present context, we should note the increased effect on V_A due to the faster drift. Thus, we should probe into an explanation that involves speed or, in essence, time—unless we want to abandon the homogeneity of Wright–Fisher processes in characteristic time (Section 2.5 on page 41). Recombination seems a prime candidate. Random genetic drift causes loss of variation and a concomitant buildup of linkage disequilibrium. Linkage disequilibrium is eroded by recombination, so drift during a short time can build more intense linkage disequilibrium than an equivalent amount of drift spread over a long time interval. The ensuing "chunks" of chromosome will be subject to inbreeding depression and heterosis if variation is present. QTLs of bristle number therefore have a higher probability of hitchhiking on these effects for $N_e = 2$ than for $N_e = 8$, and at the same time, the associated fitness effects will be higher for the small than for the large population size.

The main effect of such hitchhiking is that the genotypes at the QTL loci will tend to show a deficit of homozygotes compared to the expected Hardy–Weinberg proportions. This is comparable to the effect of random genetic drift in the first place, the main difference being the sign of F_{IS} in equation (5.18) on page 135. Deviations from Hardy–Weinberg proportions in the local population produce the additive variance as $V_{AF_{IS}} = (1 + F_{IS})V_{A0}$ (calculated from the father–son covariance using the formulation in equation (5.18) on page 135), where V_{A0} is the usual additive variance assuming $F_{IS} = 0$. We thus expect to estimate a lower additive variance the more the genotypic proportions among the fathers deviate from the Hardy–Weinberg proportions due to heterotic effects ($F_{IS} < 0$).

The immediate conclusion is that the hitchhiking of QTLs on inbreeding effects can account for the qualitative impact of the pace of drift. Whether their quantitative impact suffices to explain the observed difference between large and small population size is a much more difficult question to answer, because heterosis interferes with the buildup of consanguinity. Other phenomena may possibly contribute to the observed effects. The dominance variance of bristle count is small, but it is not zero, and its influence is likely to be amplified due to the heterotic effects. Dominance effects will in the analysis of Kristensen et al. (2005) be relegated to the residual variation component and therefore interpreted as a contribution to the environmental variance.

9.5   Sweeps by Selection

Random genetic drift and other stochastic effects like mutation and recombination determine the dynamical unit of selection as different from the functional unit of selection defined by the molecular, physiological, and phenotypic properties of the information carried in the DNA of the gene and in its immediate vicinity. The dynamical unit is a metaphorical expression of linkage disequilibrium, and as such its size is determined by a balance between recombination and those phenomena that create associations among sequence variations along a chromosome. A stretch of maximal linkage disequilibrium is clearly a dynamical unit until recombination breaks it. The dynamical unit is thus a rather soft entity, and it may be illustrated as the shaded object below, emphasizing it as a distribution of the strength of association along the chromosome. In

human populations the lengths of the basic dynamical units are of the order of 100kb and mainly determined by recombinational hotspots (see Section 6.5 on page 177).

A mutation in a DNA molecule is in complete linkage disequilibrium with alleles in the meiotic product at all varying loci in the population. Recombination erodes this association to unlinked and loosely linked loci within a few generations, and the mutant therefore quickly becomes associated only with alleles at nearby loci. This stochastic source of linkage disequilibrium works equally in large or small populations and is simply a consequence of the random placement of a mutation in its genetic background. For instance, due to inbreeding depression we expect a more rapid fixation of the advantageous allele by directional selection when the locus is in a region of high recombination compared to the rate in a region of low recombination. Random genetic drift produces associations among closely linked loci in any population, emerging as a squared correlation roughly inversely proportional to the population size (Section 6.3.4). Recombination modifies this to appear as heterogeneity in coalescence along a sequence (Section 6.4). The effect of such random associations of an allele with its neighbors is to make the influence of its phenotypic effects on selection uncertain and unpredictable because of associative fitness effects. When the dynamical effects are spectacular, as when a gene causing a lethal disease is carried to high frequencies, they are ascribed to hitchhiking. This is not a strict definition, however, and hitchhiking is in general used as a synonym for local associative fitness effects (see Thomson 1977).

Evolution is driven by natural selection caused by the phenotypic variation extant in the population. The contribution to this variation is determined by variation in the sequences of functional genes, which are then the functional units of selection. The phenotypic effect of the alleles of a gene is shaped in functional interactions with the rest of the genome, its genetic environment. Selective effects of the gene are defined in terms of these phenotypic effects, and the associative fitness effects caused by linkage are merely noise in the process

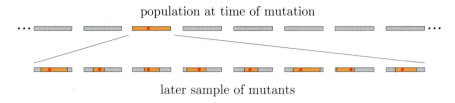

population at time of mutation

later sample of mutants

Figure 9.15: Detection of hitchhiking around a favorable mutant.

of Darwinian evolution. Selection may thus amplify the stochasticity added to the evolutionary process by random genetic drift, mutation, and recombination. Our discussions of the initial dynamics of a new favorable mutant (Section 8.5 on page 248) should therefore in an evolutionary context be extended to incorporate associative fitness effects.

Hitchhiking due to directional selection at a locus, which we may call hitchhiking *sensu stricto*, leaves a recognizable effect on the neighboring genomic region because variation is lost (Figure 9.15)—the gene genealogy becomes more shallow. In molecular population genetics this process is often referred to as a *selective sweep* (see Section 8.6.1 on page 259), because the effect is seen as a stretch of genome sequence with a low level of variation—the cause of the hitchhiking is rarely known. When the favorable mutation appears (red dot in the figure) it is in complete linkage disequilibrium with all polymorphic sites in the genome. Most of these associations fade quickly, except those on the chromosome where the mutant arose (orange). Sequences including the favorable mutation sampled shortly after, at a time when the mutant is still rare, will contain a section that is identical by descent to the homologous section of the original mutant chromosome (orange) surrounded by sections that by recombination are picked at random from the population (light shade). Analysis of this sample will reveal homogeneity next to the mutation (orange) and limited

variability in an extended stretch of the sequence (yellow). In the population the mutant will thus show absolute linkage disequilibrium with markers next to the mutant (orange area), and this fades out through the yellow area where variation among chromosomes exists. The extent to which this effect spreads out on the chromosome is determined by the local recombination frequencies. As the mutant becomes more frequent, recombination between chromosomes carrying the mutant becomes likely, and a sample may show a much more complicated variation in linkage disequilibrium, where the sections homologous to the original mutant chromosome may be noncontiguous. A similar pattern may

also be produced by gene conversion early in the sweep.

Maize (*Zea mays*) was domesticated from teosinte in a process that began

Figure 9.16: Nucleotide variation around the *tb1* gene in teosinte (green) and maize (orange). The abscissa shows the translated exons in red, the UTRs in black, and the intron in blue (see page 15). The measure H of variation is the expected heterozygosity calculated by formula (4.19) on page 99. Simplified after Wang et al. (1999).

five to ten thousand years ago. A major change is that the tassels in teosinte are placed at the tip of long branches, while those of maize have a short stalk. This character is determined by the locus of the *tb1* gene (teosinte branched). Wang et al. (1999) studied the effects of domestication on its sequence. In particular, they analyzed the nucleotide variation in a 2.9kb region containing most of the gene. Observations in varieties of maize grown in the Americas and in teosinte from Mexico are summarized in Figure 9.16. The amino acid sequences of the proteins in teosinte and maize are very similar, with no fixed differences, but the overall variation within the transcripts in maize is about 40 percent of that in teosinte—a magnitude that coincides with the differences observed in other genes. However, in teosinte the variation in the 5′ nontranscribed region is considerably higher than that within the transcribed regions, whereas the maize sequences show very little variation. Wang et al. (1999) interpreted this difference as an effect of a selective sweep due to a preference for a more effective harvest of short-stalked tassels during domestication, and they concluded that the morphological difference is due to a difference in gene expression mediated by a sequence motif upstream from the gene.

Nielsen et al. (2005) scanned the human chromosome 2 for selective sweeps using the HapMap resource (see Box 25 on page 178). The clearest evidence of a sweep was found around the *lactase* gene. Humanity is polymorphic in adult lactose tolerance and therefore in the ability to digest milk (Box 36). When the putative agent of the sweep is still polymorphic, the characteristic feature is a high frequency of a single haplotype—much like the Adh^F allele in *Drosophila melanogaster* (see Table 8.9 on page 260). Further discussions of methods and complications in genome scans for selective sweeps are offered by Sabeti et al. (2006) and Teshima et al. (2006).

Williams and Hurst (2000) provided evidence of more general effects of hitchhiking. The sequences of orthologous functional genes in the mouse and rat genomes were compared, and the synonymous and nonsynonymous codon differences counted. The data were expressed as the per codon evolutionary rates, K_s and K_a respectively. They found that loci situated with a distance of less than 1cM on the linkage map of the mouse are more alike in their divergence

---

Box 36: Variation in lactose tolerance

Lactose tolerance is the rule in milk-drinking cultures and rare outside these, and this differentiation has emerged since the Neolithic domestication of cattle. Lactose tolerance is a genetic trait, whereas dairy farming is cultural, and the evolution of lactose tolerance should therefore be addressed as an interplay of genetic and cultural transmission (Feldman and Cavalli-Sforza 1989). In Europe, for instance, a close correspondence exists between the geographical distribution of lactose tolerance and that of Neolithic archeological sites with evidence of cattle farming (Beja-Pereira et al. 2003). Tutsi populations in Africa show as high a frequency of lactose tolerant individuals as Scandinavian populations (*viz.* about 0.9). Tishkoff et al. (2007) investigated the genetics of tolerance in East Africa and identified a SNP associated with persistence of lactase in adults and two SNPs that showed weak association. One allele at each SNP marks tolerance, and these alleles occur in different characteristic haplotypes and are all different from the haplotype of the fourth SNP allele that is highly associated with lactose tolerance in Europeans. Lactose tolerance thus developed at least twice.

---

than loci further apart. Closely linked loci are rarely functionally related, so the cause of this observation must be their higher probability of association. The mean rate difference of protein evolution K_a between closely linked loci is lower than the mean K_a difference in each of 10,000 similar data assembled randomly from all the compared genes. The contrast for synonymous codons is less extreme, but still significant. The fraction $\frac{K_a}{K_s}$ of amino acid differences between mouse and rat, calculated within a functional gene, is also more similar for closely linked loci—again more extreme that any of those seen in the 10,000 random data. These are precisely the kind of phenomena expected for closely linked loci when hitchhiking is a common phenomenon. For Y or W chromosomes these phenomena are also observed but the heterogeneity is more extreme (Berlin and Ellegren 2006).

Deleterious mutation seems to produce associated fitness effects of a quite different nature in the phenomenon of *background selection* (Charlesworth et al. 1993*a*, Hudson and Kaplan 1995, Nordborg et al. 1996). Such mutants cause increased mortality or infertility, and they can therefore be viewed as a burden or a *load* on the population. The load is an expression of the mortality and infertility that causes the elimination of deleterious alleles. In this sense the dynamical unit may be viewed as if it was one functional gene, since the elimination of a deleterious allele in the unit causes elimination of associated genes. This causes a mortality rate of a gene equal to the deleterious mutation rate of all the genes in its particular unit (see Exercise 8.4.1 on page 262). The unit is of course rarely a fixed entity, but the mortality associated with a particular gene equals the sum of the mutation rates of itself and the neighboring genes weighted by the strength of the association.

This is an almost trivial assertion, but the interest derives from the variation in the size of the dynamical unit across the genome. A higher mortality rate means a smaller population size, all else being equal, and all else is equal because we are talking about the same population. For regions with a high deleterious mutation rate N_e is low because of the high mortality, and in ones with a low rate N_e is comparatively high. The dynamical unit, on the other hand, is large when recombination is low and small when it is high. Low recombination around a gene therefore means a high deleterious mutation rate, and vice versa for high recombination. At a particular locus with the neutral mutation rate of μ, genetic variation is given by a balance between mutation and drift characterized by $N_e\mu$. Thus, we expect to find more neutral variation in genomic regions with a high density of recombination events than in ones with a low density.

The dynamical unit may be measured in number of DNA bases or in number of functional genes. The deleterious mutation rate is a function of the number of genes, so the physical measure of unit size seems more appropriate, but then we have to count the number of genes on the sequence map in various regions of the genome. This should be possible, but such detailed data is not necessary: the initial observation of the effect of background selection was done on the X chromosome of *D. melanogaster* (Aguadé et al. 1989). At the tip of the X chromosome (around y and w, Figure 6.5 on page 177) we expect a low amount of recombination per base pair, given that distance on the salivary gland chromosomes is a simple reflection of distance in the DNA sequence. Therefore the area of linkage disequilibrium around a locus A is large. Further into the chromosome

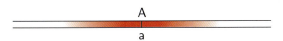

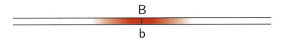

(around w and N) recombination per base pair is higher and the physical size of the dynamical unit around a locus B becomes smaller. Thus, if the densities

of functional genes are fairly constant, we expect many per dynamical unit at the tip of the chromosome and therefore comparatively low levels of variation at neutral loci in this region. This is precisely what Aguadé et al. (1989) observed. The phenomenon was later observed in other regions of the *Drosophila* genome (see Charlesworth 1996). It was also investigated in other species with results ranging from clear evidence to negative reports (see Kraft et al. 1998, Nachman et al. 1998, Wright et al. 2006). An alternative explanation of the correlation between recombination rate and genetic variation is that the rates of recombination and mutation are correlated (Hellmann et al. 2003)—recombination requires a fair amount of DNA synthesis and repair in the formation and resolution of the Holliday structure. Nevertheless, both background selection and higher mutation rate are expected to cause weak associative overdominance because higher association to recessive deleterious mutants is expected in regions with lower probability of recombination.

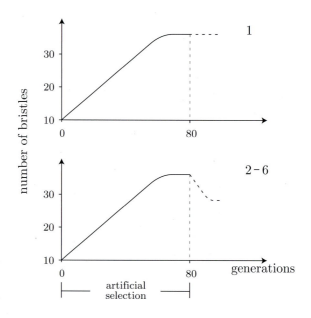

Figure 9.17: Sketch of results in Yoo's (1980a) experiments.

Hitchhiking and background selection affect variation at linked loci in similar ways. The probability that a neutral allele hitches a ride is likewise determined by the extent of linkage disequilibrium in its neighborhood. Both phenomena may cause correlation between recombination and variation in the way observed by Aguadé et al. (1989) and others. To separate their effects, note the more localized effects of hitchhiking, which should produce larger variance in the correlation than the general effects of background selection.

9.5.1   Selection on a bristle character

Selection on the number of bristles on a given segment of the abdomen of *Drosophila melanogaster* was described in Section 9.3.1 (page 289), and we can now discuss more aspects of such experiments based on our discussions of the genomic effects of selection. Data will be given by very simplified representations to facilitate focus on the generic trends seen in many similar experiments.

Figure 9.17 outlines the change over time in the average number of bristles in large laboratory populations where truncation selection favors a high number of bristles (in each generation the 20 percent flies with the highest number of bristles are selected to be the 50 parents of the next generation). The average number of bristles is shown for each generation through 80 generations of selection over the period marked "artificial selection" (the curves outline the course of the response, and the usual experimental fluctuations have been smoothed out). The populations show a typical response to selection. The average number

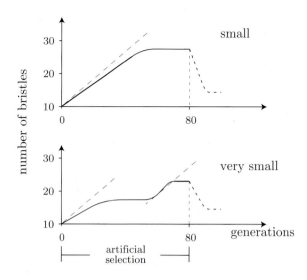

Figure 9.18: Sketch of results in smaller laboratory populations.

of bristles increases approximately linearly in the beginning, and by the end of the experiment the effect of selection ceases.

The responses to selection in these populations are as expected for a character reflecting variation in the loci that influence the character. A simple description of the course of the experiment is that selection of bristles induces selection on the genetic variation (see Figure 8.3 on page 218). This in turn changes the genetic composition of the population. The final lack of response may be caused by fixation of the alleles that cause a high number of bristles. If this simple description is correct, the cessation of selection after generation 80 will not cause a change in the average number of bristles in the population. Experiment 1 meets this expectation because the bristle number stays constant after selection stops (shown in Figure 9.17 as the dashed line continuing the response curve). This nice result was obtained by Yoo (1980a) in one of six experiments. The responses to selection in the other experiments are virtually indistinguishable from that in the first. When selection ceased, however, the average number of bristles decreased—the response reverted. This phenomenon is commonly observed in selection experiments and is evidently due to selection opposing the experimentally induced viability differences—often described as "natural" selection running counter to artificial selection. The reason is probably deleterious genes hitchhiking on QTLs that influence bristle number. In four of the laboratory populations showing reverted response, Yoo (1980b) showed the total frequency of recessive lethals on chromosome 3 to be strongly elevated (0.4–0.5) in generation 79. Experiment 1 showed the same total frequency of lethals as in the founding population (0.14), and the final experiment showed about double that.

Jones et al. (1968) conducted similar experiments in both large and small populations (see Figure 9.11 on page 290). In large populations their results resemble those of Yoo. Their results for smaller populations are outlined in the top drawing in Figure 9.18 in a way that allows immediate comparison to Yoo's results (the initial response in the large populations is indicated by the thin dashed line). This raises the questions asked in Exercise 9.5.1.

In very small populations the response to selection can deviate considerably from those seen so far (see, e.g., Thoday 1955). An example is shown in the lower drawing in Figure 9.18 (again drawn such that immediate comparisons to the previous drawings are possible). This raises the questions asked in Exercise 9.5.2. Yoo (1980a) saw similar phenomena in three of his experiments, although on a smaller scale. The renewed response is referred to as an *accelerated response*.

Exercises

Exercise 9.1.1 * Find the linkage disequilibrium at the high complementarity equilibria (Box 33) given they exist. Show that as r varies in the interval where the high complementarity equilibria exist, the linkage disequilibrium varies between 0 and $\frac{1}{4}$.

Exercise 9.1.2 Show that in the symmetric viability model (Table 9.7) the condition for stability of the monomorphic equilibria is $\alpha < \beta$, $\alpha < \gamma$, and $r > \alpha$.

Exercise 9.4.1 * Show that the contribution of a rare recessive allele to inbreeding depression in a population with a high inbreeding coefficient is approximately

$$W(f) - W(F) \approx -\frac{f - F}{F}\,\hat{q}(1)(1 - w_{\mathsf{aa}}).$$

The contribution is thus proportional to the relative increase in the inbreeding coefficient.

Exercise 9.5.1 Describe the differences in the response to selection in the large populations (Figure 9.17) and the small population in the upper diagram in Figure 9.18. Discuss possible population genetic explanations of these differences.

Exercise 9.5.2 Describe the differences in the response to selection in the lower diagram in Figure 9.18 and those in the upper diagram in Figure 9.18 and in Figure 9.17. Discuss possible population genetic explanations of these differences.

Chapter 10

Population Structure

Natural selection can vary with the local environment—the obvious example is selection on the morphs of *Biston betularia* in polluted and undisturbed forest. As a first approximation the changes in polluted forest populations may be understood by neglecting exchange of individuals among different habitats, but as the *carbonaria* allele becomes more common immigration may play a role. Surely, the incidence of *carbonaria* individuals in pristine forest must be higher than expected by mutation alone, and we may inquire about the properties of such an immigration–selection balance, which should resemble mutation–selection balance when the immigration rate is low. Associative fitness effects must, however, affect the dynamics of an immigrated *carbonaria* allele because individuals, not alleles, migrate and the observed allele thus becomes a marker of the genomes in a population of a polluted forest. In this example the main effect of selection is clearly to limit immigration from polluted areas to undisturbed forest because the dominance of the *carbonaria* allele immediately expresses the highly deleterious trait.

Selection will, however, interfere with migration whether or not selection has a direct influence on local differentiation. Random genetic drift will strongly influence the local frequency of recessive deleterious alleles. Differentiated populations may hence differ in the genetic makeup of their load of those. Hybrids may therefore on average show increased fitness because more of the deleterious alleles will be carried in heterozygotes, increasing the effect of immigration. On the other hand, as in the above example, immigration is diminished when hybrids show low fitness because they are not adapted to the local environment, an effect that is necessarily related to the phenotype. The selection effects on immigration are inherently genomic in that variants at every locus will be affected, and they are of particular importance when highly isolated populations, subspecies, and species are compared.

10.1   Polymorphism

Geographical structure has a fundamental influence on the possibilities for natural selection to maintain polymorphism in a population. In an unstructured randomly mating population we can study most aspects of the genetic impact of selection without reference to a population dynamic specification of how selection acts. That is, we need not worry about changes in absolute population size caused by changes in mortality and fecundity. Of course, a dynamical description becomes necessary if the characteristics of selection change as a function of the population density, but at any given point in time the changes in genetic composition of the population are mediated by current selection on current variation. Density-dependent selection has little impact on the qualitative aspects of the genetic dynamics (Christiansen 2004). On the other hand, in a structured population we compare the size of subpopulations in each generation even though this comparison is implicit in most models because they are formulated in terms of backward migration. Migration is in terms of number of individuals, while the genetic effect is fraction of immigrants.

Individuals may live in isolated patches during their development and still collect as one randomly mating population as mature adults. Insect larvae, for instance, often cover a limited area in search of food, but winged adults roam over large areas in search of mates and egg-laying opportunities. The main

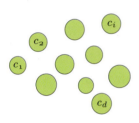

intraspecific competition can therefore occur in patches, which dynamically partition an otherwise randomly mating population. Levene (1953) analyzed a simple model of such a population for a locus with two alleles. Assume that the viabilities of genotypes AA, Aa, and aa in patch i are w_{iAA}, w_{iAa}, and w_{iaa}. The gene frequencies in the breeding population are p and q, and with Levene we will assume random mating, so that each patch is initiated by a random sample of offspring in Hardy–Weinberg proportions. The gene frequency among adults emerging from patch i is then

$$p_i^* = p\frac{W_{iA}}{W_i}, \quad W_{iA} = pw_{iAA} + qw_{iAa}, \quad W_i = p^2w_{iAA} + 2pqw_{iAa} + q^2w_{iaa},$$

and the gene frequency among the mature adults in the offspring population is

$$p' = \sum_{i=1}^{d} c_i p_i^*,$$

where $c_1, c_2, \ldots, c_d$ are the relative sizes of the contribution of adults from the d patches.

Even this simple model is quite complicated to analyze, and Levene therefore only analyzed the external stability of the two monomorphic equilibria. These are stable when

$$\sum_{i=1}^{d} c_i \frac{w_{i\text{Aa}}}{w_{i\text{aa}}} < 1 \quad \text{and} \quad \sum_{i=1}^{d} c_i \frac{w_{i\text{Aa}}}{w_{i\text{AA}}} < 1.$$

He interpreted this result in terms of the weighted harmonic means of the homozygote fitnesses relative to those of the heterozygote: a monomorphic equilibrium is stable when the harmonic mean of the relative homozygote fitnesses is larger than that of the heterozygote fitness, or

$$\left(\sum_{i=1}^{d} c_i \frac{1}{v_{i\text{aa}}} \right)^{-1} > 1, \quad \text{where} \quad v_{i\text{aa}} = \frac{w_{i\text{aa}}}{w_{i\text{Aa}}}.$$

If both monomorphic equilibria are unstable, the population is in a state of protected polymorphism where neither allele can be lost due to the action of selection. When both alleles are present the population is therefore expected to remain polymorphic, but random genetic drift may still engender loss of variation.

The surprise in Levene's result was that polymorphism could be maintained even when the probabilities of survival of the genotypes in the population[1] did not exhibit overdominance. Dempster (1955) recognized this result as due to a lack of effect of selection on the size of the demes. He contemplated a model where $c_1, c_2, \ldots, c_d$ are the relative sizes of the initial population in the d patches, and the contribution of adults in patch i is then

$$\frac{c_i W_i}{W},$$

where $W = W_1 c_1 + W_2 c_2 + \cdots + W_d c_d$. The dynamics of this model are entirely determined by the probabilities of survival of the genotypes in the population, and it is therefore a classical model of random mating and fixed viabilities.

The two models are widely used in population genetics to model extremes in density regulation of population size. The regulation in Levene's model is known as *soft selection*, and that in Dempster's model as *hard selection* (Christiansen 1975). They are not, however, simple extremes of population dynamic models (Walsh 1984, Christiansen 1985). Dempster's model is a simplified but general formulation of global population regulation. Levene's and Dempster's results, however, rely on a formulation in terms of fitnesses specified relative to the heterozygote.

The patches in Levene's (1953) and Dempster's (1955) models are not subpopulations in the sense of local breeding populations (Chapter 5). The only local phenomenon is selection, but to study selection in a population where the breeding population is also subdivided we need to consider limited migration among the patches as described by the migration matrix.

---

[1] By the law of total probability (A.6) these are equal to the weighted arithmetic mean of the genotypic viabilities.

10.1.1   Local and global effects of selection

A general model describing viability selection in a population subdivided into d local populations is very complicated, not least because of the many parameters required to specify the model even with only two alleles ($2d$ relative viabilities and d^2 immigration fractions). However, specific very simplified models may draw our attention to fundamental phenomena caused by the combined action of selection and immigration in interacting populations.

The simplest model assembled from local breeding populations is Deakin's (1966) model (5.8) on page 120. He originally formulated it as an extension to Levene's model,

$$m_{ii} = (1 - m) + mc_i^* \quad \text{and} \quad m_{ij} = mc_j^*, \; j \neq i,$$

in that the fraction $1 - m$ in each patch does not migrate. The relative size of subpopulation i at breeding is c_i, and this also reflects the relative magnitude of the number of offspring zygotes initiating the local population. The relative size c_i^* of population i at the time of migration reflects the influence of selection: soft selection is $c_i^* = c_i$ and hard selection is $c_i^* = W_i c_i$.

The condition for external stability of a monomorphic equilibrium is very simple in this model, and it is given by a series of local conditions and one global condition (Christiansen 1974). The local conditions make more sense when formulated as a condition for the initial increase in frequency of allele a when introduced as rare in a population monomorphic for AA ($\hat{p}_i = 1$, $i = 1, 2, \ldots, d$). With soft selection allele a increases in frequency in population i when

$$(1 - m)w_{i\mathsf{Aa}} > w_{i\mathsf{AA}}, \tag{10.1}$$

no matter what happens in the rest of the populations. Even when genotype Aa is deleterious in the rest of the population, allele a increases in subpopulation i when condition (10.1) is satisfied. The condition requires the viability of genotype Aa to be sufficiently superior relative to AA to allow the increase of the frequency of allele a when all immigrants are of genotype AA.

With soft selection the condition for external stability of the equilibrium $\hat{p}_i = 1$, $i = 1, 2, \ldots, d$, is, first, that the condition for local increase of allele a is violated in each deme, that is,

$$(1 - m)w_{i\mathsf{Aa}} \leq w_{i\mathsf{AA}} \quad \text{for all } i = 1, 2, \ldots, d. \tag{10.2}$$

These are augmented by a global condition:

$$\sum_{i=1}^{d} \frac{mc_i w_{i\mathsf{Aa}}}{w_{i\mathsf{AA}} - (1 - m)w_{i\mathsf{Aa}}} < 1. \tag{10.3}$$

The conditions (10.1) for local allele protection holds in general, in that for any migration matrix the frequency of allele a will increase in subpopulation i when $m_{ii}w_{i\mathsf{Aa}} > w_{i\mathsf{AA}}$. The simplicity of Deakin's model allows the formulation of the full external stability condition as just these local conditions and one global condition. If the opposite of condition (10.3) holds,

$$\sum_{i=1}^{d} \frac{mc_i w_{i\mathsf{Aa}}}{w_{i\mathsf{AA}} - (1-m)w_{i\mathsf{Aa}}} > 1,$$

then allele a is protected and increases in the population when at a low frequency. If equality holds in condition (10.3), as happens when allele A is dominant, then we need further investigations to decide whether the equilibrium is externally stable or unstable. With dominance the probabilities of survival of genotype aa have to be inspected—just as the direction of selection is decided by a comparison of $w_{\mathsf{A-}}$ and w_{aa} in an isolated randomly mating population.

With hard selection the conditions for external stability in Deakin's model are again given by equations (10.2) and (10.3) if we substitute the local relative viabilities of the homozygotes with

$$(1-m)\frac{w_{i\mathsf{AA}}}{w_{i\mathsf{Aa}}} + m\sum_{j=1}^{d} c_j \frac{w_{j\mathsf{AA}}}{w_{i\mathsf{Aa}}} \quad \text{and} \quad (1-m)\frac{w_{i\mathsf{aa}}}{w_{i\mathsf{Aa}}} + m\sum_{j=1}^{d} c_j \frac{w_{j\mathsf{aa}}}{w_{i\mathsf{Aa}}}$$

(Christiansen 1975). For $m \to 1$ the stability conditions approach that of Dempster, and for very small migration rates ($m \to 0$) they approach the one-locus conditions in isolated populations. For much migration hard and soft selection give rise to different conditions, but for low migration rates the distinction is unimportant. For two loci the external stability conditions of a monomorphic equilibrium in Deakin's model are modified in the same way as in a homogeneous population, and as expected, recombination increases the possibility for external stability of an equilibrium (Christiansen and Feldman 1975).

10.1.2 Transient polymorphism

The example of three-allele directional selection in *Biston betularia* in Figure 8.5 on page 225 may seem a bit odd with an *insularia* gene frequency of 0.01 and a *carbonaria* allele frequency of 0.001. The *carbonaria* allele emerged first. The reason is that even with a viability of the *insularia* morph of 0.9 relative to that of the *carbonaria* morph, it is hard to make a realistic example where *insularia* reaches appreciable frequencies (Figure 10.1).

An *insularia* mutant has little opportunity to increase to a frequency where it may be detected in a population where the *carbonaria* allele is already present. The most likely scenario is therefore that *insularia* emerged in a monomorphic *typica* population. Its probability of establishment and rate of increase are similar to those for the *carbonaria* allele. A rare but established *carbonaria* allele in an *insularia* population will eventually displace the *insularia* phenotype, but the process is considerably slower. In the time period where melanics replaced the *typica* moths in polluted areas, the two melanic types could well take over different local populations if dispersal was limited—even if they originated at quite different times. The rate of increase of melanic forms in a local population is fairly independent of the alleles present (Figure 10.2).

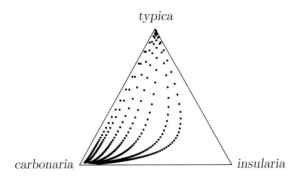

typica

carbonaria   *insularia*

Figure 10.1: Gene frequencies of the three alleles at the *carbonaria* locus in *Biston betularia* during a transient melanic polymorphism. The rightmost curve shows the situation in Figure 8.5 on page 225 starting with 0.01 *insularia* and 0.001 *carbonaria* alleles. All curves have the same initial *insularia* frequency. Initial *carbonaria* allele frequencies are 0.1, 0.05, 0.02, 0.01, 0.005, 0.002, and 0.001, from left to right. Explanation of the coordinate system in the illustration is given in Box 22 on page 153.

10.1.3   Migration–selection balance

The two-island model is the simplest model of the Deakin family, and it is convenient for exemplifying genetic phenomena in subdivided populations. Its parameterization is simplified as follows. The fraction of immigrants in population 1 is $m_1 = m_{12}$ and that in population 2 is $m_2 = m_{21}$, with $m_{11} = 1 - m_1$ and $m_{22} = 1 - m_2$. To be a Deakin model, $m_1 + m_2 < 1$. The recurrence equations with soft selection then become

$$p_1' = (1 - m_1)\frac{p_1 W_{1A}}{W_1} + m_1 \frac{p_2 W_{2A}}{W_2} \quad \text{and}$$

$$p_2' = m_2 \frac{p_1 W_{1A}}{W_1} + (1 - m_2)\frac{p_2 W_{2A}}{W_2},$$

where $W_{iA} = p_i w_{iAA} + q_i w_{iAa}$ and $W_{ia} = p_i w_{iAa} + q_i w_{iaa}$, $i = 1, 2$ are the average allele viabilities in the two subpopulations. The qualitative properties of this model were described by Maynard Smith (1970).

Suppose that population 1 is very large compared to population 2. We can approximate this by assuming that $m_1 = 0$ to get the classical island model. Population 1 is then isolated and we can assume that it is at the equilibrium $\hat{p}_1$, $0 \le \hat{p}_1 \le 1$. Further, assume directional selection in population 2 such that allele a is expected to fix had the population been isolated, that is, $W_{2A} \ll W_{2a}$. With these assumptions we can expect allele A to be present at a low frequency in population 2. For allele A rare we

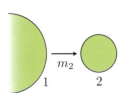

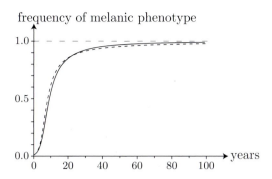
frequency of melanic phenotype

Figure 10.2: Example of the development of the frequency of melanics in *Biston betularia*. The solid curve shows the increase with three alleles, starting with 0.01 *insularia* and 0.001 *carbonaria* alleles (Figure 8.5 on page 225). The dashed curve shows the increase with only *typica* and *carbonaria* present, starting with 0.01 *carbonaria* alleles (Figure 8.4 on page 224).

get the approximate recurrence equation

$$p_2' \approx m_2 \hat{p}_1 + (1 - m_2)p_2(1 - s_2) \quad \text{where} \quad s_2 = 1 - \frac{w_{2\text{Aa}}}{w_{2\text{aa}}}$$

is the selection coefficient of the heterozygote relative to the common homozygote in population 2. This provides the *migration–selection balance equilibrium*

$$\hat{p}_2 = \frac{m_2 \hat{p}_1}{1 - (1 - m_2)(1 - s_2)} \approx \frac{m_2 \hat{p}_1}{s_2} \,, \tag{10.4}$$

where the last approximation holds for m_2 sufficiently small. The parameter m_2 expresses the level of "disturbance" caused by immigration. Unlike the situation for mutation–selection balance, we cannot a priori assume it to be small. This description holds even when population 2 is expected to be polymorphic due to overdominant selection in the absence of immigration. For m_2 sufficiently small we get

$$\hat{p}_2 \approx \tilde{p}_2 + m_2 \hat{p}_1 \frac{\tilde{W}_2}{1 - \tilde{W}_2} \,,$$

where $\tilde{p}_2$ is the stable equilibrium in population 2 for $m_2 = 0$, and $\tilde{W}_2$ is the corresponding equilibrium mean fitness.

Example 10.1.1 On the island suppose the one-locus multiplicative selection model where $w_{2\text{AA}} = w^2$, $w_{2\text{Aa}} = w$, and $w_{2\text{aa}} = 1$. Directional selection for allele a corresponds to $w < 1$. The mainland only supplies allele A, that is, $\hat{p}_1 = 1$. In this case the allele fitnesses simplify to $W_{2\text{A}} = w(wp_2 + q_2)$, $W_{2\text{a}} = wp_2 + q_2$, and $W_2 = (wp_2 + q_2)^2$. The balance equilibrium approximation (10.4) is now exact, namely

$$\hat{p}_2 = \frac{m_2}{1 - w} \,.$$

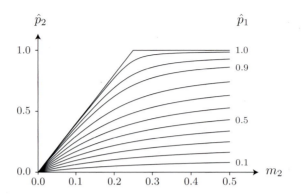

Figure 10.3: Gene frequencies at migration–selection balance equilibrium in a haploid two-island model ($w = 0.75$) with unidirectional migration, that is, $m_1 = 0$. The curves show $\hat{p}_2$ as a function of m_2 for values of $\hat{p}_1$ in increments of 0.1; curves for $\hat{p}_1 = 0.95$ and 0.99 are also included.

This equilibrium is valid and stable for $m_2 < 1 - w$. Otherwise the frequency of allele a keeps on decreasing and the only equilibrium is $\hat{p}_2 = 1$.

The dynamics of the selection model in this example correspond to selection in a haploid one-locus two-allele model. With polymorphism in population 1, $\hat{p}_1 < 1$, the qualitative behavior of the migration–selection balance equilibrium $\hat{p}_2$ as a function of m_2 is similar, although smoother (Figure 10.3). □

10.1.4   Environmental homogeneity

The study of the effect of selection when it varies among subpopulations is immediately appealing, but interesting phenomena emerge in a structured population even when the genotypic viabilities are the same in all populations. A simple 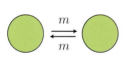 symmetrical model of such environmental homogeneity assumes that the genotypic fitnesses in both populations are given by $w_{\mathsf{AA}} = w_{\mathsf{aa}} = w$ relative to the heterozygote $w_{\mathsf{Aa}} = 1$, and we extend the symmetry to migration, $m_1 = m_2 = m$.

With overdominance ($w < 1$) the gene frequencies evidently converge to the stable equilibrium $\hat{p}_1 = \hat{p}_2 = \frac{1}{2}$ independently of the amount of migration. With underdominance ($w > 1$) the monomorphic equilibria $\hat{p}_1 = \hat{p}_2 = 0$ and $\hat{p}_1 = \hat{p}_2 = 1$ are, of course, stable, but more equilibria may exist. With no migration ($m = 0$) the subpopulations are isolated and monomorphic at equilibrium. The equilibria with $\hat{p}_1 = 0$ and $\hat{p}_2 = 1$ or $\hat{p}_1 = 1$ and $\hat{p}_2 = 0$ are therefore also stable. With a low amount of immigration the situation resembles that discussed in Section 10.1.3, and viewed from within a subpopulation the disturbance results in a slight perturbation of the population away from the stable equilibrium. Selection is therefore expected to produce a balance

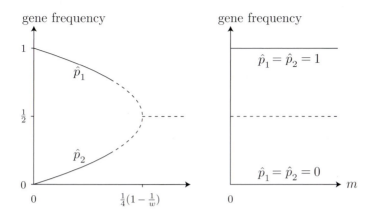

Figure 10.4: Equilibrium gene frequencies in a two-island model where selection is underdominant, symmetric, and the same in both subpopulations. The right-hand drawing shows the trivial homogeneous equilibria $\hat{p}_1 = \hat{p}_2 = 0$ and $\hat{p}_1 = \hat{p}_2 = 1$, which are locally stable, and the unstable equilibrium $\hat{p}_1 = \hat{p}_2 = \frac{1}{2}$. The left-hand drawing shows one of the heterogeneous antisymmetric equilibria. Solid lines indicate locally stable equilibria, dashed lines unstable equilibria.

equilibrium because the monomorphic equilibrium is stable without the disturbance. Immigration disturbs both populations in this way. The antisymmetric equilibrium with $\hat{p}_1 = 0$ and $\hat{p}_2 = 1$, stable for no migration, is therefore for small m turned into a pair of migration–selection balance equilibria, one in each population (Figure 10.4). As m increases the gene frequencies in the two populations become closer, and eventually the two migration–selection balance equilibria fuse with the unstable equilibrium $\hat{p}_1 = \hat{p}_2 = \frac{1}{2}$. However, the balance equilibria become unstable before m reaches the fusion point (Karlin and McGregor 1972, Christiansen 2000).

Any genetic system that allows multiple stable equilibria in an isolated population may produce such heterogeneity in a geographically structured population. Multiallelic loci under selection often allow this, as for instance in the four-allele model in Example 8.2.2 on page 232. This example corresponds to absolute linkage in the two-locus multiplicative viability model considered in Example 9.1.2 on page 273. The high complementarity equilibria, stable for tight linkage, may produce geographic heterogeneity in the sign of the linkage disequilibrium (Christiansen and Feldman 1975). In the four-allele version the model produces equilibria where both populations segregate four alleles, but at different frequencies. The two-island model produces genetic heterogeneity based on two simultaneously stable equilibria. In the same way a d-island model in a homogeneous environment may exhibit a geographically heterogeneous equilibrium based on up to d equilibria that are simultaneously stable in an isolated population.

Persistent genetic heterogeneity in a geographically structured population in a homogeneous environment is thus consistent with these models, but they do not explain the origin of the heterogeneity. The state of genetic homogeneity is also stable, so we are in need of a phenomenon that can produce heterogeneity. Random genetic drift is an obvious candidate if selection is weak, but the most spectacular and widespread heterogeneity in natural populations is karyotypic variation (see Section 10.2.1 below), which is often associated with quite strong selection. The local population size must therefore be small and migration rare—conditions that infringe on the permanence of the local population. The models we discussed in Chapter 5 are therefore no good; we need small ephemeral local populations. Such a model belongs to the class of *metapopulation models* originally introduced into ecology by Levins (1970) and Levin and Paine (1974) (see, e.g., Hanski and Gilpin 1997). The local (virtually isolated) populations undergo stochastic extinction and recolonization, and the colonization event is viewed as a *population bottleneck* where the population size is particularly small, giving an episode of strong genetic drift. In addition, the inoculum is a small random sample of the source population. In population genetics this process is known as the *founder effect*, and it is quite potent in a metapopulation. One of the best examples of metapopulation effects is seen in *Cepaea nemoralis*, where populations in homogeneous regions of habitat tend to vary in gene frequencies, but with constant gene frequencies within rather large areas (see Jones et al. 1977, Ochman et al. 1983). This phenomenon is called an *area effect*.

10.1.5   A butterfly metapopulation

Some populations of the African nymphalid butterfly *Acraea encedon* have a very low frequency of males. Others have a more even sex ratio. Chanter and Owen (1972) observed two types of females. In populations with an even sex ratio females produced offspring showing an even sex ratio. Those populations with a very skewed sex ratio in addition harbored females that produced only female offspring. Chanter and Owen suggested meiotic drive of the sex chromosomes in this kind of females as a reason (Box 37). The odd females were assumed to carry a driving W chromosome that caused all their egg cells to carry this chromosome. The homogametic sex would therefore never be formed. The only remaining question was to explain how *A. encedon* could persist while staying polymorphic for such a driver, because it is expected to fix in the population and thus drive it to extinction (Box 37).

Heuch (1978) provided an explanation by proposing that the butterfly occupies a collection of suitable habitats virtually isolated from each other. A population with the driver will eventually go extinct, and colonization of empty habitats is only possible by females carrying the "healthy" W chromosome. The driving W chromosome survives by rare infections of "healthy" populations. In this metapopulation model extinction is mediated by the genetic constitution of the local population.

The driving W chromosome acts like an infectious agent, and indeed, the maternally transmitted male-killing bacterium, the endosymbiont Wolbachia,

---

Box 37: Meiotic drive of chromosomes

The viability of the four products of the meiotic division may vary as a function of their genetic constitution— the survival may even be influenced by the tissue of the mother. This is an instance of gametic selection. In higher organisms, such viability effects are often seen as deviations from Mendelian segregation in heterozygotes, described as segregation distortion. A chromosome with a genetic element subject to strong directional gametic selection is said to be subject to meiotic drive.

Meiotic drive of an autosomal element produces directional selection as simple as gametic selection (see Exercise 8.2.1). The evolutionary dynamics of a driving element on an X or a Y chromosome are more complicated. With an uneven population ratio of the sexes a male that produces more offspring of the minority sex is expected to have more grandchildren (Fisher 1930*b*). The reason is that every offspring originates from a female and a male, and an individual of the minority sex will therefore on average have more offspring. The sex ratio should therefore be even—a parallel to the even frequencies of self-incompatibility alleles in plants (see page 257). This expectation is however valid only for autosomal genes that influence the sex ratio. For a driver on one of the sex chromosomes functioning in the heterogametic sex, the transmission of the gene is in addition influenced by the deviation from Mendelian segregation. It may therefore cause deviations from evenness of the sex ratio. Odd results may emerge. An absolute Y driver on the Y chromosome will cause the frequency of males to keep increasing in the population until it goes extinct for the lack of females.

---

is the cause (Jiggins et al. 1998, Jiggins et al. 2003). In butterflies a maternally transmitted pathogen and the W chromosome show the same pattern of inheritance.

10.1.6 Geographical variation in another butterfly

E. B. Ford (1971) for many years studied variation in the number of spots on the hind wings of the brown meadow butterfly *Maniola jurtina* (Satyridae) in British populations. Individuals possess between 0 and 5 spots.

In a population in East Cornwall most males had two spots, while females exhibited polymorphism with a wider range of variation (Figure 10.5). This sex difference is observed in most populations of *M. jurtina*. We will in the following only discuss the variation in females. The variation is heritable, but the genetic details are unknown.

M. jurtina is present on the Isles of Scilly off the coast of Cornwall, and Ford investigated the variation on three large and five small islands. The figure on the next page sketches the Scilly Isles, but omits several smaller islands. The distance to the coast of Cornwall is large compared to the distances between the islands.

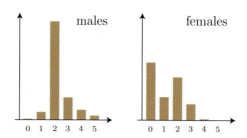

Figure 10.5: Variation in the number of hind wing spots in a population of *Maniola jurtina* in Cornwall. Abscissas show the number of spots, ordinates the frequency of individuals.

Data from the three large islands is shown in diagrams 1, 2, and 3 in Figure 10.6. The differences between these graphs are likely due to the limited number of individuals counted, that is, a statistical analysis leads to acceptance of homogeneity among the three distributions. In addition the distributions resemble the distribution among females collected in mainland Cornwall.

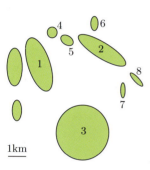

The distributions of spots vary a lot more among the five small islands (diagrams 4–8 in Figure 10.6). The differences between these graphs are larger than expected from the number of individuals in the samples. Hence a statistical analysis shows significant heterogeneity among these five distributions. The conclusion is thus homogeneity among the populations on the large islands and heterogeneity among the populations on the small islands. These observations raise many questions as outlined in Exercise 10.1.2.

10.2 Introgression

A low frequency of immigration disturbs an isolated population in much the same way as mutation when the variation of only one gene is considered. In the classical island model the distributions of gene frequencies for neutral variation are very similar. For instance, the mutation frequency μ is simply replaced by the immigration frequency m in Wright's formula (equation (4.6) on page 80

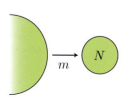

to obtain its migration counterpart in equation (5.4) on page 110). The one-locus mutation–selection balance and the migration–selection balance equilibria are very similar. The mutation rate μ is just replaced by the rate of immigration disturbance.

The whole genome is disturbed by immigration, not just the gene whose variation we happen to observe.

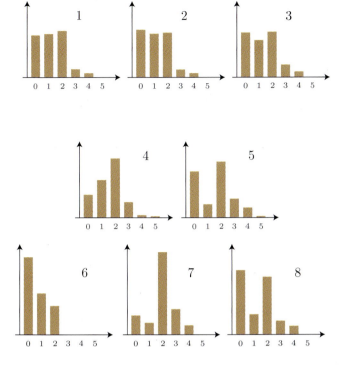

Figure 10.6: Variation in the number of hind wing spots among females of *Maniola jurtina* on the three large islands (1, 2, and 3) and the five small islands (4, 5, 6, 7, and 8). Axes as in Figure 10.5.

Mutation is a local change in the genetic material, and due to the initial high linkage disequilibrium the dynamics of the mutant allele are modified by its genomic neighborhood, and possibly vice versa. Immigration from a highly differentiated population amounts to a major disturbance referred to as an *introgression* of genetic material. The immigration frequency quantifies the disturbance equally for all autosomal loci, but differences in migration between sexes require special attention to variation at sex-linked loci. We therefore only consider autosomal loci in the following discussions.

The inclusion of more than one locus introduces stable monomorphic equilibria of another kind, namely those discussed in Section 9.1 (page 270). In these the effect of recombination hinders the increase of a rare advantageous gamete. If this gamete is recurrently introduced, then a migration–selection–recombination balance equilibrium emerges for sufficiently low immigration rate (Christiansen 2000). Hybrids between highly diverged populations usually have lower fitness than either parent, but occasionally plant hybrids are highly vigorous and fertile and therefore seem to be superior in fitness (Stebbins 1950, Grant 1963). Higher hybrid fitness is exactly what is expected at a migration–selection–recombination balance equilibrium, and the phenomenon is expected from multilocus theory (Zhivotovsky and Christiansen 1995, Pylkov et al. 1999).

Divergence of populations in a sexually reproducing species is a process that may ultimately lead to fission of a species and thus to *speciation*. Migration retards this process and has to be sufficiently low for significant separation to emerge. The ultimate test of whether speciation has occurred is to check if the two populations can coexist in the same habitat without losing their genetic integrity. The simplest mechanism to avoid introgression is a prezygotic barrier that prevents individuals of the two species from mating. Examples are plentiful: individuals do not recognize each other as potential mates, they are not in rut at the same time, they do not find mates in the same places, and technically they cannot perform coitus. In addition to such differences, closely related species or diverged populations within a species also show postzygotic barriers. These emerge after matings between individuals from different populations and show as decreased viability or fertility of hybrids and their offspring. If both pre- and postzygotic barriers are weak, then mating is possible in nature and hybrids are fertile. Introgression may then occur with the potential to disturb the divergence of the populations. This process is studied within a population genetic framework.

The basis of postzygotic barriers is genetic incompatibilities, also called genetic barriers. The archetypical postzygotic barrier giving low hybrid fertility is a translocation difference in the karyotype of the two populations, and such differences abound. A classical question is to what extent such genetic barriers contribute to population divergence.

10.2.1   Karyotypic variation

Karyotypes vary among species, even among subspecies, and karyotype changes may play a role in speciation (White 1978). A prime example is the subspecies of the house mouse (*Mus musculus*) in Europe. The standard karyotype of *Mus musculus* shows 20 pairs of very similar telocentric chromosomes, but isolated populations in southern Europe have a lower number. For instance, a population in the Apennines has only four of the telocentric chromosomes, but eight metacentric chromosomes not present in the standard karyotype. Such karyotypic races arise by local fixation of changes in the chromosomes.

Errors in the segregation of homologous chromosomes may occur during meiosis. This may lead to duplication or deletion of entire chromosomes in the resulting gametes. The phenomenon that two homologous chromosomes end up in the same daughter cell after the first meiotic division is called *nondisjunction*. Nondisjunction often has very detrimental effects for the resulting zygote because one of the chromosomes exists in either one or three copies rather than the usual pair in diploid organisms. Such a condition is called *aneuploidy*. In humans most such zygotes die early in development, often without noticeable signs of pregnancy, but some survive. Down's syndrome, for example, is caused by trisomy of the small chromosome 21. The lack of entire chromosomes is usually lethal for the zygote. In humans about half of all fertilized eggs die so early that they are not even recognized as abortions, and many of these zygotes are aneuploid or have other severe aberrations of the karyotype.

Sex chromosomes are special in this regard in that aneuploidy in these has less grave effects. Nondisjunction in males may produce XXY zygotes suffering from Klinefelter's syndrome or X zygotes suffering from Turner's syndrome. Klinefelter individuals develop into sterile males with feminized traits, and Turner individuals are sterile masculinized females.[2] Nondisjunction in females may also produce Klinefelter zygotes and, in addition, XXX zygotes that are females.[3]

Nondisjunction causes aneuploidy, but other errors influencing the transmission of chromosomes occur. Errors like base insertions or deletions during DNA replication may also occur at the chromosome level, resulting in duplication or deletion of a piece of the chromosome. Mispairing and crossover produce a duplication and a corresponding deletion.[4] This characteristic duplication is called a

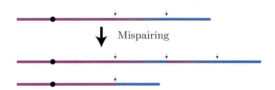

tandem duplication. Duplication of a large piece of chromosome has effects similar to aneuploidy, but some small duplications seem not to have any deleterious effects. For instance, populations of bank voles (*Clethrionomys glareola*) show variation in the number of genes coding for salivary amylase (Nielsen 1977), and *Drosophila melanogaster* has two closely linked genes coding for α amylase (Bahn 1968). The mammalian genes for salivary and pancreatic amylase seem to be a recent duplication, whereas the split between the α and β hemoglobin occurred in the common ancestor we share with fish (Figure 9.2 on page 267). Repeated duplications followed by differentiation of the genes produce gene families, which are evolutionarily closely related genes in the genome (see page 267). These are often referred to as paralogous genes to distinguish them from orthologous genes seen in comparisons between species (Fitch 1970, Koonin 2005).

Polymorphisms for small (submicroscopic) duplications and deletions are quite numerous in the human genome, and their frequencies vary among populations. Tuzun et al. (2005) compared the human genome reference sequence with a second genome using a technique capable of finding structural variants longer than 8kb. They found 139 insertions and 102 deletions. Redon et al.

---

[2]The presence of a Y chromosome produces a male-like appearance in humans. In *Drosophila* the number of X chromosomes is more important in that XXY flies are fertile females and X zygotes resemble males.

[3]XXX women have minor aberrations. The reason is that females only have one active chromosome. In the cells of normal XX females an inactivated X is present as a visible chromatin body, the bar body (see Chow et al. 2005). XXX women have two chromatin bodies in their cells. This condition is shared by all mammals.

[4]First described by Sturtevant (1925) in an analysis of recombination at the *Bar* locus of *Drosophila melanogaster.* The bar allele *B* is a small duplication on the X chromosome, and *BB* females produce a low frequency of wild-type male offspring by recombination.

(2006) surveyed the HapMap collection (see Box 25 on page 178) and found about 1500 structural variants covering a combined 12 percent of the human genome. The combined aberrations covered about 2000 genes, which was fewer than expected in a randomly chosen collection of sequences of similar size. Comparing these genes to those in OMIM (2006) revealed 285 loci for disease genes, suggesting a possible role for duplications or deletions in the expression of the disease.

The gene *PgiC2* coding for the enzyme phosphoglucose isomerase in the grass *Festuca ovina* is present as a tandem duplication (Ghatnekar et al. 2006), and it reveals a general phenomenon in the evolutionary destiny of tandem duplications of enzyme loci. Mutations that destroy the proper function of the gene are recessive or nearly recessive in many enzyme loci, because one functional gene is sufficient for the organism to function—to be homozygote for a nonfunctional allele is a highly deleterious condition. Such a mutation will have virtually no effect when hitting one of the two functional genes in the tandem, and the single-mutant tandem will therefore be a neutral, or nearly neutral, allele.

This situation was initially analyzed by Fisher (1935). The functional alleles in the two genes of the tandem are A and B, and the nonfunctional alleles are designated by lowercase letters. Only the genotype aabb is assumed to be deleterious. The mutation rates to a and b are μ and ν, respectively. The changes in gene frequencies of the functional alleles are

$$p'_A = \frac{1-\mu}{W} p_A \quad \text{and} \quad p'_B = \frac{1-\nu}{W} p_B,$$

where W is the average fitness that depends on the frequency and fitness of the double homozygote ab/ab. The change in allele frequencies therefore has the simple property

$$\frac{p'_A}{p'_B} = \frac{1-\mu}{1-\nu} \frac{p_A}{p_B} \quad \text{and therefore} \quad \frac{p_A^{(t)}}{p_B^{(t)}} = \left(\frac{1-\mu}{1-\nu} \right)^t \frac{p_A}{p_B}.$$

The functional allele with the highest mutation rate will thus ultimately be lost from the population, but the convergence is slow, that is, it proceeds at a rate on the order of magnitude of the difference between the mutation rates. Random genetic drift will therefore play a decisive role in the evolution of this system.

For $\mu = \nu$ the ratio of gene frequencies is not expected to change, and the population settles on a mutation–selection balance equilibrium dependent on the initial ratio of the gene frequencies p_A and p_B, and the frequency of recombination between the two genes (see Christiansen and Frydenberg 1977). This defines a continuum of mutation–selection equilibria stable for a given gene frequency ratio, but a random change in gamete frequencies will make the population adhere to an equilibrium possibly corresponding to a new gene frequency ratio. This drift of gene frequencies will rather quickly create a state of the population where one gene shows a typical mutation–selection–balance gene frequency and the other a polymorphism with the nonfunctional allele at a high frequency—exactly the situation observed by Ghatnekar et al. (2006) at the *PgiC2* locus.

Duplications of genetic material, from small local duplications over aneuploidy to duplication of a whole genome, seem to have played an important role in evolution—from prokaryotes to eukaryotes and further. The two versions of the genetic material can maintain the original function while also specializing in alternative tasks (Ohno 1970). The paralogous *Hox* gene clusters in mammals provide an excellent example of the duplication–specialization process, and the 12 gene *Rhox* cluster recently found in the mouse actually contains many more paralogs (MacLean et al. 2006). These are again expressed in testis and placenta. Sequence similarities suggest recent duplications with no changes in function.

Chromosomes may break, and a chromosome piece without a centromere is regularly lost during cell division, causing a terminal piece of chromosome to be deleted. Zygotes lacking large pieces of chromosomes will suffer phenotypic degradation like that caused by aneuploidy. However, broken chromosomes can be repaired with no further problems, but in case of multiple breaks errors in the repair may cause reproductive aberrations. Three types of error occur: deletion, inversion, and translocation. Such structural mutations may occur by spontaneous errors or they may be induced by environmental influence. Particle radiation is a particularly potent agent of chromosome breaks. Such breaks also occur in the soma—cancer cells are usually aneuploid and have deletions and duplications.

If two breaks occur in the same chromosome, a deletion may result because the orphan chromosome piece is lost. The effects of this deletion are indistin-

guishable from those produced by mispairing and recombination. A less deleterious result is that both breaks are repaired. If joined correctly, the trouble is over. If not, the result is a chromosome *inversion*. For most inversions the

chromosome is restored and no harm is done—except, of course, when one of the break points is situated within a gene, but this has a low probability.

A person carrying an inversion is thus commonly unaffected, and a homozygote for an inversion produces balanced gametes. Problems start when an individual heterozygous for an inversion breeds. The chromosome pairing of an inversion during meiosis looks as shown below. If a chiasma with crossover is

formed within the inversion, then a chromosome with two centromeres and an acentric chromosome are formed. The acentric piece of chromosome is lost.

The centromeres go to opposite ends of the cell during the first division and a chromosome bridge is therefore formed between the two putative daughter cells. The subsequent division of the cell breaks the bridge at a random point, leaving each cell with an odd chromosome with a terminal deletion and possibly a duplication. An individual heterozygous for a large inversion is therefore expected to produce few functional gametes. Zero crossovers within the inversion produce balanced gametes, as do two crossovers if both involve the same two chromatids.

The chromosome bridge is formed only by *paracentric* inversions, where the centromere is outside the inversion. In *pericentric* inversions, where the centromere is inside, meiotic pairing produces the configuration shown below. The

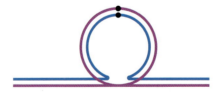

meiotic division proceeds normally, but any recombination within the inversion causes large duplications or deletions in the meiotic products. Again, only double and even numbers of crossovers within the inversion produce balanced gametes. An inversion heterozygote may thus be viewed as having lowered fertility. Inversion variation in a population is thus subject to underdominant selection, and fixation of one of the chromosome structures is expected. Variation in a geographically structured population is still possible, however (Section 10.1.4).

By comparing two human genome sequences, Tuzun et al. (2005) found 56 inversion breakpoints and estimated that most corresponded to small inversions (8–40kb; only inversions of length < 2Mb were considered). These inversions rarely cause meiotic problems because of the low probability of recombination within the inversions. *Drosophila* species are commonly polymorphic for large paracentric inversions—Theodosius Dobzhansky, in particular, made thorough investigations of these in *D. pseudoobscura* (see Dobzhansky 1970). These are possible because of the absence of crossovers in males. Female meiosis in higher animals produces only one gamete, the egg cell, and the three remaining products, the polar bodies, are nonviable cells with little more than a nucleus. The four daughter cells after a meiosis are initially placed in close proximity, more or less like in Figure 6.1 (on page 142), and a chromosome bridge is caught in the two middle cells, which always become polar bodies. The egg cell is therefore rarely disturbed by the aberrant meiosis and the inversion heterozygote has normal fecundity. This peculiarity of *Drosophila* has allowed the use of paracentric inversions as crossover suppressors in genetic experiments. Herman Muller championed these techniques in his investigations of mutation. The *Ba gl* chromosome mentioned in Box 35 is an example.

The effect of crossover suppression allows inversions to be advantageous when rare due to heterotic effects (Bengtsson and Bodmer 1976)—see equation (9.1) on page 270. Inversion polymorphism is thus likely in *Drosophila* species, but in general the heterotic effects in an inversion heterozygote could dwarf its fertility effects (for further discussions, see Kirkpatrick and Barton 2006).

Repair of chromosome breaks in nonhomologous chromosomes can create a new kind of structural mutation by joining the orphan pieces to the wrong chromosome. Such an interchange of chromosome pieces between two chromosomes

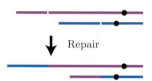

is called a *translocation*.

The translocation haplotype is balanced in that nothing is lost and nothing is duplicated. The translocation homozygote should therefore be indistinguishable from the normal homozygote. The translocation heterozygote is balanced in the

same way. The meiosis in the translocation homozygote is also completely normal and the four meiotic products carry a balanced haploid set of chromosomes. Normal meiotic chromosome pairing, however, is not possible in the translocation heterozygote because the homologous regions of one of the chromosomes are spread over two chromosomes. This gives rise to a characteristic quadrivalent pairing (Figure 10.7, left). In the meiotic reduction division (the first), the homologous centromeres segregate and the result is one of the four gametes

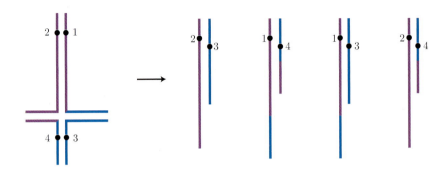

Figure 10.7: Meiosis in a translocation heterozygote. Left, profase quadrivalent; right, meiotic products.

shown in Figure 10.7. If centromeres 2 and 3 go to the same pole, the gamete carries the normal chromosomes and the complementary gamete (1 and 4) carries the translocated chromosomes. Both these gametes are balanced. If 1 and 3 go to the same pole, the gamete is unbalanced, with a large duplication of blue material and a large deletion of purple material. The complementary gametes (2 and 4) are equally unbalanced. Such unbalanced gametes are likely to convey strong deleterious, and often lethal, effects upon a zygote, and a translocation heterozygote therefore has low fertility, resulting in underdominant selection.

A particular kind of translocation is that observed in *Mus musculus* in isolated populations in southern Europe. Here two telocentric chromosomes participate in the translocation at the centromere, producing one metacentric chromosome (and maybe a very small chromosome that is subsequently lost). Such a translocation bears the name *Robertsonian translocation*, or centric fusion, referring to the apparent fusion of the centromeres. The opposite process is, of course, also called a Robertsonian translocation, or centric fission. The observation of local differentiation for translocations is consistent with our results in Section 10.1.4.

A quite different class of originally autonomous insertions exists having sizes between those of small insertions produced by local errors in DNA replication and larger duplications caused by recombination in mispairing chromosomes. Viruslike elements are quite abundant, and form a large part of the human genome, for instance the endogenous retrovira mentioned on page 267 and the insertions and deletions studied by Langley and Aquadro (1987) (page 18). Some, the *transposable elements*, can occasionally shift position in the genome—they transpose and are inserted at a new place. If a transposable element is inserted into a gene its function may be severely disturbed, often with detrimental effects. This shows as a deleterious mutation of the gene. Some elements can excise and carry an extra piece of DNA when they transpose. If this piece contains a gene, it appears as if having jumped to a new position on the linkage map.

Bacteria may also supply insertions in eukaryotic genomes. Insertions originating from endosymbiont Wolbachia has been found in insects and nematodes (Dunning Hotopp et al. 2007). In insects these bacteria may kill male offspring, as in *Acraea encedon*. Genes in the insertions may be transcribed even when the endosymbiont is not present in the host—a parallel to the expression of genes in old endogenous retrovira in the human genome (see page 267). Transfer of bacterial genes may therefore be a source of new functions in eukaryotic cells, given, of course, that germ line cells are transformed. This phenomenon is also seen for what is perhaps the oldest endosymbiontic bacterium in eukaryotic cells, the mitochondrion. It still functions as an autonomous entity within the cell, but a sizable fraction of genes important for its function now resides in the eukaryotic genome.

Whole genomes may be duplicated. A diploid species may then give rise to tetraploid individuals, in which cell nuclei have four homologs of each autosome. Tetraploid or other polyploid species are widespread in the plant kingdom, and Salmonid fishes show signs of originating from a tetraploid ancestor. Physical or chemical disturbances of normal meioses or mitoses may give rise to errors that generate such products.

Mitoses in polyploid individuals are problem free, whereas meioses are more complicated. Chromosome pairing occurs in the tetraploid meiotic prophase, but pairing does not progress from a unique place on the chromosomes, so the usual divalent pairing in diploids may turn up as quadrivalent pairing, in which local divalent pairing entangles all four homologs. The centromeres pair and disjoin in the first division, producing four balanced diploid cells, and so far reproduction proceeds happily. The trouble starts in mating. A newly arisen tetraploid individual is surrounded by diploids, and matings with these produce triploid zygotes, which are usually balanced and viable, but their meioses go haywire—paired centromeres disjoin, while singleton centromeres go to random poles. The result is unbalanced gametes that give rise to aneuploid zygotes. The only individual available for mating is thus itself, and selfing is indeed frequent in many plants. A single tetraploid individual may therefore be the source of a population of plants that is reproductively isolated from the diploid parent population, and by the usual definition the result is a new biological species. This is speciation by polyploidy—a form of speciation by saltation—really speciation by *autopolyploidy*.

The alternative is speciation by *allopolyploidy*, where the chromosomes originate from different species. The mechanism is here the formation of hybrids, which often have meiotic problems, followed by polyploidization, making a functional diploid with full chromosome complements from two species—again, saltatoric speciation by polyploidy. Examples of this are found in many plant genera. One is the genus *Brassica*. Among the many vegetables and crops of this genus, three species are polyploid and all three are allotetraploid (Table 10.1).

Table 10.1: Allopolyploidy in *Brassica*

Brassica species	haploid number	genome	cultivar examples
carinata	17	BC	Ethiopian mustard
juncea	18	AB	leaf mustard, brown mustard
napus	19	AC	canola, rape, rutabaga, swede
nigra	8	B	black mustard
oleracea	9	C	broccoli, cabbage, cauliflower
*rapa**	10	A	field mustard, oriental cabbages, turnip

*synonym *B. campestris*.

10.2.2   Genetic barrier

The basic model for studying genetic barriers is the classical island model. The focus is on the recipient population $\mathcal{R}$ which receives immigrants from a donor population $\mathcal{D}$. We study the fate of a small proportion of individuals in $\mathcal{R}$ immigrated from $\mathcal{D}$. To keep arguments simple we assume no prezygotic barrier, that is, immigrants mate at random within the recipient population. As they are rare, we may assume that they only mate with residents. All offspring of immigrants are therefore hybrids and we focus on these hybrids and their descendants—the hybrid swarm following the introgression. The population of hybrids is designated $\mathcal{H}$ and their origin may be symbolized by the cross $\mathcal{H} = \mathcal{D} \times \mathcal{R}$. The development of the hybrid swarm is followed for a while, still assuming it to be small and formed by backcross to the resident population:

$$\mathcal{H}_t = \mathcal{H}_{t-1} \times \mathcal{R}, \quad \text{with} \quad \mathcal{H}_0 = \mathcal{D}.$$

This description of introgression corresponds to the branching process description of the initial increase of a favorable mutant (page 250) and to any analysis of external stability of a monomorphic equilibrium in a randomly mating population.

   The donor population $\mathcal{D}$ is monomorphic for allele B and the recipient population $\mathcal{R}$ is monomorphic bb. Assume that the heterozygote Bb has the viability w in population $\mathcal{R}$ relative to that of the homozygote bb, and that $w < 1$. We follow a small fraction m of immigrants, and the initial frequency of allele B in population $\mathcal{R}$ is then $p = m$. All offspring of immigrants are heterozygotes because they mate with individuals of genotype bb, and the gene frequency among mature adults in the next generation is then $p' \approx mw$. These hybrids again mate with individuals of genotype bb and the next generation in the hybrid swarm therefore consists of $\frac{1}{2}$ heterozygotes and $\frac{1}{2}$ homozygotes bb. The homozygotes mingle with the residents, and only the heterozygotes—the "real" hybrids at locus B—are interesting. The ultimate result is a gene frequency in population $\mathcal{R}$ that decreases to zero as $p^{(t)} \approx mw^t$.

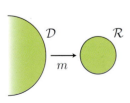

Bengtsson (1985) extended this argument to describe the genomic influence of the genetic barrier provided by the alleles B and b described above. He followed the two neutral alleles A and a at a linked locus. Population $\mathcal{D}$ is monomorphic for allele A and $\mathcal{R}$ for a. Allele B will eventually disappear from population $\mathcal{R}$, whereas the neutral marker allele A has the potential to stay.

Immigrants thus have the genotype AB/AB and first-generation hybrids are *cis* double heterozygotes AB/ab. These hybrids have the probability w of surviving to maturity. Their frequency in population $\mathcal{R}$ at breeding is therefore $2mw$ and the frequency of AB is $x_1' = mw$. The parental gametes produce offspring of genotypes AB/ab and ab/ab. The double heterozygotes therefore have the frequency $2(1-r)wx_1'$ among breeders in generation 2, and the frequency of the immigrant gamete is $x_1'' = (1-r)wx_1'$. Thus we have the recurrence equation

$$x_1^{(t+1)} = (1-r)wx_1^{(t)},$$

giving the frequency

$$x_1^{(t)} = m\big((1-r)w\big)^t$$

of gamete AB in generation t because $x_1^{(0)} = m$, and the frequency of the hybrid genotype in that generation is then $2m\big((1-r)w\big)^t$.

The double homozygotes have a resident phenotype, and we need not keep further track of them. The two recombinant types give offspring that behave like hybrids in one locus and residents in the other. The genotype aB/ab is a one-locus hybrid with respect to the locus with an effect on fitness. Its frequency decreases to zero as w^t, and it eventually disappears without leaving a trace in the population. The genotype Ab/ab, however, is like the resident ab/ab except for a neutral allele at the A locus. The gamete Ab occurs in the frequency $x_2'' = rx_1'$ in the second generation, and this contribution of Ab is expected to be conserved in population $\mathcal{R}$ in the future. Allele A has therefore introgressed into population $\mathcal{R}$. The fate of the offspring of the hybrids is therefore as described in Table 10.2.

Table 10.2: Segregation from hybrids

frequency	gamete	phenotype
$\frac{1}{2}(1-r)$	AB	hybrid
$\frac{1}{2}r$	Ab	introgressed
$\frac{1}{2}r$	aB	eliminated
$\frac{1}{2}(1-r)$	ab	resident

The description of the introgression of allele A is therefore fully covered by the description of change in the frequency of the immigrant gamete AB. The frequency of gamete Ab produced by the hybrid genotype in generation t is $rwx_1^{(t)}$, and summing these contributions provides the ultimate frequency of

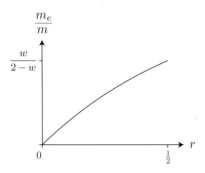

Figure 10.8: The reduction in migration rate at a neutral locus due to a one-locus selection barrier.

gamete Ab in population $\mathcal{R}$ as

$$x_2^{(\infty)} = rw \sum_{t=1}^{\infty} x_1^{(t)} = rw \sum_{t=0}^{\infty} m\big((1-r)w\big)^t.$$

The frequency of allele A in population $\mathcal{R}$ after this immigration event is therefore

$$m_e = m \cdot \frac{rw}{1 - (1-r)w}$$

(Bengtsson 1985). The immigrant frequency m will therefore result in the effective migration frequency m_e at a neutral marker locus with the recombination frequency r to the locus, which provides a one-locus genetic barrier between the populations. Recurrent immigration at a low rate m thus corresponds to recurrent immigration at the rate $m_e < m$. A barrier exists even for free recombination in that

$$m_e = \frac{mw}{2 - w},$$

and the effective migration rate decreases to zero for absolute linkage because at that point the selection effects reflect directly on the genetic marker (Figure 10.8).

An extreme model of a translocation difference between the populations has $w = \frac{1}{2}$. This gives $m_e = \frac{1}{3}m$ for unlinked loci, clearly a rather small reduction in the effective migration rate. One locus in general causes a barrier of minor significance, and introgression is markedly reduced only for very closely linked loci (Bengtsson 1985, Bengtsson and Bodmer 1976). Neighboring *Mus musculus* populations that differ by Robertsonian translocation lend support to this result. Britton-Davidian et al. (1989) investigated the genetic differentiation of such populations by surveying 34 loci by protein electrophoresis. They found very little differentiation between populations with different karyotypes and no evidence that the differentiation was larger than for comparable populations with similar karyotypes.

The isolation effect becomes stronger for a genetic barrier due to multiple selected loci. If both the donor and the recipient populations are monomorphic, some effects of linkage may be conjectured. If the marker locus is situated as loosely linked to all of the selected loci, we expect a situation much like the one-locus situation. However, a marker locus placed internally on a chromosome between selected loci needs at least two recombinations in order to be freed from the deleterious alleles of the donor population. The effect of linkage thus becomes a second-order effect, and close linkage has a much stronger effect on the effective migration rate than the almost linear effect seen in Figure 10.8. This was argued by Barton and Bengtsson (1986). Christiansen (2000) considered an example with variation at two loci B and C at a distance of 5cM subject to selection, as modeled by a two-locus symmetric viability model (Table 9.7 on page 278, with the viability of single and double heterozygotes equal and half that of the homozygotes). The effective migration rate for locus A is a function of its position relative to the two loci that constitute the genetic barrier. The rate is greatly reduced between the loci B and C, with the maximum m_e for locus A in the middle of the interval between the selected loci. This maximum value corresponds to that of a neutral locus placed outside the interval and about 0.1cM from one of the selected loci. The dependence of the effective migration frequency on an outside marker is similar to that in the situation shown in Figure 10.8.

More loci are thus much more effective in preventing introgression than a single-locus barrier. Of course, the magnitude of the effect depends on the linkage among the barrier loci. Loosely linked loci show very little interaction in their effect on m_e. Nevertheless, barrier loci will cause heterogeneity among neutral loci in their ability to introgress, and the more loci, the better the chance of observing such heterogeneities in a study of chosen marker loci. An example of such heterogeneities is found in the study by Szymura and Barton (1991) on the hybrid zone between the fire-bellied toads *Bombina bombina* and *B. variegata* (discussed in Section 6.3.2 on page 159). They provided evidence that alleles at the six marker loci indeed introgress across the zone and that the pattern of introgression is very heterogeneous. This heterogeneity is most likely due to close linkage between some of the markers and the barrier loci, implying a high number of such loci. The marker loci act like markers for QTLs that determine the hybrid fitness, and as for the "real" QTLs considered in Section 7.4 (page 202), the effect of the QTL on fitness and its distance to the marker locus are confounded in the determination of the association.

In their investigation of phosphoglucose isomerase in fescues (discussed in Section 10.2 on page 324) Ghatnekar et al. (2006) found *Festuca ovina* to be the only species having two loci coding for the cytostolic enzyme, whereas others had only one. The DNA sequences of the two, *PgiC1* and *PgiC2*, were found to be highly diverged and the two loci show independent assortment. Comparison to other species revealed that the *PgiC* gene of *F. polesica* in particular is closely related to the five alleles of *PgiC1* observed in *F. ovina* in Sweden. The functional and nonfunctional *PgiC2* genes of *F. ovina*, however, were found to be closely related to *PgiC* genes of species in the genus *Poa*, suggesting that this is their origin. This observation suggests a recent introgression, as only up to 10 percent of the plants have active genes at the second locus in southern

Sweden (Bengtsson et al. 1995). The introgression is limited, as plants lacking active genes do not have a related sequence at the *PgiC2* locus.

In a sample of sequences a clear sign of a very recent introgression is a very heterogeneous distribution of alleles at SNP sites—a number of rare alleles is expected to occur in strong linkage disequilibrium. The genealogy of such a sample will deviate clearly from Kingman's coalescent. The effects of this are numerous. For instance, we may elaborate on the examples on page 103, where θ is estimated in a sample of n sequences with $\mathcal{S}_n = 10$. If $n-1$ of the sequences are equal but differ from the last at 10 sites, then Tajima's estimator is smaller than Watterson's estimator. The immediate reaction would of course be that the sample was contaminated by a sequence of unknown provenance—a straightforward expression of introgression. As time passes the linkage disequilibrium degrades and the deviant pattern emerges as restricted genomic islands where strong linkage disequilibrium persists among rare alleles.

10.2.3   Introgression of a microcephalin allele

The divergence of the two microcephalin alleles considered in Section 8.6.1 (page 259) is very large. Based on a molecular clock rate estimated by Tang et al. (2002), Evans et al. (2006) judged the coalescence time for the D allele to be about 40,000 years, that for the non-D allele 990,000 years, and that of the whole sample to be 1,700,000 years. The coalescence time of the whole sample is thus a bit less than twice that of the most recent non-D allele, and the coalescence times of the two oldest sequences thus seem to be of the right order of magnitude. The surprise is really that the ancestral sequence where the D allele mutation arose is also closest to the common ancestor of the non-D allele.

Evans et al. (2006) argue that this scenario is highly unlikely and suggest that the D allele introgressed from another hominid line. The separation of this line from the human ancestral line should be sufficiently long to allow the supply of a gene separated from the common ancestral gene by about 2 million years. In this context introgression is reminiscent of the phenomenon, discussed in Section 5.1.2 on page 115, that a low probability of immigration from a "neighbor" population may occasionally produce very long coalescence times. When it transgresses the time since the populations diverged into separate species, immigration becomes introgression, and in the description of genetic relations the coalescent is replaced by phylogenetic descriptions of relations among species. In this transition population genetic analysis is replaced by analyses of molecular evolution, but during the specific population divergence the fields fuse into an analysis of the genetics of the speciation process.

The Neanderthal (*Homo neanderthalensis*) lineage diverged from humans between 0.3 and 0.8 million years ago (Green et al. 2006). The D allele thus seems to have introgressed from a hominid that was at least as diverged from humans as the Neanderthal lineage was. Modern humans expanded into the realm of the Neanderthals about 45,000 years ago (Mellars 1992), which happens to be around the time of the introgression.

Assuming that the D allele introgressed into *Homo sapiens* about 40,000 years ago, we may again ponder the explanation of its rapid increase. The introgression occurred around the time when our species expanded into Asia and Europe. This expansion happened along the border of the distribution because migrants from the presumably small settlements in the border area founded new populations. In this process the D allele might have been fixed in an area of the border close to a large area of suitable habitats. The expansion could in this way carry the D allele to a high frequency due to reasons entirely unrelated to its physiological properties—really a series of founder effects. This would look like selection for the D allele and amounts to a kind of hitchhiking on the dynamics of the human population that works for a neutral or even deleterious allele.

Is this a credible evolutionary anecdote? As a scientific hypothesis it should be tried against our knowledge of evolutionary dynamics. Assuming a neutral D allele, the observed pattern of haplotypes is explained entirely in terms of migration events within the human population. Migration events, however, involve the whole genome, and so we would expect to observe similar patterns at other loci. If we fail to do so, the hypothesis becomes an anecdote, or rather a tale of learning.

A related hypothesis could be that introgression was a recurrent phenomenon in an area of coexistence of two species of *Homo*. When introgression is presumed possible, this is not a radical extension. The trouble is, as before, that genomes introgress even though genetic barriers may limit the genomic extent of the introgression. The real trouble is, however, that introgression originates in a population virtually monomorphic for the high-frequency haplotype in allele D. The explanation is thus at variance with the observed diversity in the non-D allele.

Selection for the D allele seems to be the parsimonious explanation of Evans et al.'s (2006) observations, but whether the physiological effects of allele D are the cause of selection remains to be seen. The allele may merely act as marker for a closely linked locus. Yu et al. (2007) subjected variation in and around the microcephalin gene to further examination based in part on the HapMap data (Box 25 on page 178), and they found the extent of linkage disequilibrium around the locus to be smaller than expected based on the interpretations of Evans et al. (2006).

10.2.4 Vertical transmission of infection

Molecular genetic techniques may be used in the further illumination of the Wolbachia infection in *Acraea encedon* mentioned in Section 10.1.5 on page 318. Jiggins (2003) made a comparative analysis of variation in Wolbachia and mitochondria in individual butterflies and included the coexisting and closely related *A. encedana* in the investigation. Mitochondria are bacteria-like organelles in the cell cytoplasm that possess their own genomes; they are also maternally transmitted.

Two Wolbachia strains were designated Ug and Tz after Uganda and Tanzania, where they were first found. *A. encedana* had only Ug parasites, whereas both were found in *A. encedon*. A short mitochondrial sequence of 430 base pairs was sequenced for each butterfly. In *A. encedana* all had the same sequence, designated *Ug*. Infected *A. encedon* from Uganda and Ghana also had the Ug infection and the *Ug* sequence. A sample from Tanzania revealed both Ug and Tz infected butterflies. Those with Ug infection had the *Ug* sequence and those with Tz had a different sequence, designated *Tz*. Uninfected *A. encedon* were polymorphic, however, as seven new mitochondrial sequences were found in addition to one individual that carried the *Ug* sequence (the Wolbachia infection is rarely lost). The complete association between infection type and mitochondrial sequence suggests that the mitochondrion genome hitchhiked on the advance of the Wolbachia infection.

Jiggins drew the phylogenetic tree of the mitochondrial sequences and rooted it using two distant species from the genus *Acraea*. The split between the *Ug* and *Tz* sequences were shown to be significant, as is the one from the sequences in uninfected butterflies. Mitochondria in infected *A. encedon* are thus more closely related to those of *A. encedana* than to those of uninfected *A. encedon*. This suggests an introgression of infected *A. encedana* into

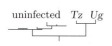

A. encedon populations. As expected, he found no indication of this in autosomal nuclear genes because as sons of uninfected females, males contribute half the genome of infected females through mating.

He was able to date the infection of *A. encedana* to within the last 16,000 years, thus the introgression and infection of *A. encedon* must be more recent than this. The origin and date of introgression of *Tz* are still to be determined.

10.2.5   Emerging barrier

Bengtsson's introgression model assumes specific genetic differences like inversions of translocations between the donor and the recipient population as a reason for the low fitness of hybrids. Zhivotovsky and Christiansen (1995) treated the process of introgression from a different perspective. We studied the intrinsic barrier against introgression of genetic material from the donor population $\mathcal{D}$ into the recipient population $\mathcal{R}$ due to genetic differentiation between the two populations. In other words, we studied the buildup of a genetic barrier rather than its effect. Introgression happens after immigration of whole genomes from the donor, and the intrinsic barrier should therefore be studied as a multiple locus problem.

We assumed that the recipient population was at a stable equilibrium with the average fitness $\hat{W}$ and that the performance of the hybrid population after t backcrosses was described by the average viability $W_{\mathcal{H}}^{(t)}$. Except for the trivial case in which the hybrids all die, the introgression corresponds to a small perturbation of population $\mathcal{R}$ away from its stable equilibrium state. At the loci subject to selection, the hybrid population therefore ends up resembling the recipient population, that is, $\lim_{t \to \infty} \mathcal{H}_t = \mathcal{R}$. The average viability of the hybrid descendants will ultimately be the same as that of the residents, $\lim_{t \to \infty} W_{\mathcal{H}}^{(t)} = \hat{W}$. In the limit the combined hybrid and recipient population will resemble the original recipient population with respect to the loci under selection. At other loci, however, the original immigrants will have left a trace in the population. To evaluate this we again added a neutral locus A to the model.

In general, the calculation of $W_{\mathcal{H}}^{(t)}$ is quite complicated, but with a marker locus unlinked to the selected loci and monomorphic in both populations, we got the recurrence equation for the introgression of the marker as approximately

$$p_{\mathsf{A}}^{(t+1)} \approx \frac{W_{\mathcal{H}}^{(t)}}{\hat{W}} \, p_{\mathsf{A}}^{(t)},$$

where $p_{\mathsf{A}}^{(0)} = m$ is the fraction of introgressed individuals in population $\mathcal{R}$. The final impact of the introgression across the selection barrier is therefore $p_{\mathsf{A}}^{(\infty)} \approx m_e$, where

$$m_e = m \prod_{t=0}^{\infty} \frac{W_{\mathcal{H}}^{(t)}}{\hat{W}}$$

measures the strength of the genetic barrier that retards the change of the gene frequencies in the recipient population at an unlinked neutral locus.

For a one-locus barrier the convergence of $W_{\mathcal{H}}^{(t)}$ to $\hat{W}$ is always monotonous and increasing. The mean fitness principle is not valid for two loci, and the hybrid swarm may for a while show a higher mean fitness than the resident population. An example was given by Christiansen (2000) using a simple two-locus symmetric viability model in which the monomorphic equilibria are stable for loose linkage. Suppose population $\mathcal{R}$ is monomorphic bc and $\mathcal{D}$ monomorphic BC. The immigrants and

	CC	Cc	cc
BB	1	0.5	1
Bb	0.5	1.5	0.5
bb	1	0.5	1

the residents then have the same fitness, but the hybrids have an elevated fitness (Figure 10.9). The mean fitness in the hybrid swarm subsequently decreases and finally converges to the mean fitness of population $\mathcal{R}$ at equilibrium. Even in this simple model the hybrids show heterosis followed by hybrid breakdown, a phenomenon reported by Stebbins (1950).

The simplest model of divergence assumes allopatric populations, that is, populations that do not exchange migrants. Isolation between the populations need not be complete, just low enough to allow divergence. Divergence does, however, occur among populations distributed over a sufficiently large area with ample migration between neighboring populations (Section 5.2.4). This causes so-called parapatric divergence. In many plant species such divergence is described under the name of outbreeding depression, in which hybrid fitness decreases with the distance between the two parents (Price and Waser 1979, Waser and Price 1994, Schierup and Christiansen 1996).

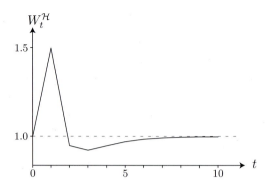

Figure 10.9: The change in mean fitness of the hybrid swarm in the simple symmetric model described in the text when a complementary gamete is introduced at a stable monomorphic equilibrium for $r = 0.4$.

10.3   Divergence of Populations

The results on genetic barriers are of practical importance for cultivars growing in areas within the range of the wild progenitor. If hybridization is possible, then gene flow between the two populations may impact both species, even though they retain their distinct appearance in those characters important for their different modes of reproduction and growth—in cereals, for instance, the ease of harvest of grains compared to the ease of dispersal in the wild population. The spread of cultivar transgenes has been of concern in such scenarios, and procedures for how to investigate such outcomes have been discussed. The conclusion based on the results in this chapter is that introgression of the transgenes is possible and likely unless it conveys detrimental properties to wild carriers or is closely linked to genetic elements that give rise to such effects. It is difficult to test whether introgression can proceed. With recurrent immigration introgression can on the one hand proceed even in the face of a fairly strong barrier (Section 10.2.2), while it may be unlikely on the other even with fairly vigorous hybrids (Figure 10.9). The results of the experiments with *ebony* and *white* (Section 9.2.1 on page 281) complicate the interpretation of experiments even further because the observed performance of the transgene may bear little relation to its physiological effects, and hence to its long-term persistence in the natural population. In general the quantification of genetic barriers is difficult when focus is on the possibility of introgression of particular genetic elements.

Analysis of neutral molecular markers from populations that have been through a long period of hybridization and introgression may reveal marked heterogeneities in divergence among loci. In phylogenetic analyses of species the effect may even give inconsistencies, depending on the molecular markers used. Heterogeneities may show as differences in divergence rates as well as in the topology of the phylogeny. These will persist until the species have diverged sufficiently in complete reproductive isolation. If the period of mutual introgression is long enough, however, the heterogeneities in the molecular evolution of the species may persist for a long time.

Analysis of genetic variation within one species may show the effect found in Section 5.1.2 on page 115 for low migration, namely that most loci show a fairly normal coalescence time of the sample although an occasional locus shows a very deep coalescent. This phenomenon is a signature of migration between populations that are highly differentiated due to a long period of isolation from one another—or a sign of introgression as in *Festuca* (page 324).

Introgression of genetic material occurs despite genetic barriers. We measured their magnitude by considering neutral genetic markers. Our description of such introgression, however, immediately extends to an approximate description of the introgression of almost neutral variation (see Box 32 on page 251). Genetic elements that confer advantageous properties to individuals in the recipient population will have the potential to increase in frequency in their own right, but the conditions for initial establishment in the population will not be much different from those of a neutral element (see page 250). Among the candidates for such introgression are self-incompatibility alleles (see page 256). An allele absent in a population may introgress from other populations or even species and subsequently establish itself. Some aspects of the phylogeny in Figure 8.18 on page 257 may be interpreted in that way. The alleles of *Physalis crassifolia* are mainly present in one homogeneous clade (left), with four alleles in a distant but homogeneous clade (right) in the *Solanaceae* tree. The small clade might then originate  from an ancient introgression. The alleles of *Solanum carolinense* are more entangled in the tree among those of the remaining species. This may suggest either repeated mutual introgressions or transspeciation polymorphism, or most likely a combination of the two.

The biological interest in introgression hangs in the balance of speciation: free exchange of genetic material given the opportunity means conspecificity, whereas no exchange means specific difference. Between these brackets is a continuum of possibilities from minor obstacles to gene flow between mildly differentiated populations to hybridizations that rarely cause introgression. Right in the middle sits the hybrid zone between the fire-bellied toads *Bombina bombina* and *B. variegata* (see pages 159 and 333) which, as the names indicate, are considered "good species." On the other hand, good species may be shown to be an assembly of biological species, that is, the taxonomic species dissolves into a number of mutually infertile populations. A good example is found in Scheel's (1968) investigations of West African killifish (Cyprinodontidae). The two fire-bellied toads probably diverged as isolated populations in allopatry. Their subsequent contact forms one of many European hybrid zones that originated in the Pleistocene glaciation, which caught much of the biota in isolated refugia. These hybrid zones thus expose the incomplete level of isolation between the species along the contact of their ranges. The apparent stability of the contact zone suggests a low fitness of hybrids because it amounts to interference competition between the two species, causing limited overlap of the species ranges (Christiansen and Fenchel 1977). This phenomenon may also emerge when isolation is complete, but infertile interspecific matings do occur, for example the forming of precopula among coexisting *Gammarids* (Fenchel and Kolding 1979).

For the population geneticist these phenomena emerge as the questions: what is the biological unit of study? what is the demarcation of the population being sampled? Failure to ponder these questions may produce strange data and interpretations. The questions persist even in well-designed investigations because the study of current variation refers to the evolutionary past. In classical genetic investigations of variation the evolutionary aspect emerged as soon as different populations of the same species were compared. As the studied variation by molecular methods approached that of the gene, the evolutionary past gained more presence even in observations on one population. This was true for observations of selectively neutral variation, in particular. The observed patterns of electrophoretic variation in enzymes was interpreted as the result of an evolutionary process. It was thus accorded a historical explanation, which is quite respectable as long as the background for the explanation is currently observed phenomena and processes—the so-called uniformitarian principle.

Data on variation in DNA sequences provide a clearer and deeper view into the past. Inferring the genealogy of sequences immediately raises the question: do the present circumstances of the sampled population reflect those of the ancestral populations revealed in the tree? The answer to this question is implicit in the data because we may ask whether the observed genealogy is a reasonable realization of the coalescent with mutation. Coalescent-based methods thus allow inference on historical populations, and hence on the evolutionary history of the sampled population. The biological unit of study is therefore the populations containing the line of descent of our sample.

For highly diverged populations or species, phylogenetic methods provide an impression of the genealogy of sampled sequences and may ask the same question about the ancestral populations described in the tree. The ancestral populations around the time of a specific split are particularly interesting. Hobolth et al. (2007) analyzed the genetic relationship of human, chimpanzee, and gorilla using four homologous autosomal sequences covering a total of 1.9Mb from these species and the orangutan, which is used to root the human–chimp–gorilla tree. During evolution the gorilla and the ancestor of human and chimpanzee first split into separate lineages, so the relationship of the three species is as shown to the left. This tree may be described as ((HC)G) using a coalescent-like set of symbols (Section 2.6.2 on page 46). Hobolth et al. also found this relationship to be generally supported by the sequence data. Assuming that divergence from the orangutan line began 18 million years (My) ago, they estimated the time since divergence of human and chimpanzee to be about 4My and that from the gorilla to be about 6My. The effective population size of the human–chimp ancestor was estimated to be about 65,000, compared to 10,000 estimated from extant variation in the current human population.

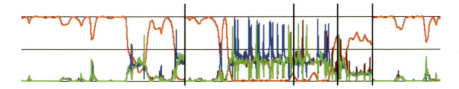

Figure 10.10: Results of the analysis of the genetic relationship of human, chimpanzee, and gorilla along a 50kb interval of the analyzed 1.9Mb sequence—for explanation see text. The curves show the posterior probability of the genealogies ((HC)G)–red, ((HG)C)–blue, (H(CG))–green, and (((HC)G))–dark red. Horizontal lines show probabilities 0, $\frac{1}{2}$, and 1, respectively. Vertical lines show places where an indel of more than 50b interrupts the analysis. From Hobolth et al. (2007).

However, Hobolth et al. found that the ((HC)G) tree was not the most likely tree for the entire 1.9Mb of the sequence. They considered the alternatives ((HG)C) and (H(CG)) and expressed the relative likelihood of the three alternatives in terms of Baysian posterior probabilities. Figure 10.10 shows these for a 50kb interval illustrating the alternating support for the three alternative genealogies. The ((HC)G) relationship (red curve) is strongly supported in long stretches where its posterior probability hovers at or just below one. These are interrupted by segments where support for it becomes low or vanishes entirely. Within these segments the stretches of strong support for any of the alternatives are short.

The human genome thus contains traces of human evolution going back across the split of the human and chimpanzee evolutionary lines. The distribution of intervals with different genealogies is, of course, subject to the population genetic processes we have considered. Recombination in particular shapes the properties of the genealogical heterogeneity. The human–chimp comparison reveals the structure of the divergence between the two species since their mutual isolation. The divergence of a sequence from each species can in two tempi be illuminated by our discussions of the coalescent of a sample of two sequences, allowing for recombination (Section 6.4 on page 161). First, the ancestors of the sampled sequences at the time of speciation are numerous and split up by recombination in the human and chimpanzee lineages. This amounts to a sample of two versions of the 1.9Mb sequence in the ancestral (HC) population spread over a great many individuals. The stretches of high support for the evolutionary tree correspond to segments where the human- and chimpanzee-derived versions coalesce before the root of the evolutionary ((HC)G) tree. Random genetic drift and recombination finally collect these ancestral segments into the intervals of the observed sequences that emerge as high support for the genetic ((HC)G) tree in the analysis.

The stretches of low support for the evolutionary tree correspond to segments where the human and chimpanzee sequences coalesce after the root of that tree. Thus, at the split of the human, chimpanzee, and gorilla lineages we have a sample of three versions of these parts of the 1.9Mb sequence. Studying the coalescence of this sample provides three immediate alternatives, namely ((HG)C), (H(CG)), and the coalescence of human- and chimpanzee-derived material first—this event we designate (((HC)G)). A parsimonious assumption is that these coalescent events are equally likely anywhere in the relevant parts of the sequence. Their likelihood hovers around a third of the probability of not ((HC)G) with prominent spikes of high support for a single of the three alternatives (Figure 10.10). The ((HG)C) tree in particular is pointed out in many intervals in the figure, and the support for this and the (H(CG)) tree shows strong negative correlation. The analysis therefore separates the alternatives rather well, but the background probability of about $\frac{1}{3}\left(1 - \text{Prob}\{((HC)G)\}\right)$ points to a lack of discriminating power.

The length of the intervals supporting the ((HG)C), (H(CG)), and (((HC)G)) genealogies are considerably shorter than those supporting the ((HC)G) genealogy. They are expected to be shorter because of the longer time available for recombination—in excess of 16My compared to 8–12My in the ((HC)G) genealogy. Even a factor of two in the time frame is not enough to produce the contrast seen in Figure 10.10. The reason is that the expected time for recombination to halve the ((HC)G) intervals is about double that of producing them (see the discussion on ancestral segment size on page 166 in Section 6.4 and the simulation results of Hobolth et al.). Mutation also breaks association tracts, and exclusion of changes associated with hypermutable CpG motifs (see Section 4.1 on page 72) do indeed produce longer support intervals in the results of the data analysis (Hobolth et al. 2007). A full understanding of the results of this population genetic analysis of phylogenetic relationships requires further theoretical insight and a deeper understanding of the properties of the tools of data analysis. The analysis does, however, convincingly illustrate the difference between the average divergence time between genetic markers and the time since speciation occurred. The former should be the longer—depending, of course, on the extent and prevalence of subsequent introgressions.

Molecular genetic variation is thus a powerful tool in the study of the transition between separated populations and species—the evolutionary process of cladogenesis. Phylogenetic analysis and molecular clocks are widely used in discussions of the systematics of extant biota. In addition, genome analysis provides windows to the genetic processes in the populations of common ancestors. In this way molecular population genetics may contribute to turning the spectrum of isolation levels described by Ehrlich and Raven (1969) into a description of cladogenesis that strictly adheres to the uniformitarian principle.

Exercises

Exercise 10.1.1 In the scenario of Example 10.1.1 consider the additive model where $w_{2AA} = 1-2s$, $w_{2Aa} = 1-s$, and $w_{2aa} = 1$, where $s \leq \frac{1}{2}$, and again suppose $\hat{p}_1 = 1$. With directional selection find $\hat{p}_2$.

Exercise 10.1.2 Discuss various explanations for the phenomenon described in Section 10.1.6: similar large islands, but different small islands. Try to use the characteristics of the observed distributions to discriminate among these explanations.

Appendix A

Probability Theory and Statistics

Mendel's law is formulated in terms of probabilities of segregation and of obtaining the various types among offspring. Probability theory is, and always has been, an important tool for the formulation of genetic rules and regularities.

The probability that an F_1 individual from the cross $AA \times aa$ is a heterozygote is one. This may be written as $Prob(Aa) = 1$, if the context is evident, and if not, we may write $Prob(Aa|F_1) = 1$. In these terms we can express Mendel's law as $Prob(Aa|F_1) = 1$, $Prob(AA|F_2) = \frac{1}{4}$, $Prob(Aa|F_2) = \frac{1}{2}$, and $Prob(aa|F_2) = \frac{1}{4}$, if we really want to be dry and formalistic. The event of an offspring of the dominant type is then

$$Prob(A\text{-}) = Prob(AA \text{ or } Aa) = Prob(AA) + Prob(Aa) = \tfrac{3}{4},$$

an expression of the general rule that the joint probability of two disjoint events equals the sum of the probabilities of each of the two events. Two events are disjoint when only one or the other may occur.

Two events are *independent* when the knowledge that one occurred does not influence the probability that the other occurs. In Mendelian segregation random combination of oocyte and pollen is expressed as

$$Prob(AA \text{ offspring}) = Prob(A \text{ in oocyte}) \times Prob(A \text{ in pollen}).$$

In general two events $\mathcal{A}$ and $\mathcal{B}$ are independent when the probability that both occur is the product

$$Prob(\mathcal{A} \text{ and } \mathcal{B}) = Prob(\mathcal{A})Prob(\mathcal{B}).$$

A.1 Discrete Stochastic Variables

Mendel's law assigns probabilities to phenotypes among individual offspring of a cross, and we validate the law by observing the number in each class. Among

n observed F_2 offspring we expect that $\frac{3}{4}n$ are A- and $\frac{1}{4}n$ are aa. According to Mendel's description the genotype of each offspring is determined independently, so the probability of first observing i A- and then $n-i$ aa offspring is $(\frac{3}{4})^i(\frac{1}{4})^{n-i}$. Alternatively, we might first observe an aa, then $i-2$ A-, then $n-i-1$ aa, and finally two A- offspring. These two alternatives have the same probability, and in both cases we observed a fraction $\frac{i}{n}$ offspring of type A-. The two observations are the same in terms of the number of A- individuals observed, so we get the probability of observing i of the dominant type by summing the (equal) probabilities for all the possible orders in which we encounter them. We observed n individuals. Among these, n may be chosen as the first, and given that choice, $n-1$ possibilities exist for the second, and so on. That is, we can order the n individuals in $n(n-1) \times \cdots \times 2 \times 1$ different ways. For any given order, i of the individuals are of type A-, and these may be ordered in $i(i-1) \times \cdots \times 2 \times 1$ different ways without changing the order of types. In the same way the $n-i$ individuals may be reshuffled in $(n-i)(n-i-1) \times \cdots \times 2 \times 1$ different ways. We may thus get exactly the same observation in

$$\frac{n(n-1) \times \cdots \times 2 \times 1}{\left(i(i-1) \times \cdots \times 2 \times 1\right) \times \left((n-i)(n-i-1) \times \cdots \times 2 \times 1\right)}$$

different ways.

These numbers occur over and over again, so we need definitions of shorter representations of them, namely

$$n! = n(n-1) \times \cdots \times 2 \times 1 \quad \text{and} \quad \binom{n}{i} = \frac{n!}{i!(n-i)!}.$$

The probability of observing i A- and $n-i$ aa individuals in F_2 is therefore

$$\text{Prob}(X = i) = \binom{n}{i} (\tfrac{3}{4})^i (\tfrac{1}{4})^{n-i}$$

where X symbolizes the number of A genes. The number of A genes varies among experiments, and our interest is to describe the variability. In this spirit X is referred to as a *stochastic variable*, or sometimes a random variable, and the specifications of $\text{Prob}(X = i)$, for $i = 0, 1, 2, \ldots, n$ is its *distribution*. The variable X takes only integer values, and such a variable is called a *discrete stochastic variable*.

The distribution of the number of A- individuals in F_2 is a *binomial distribution* and the coefficient $\binom{n}{i}$ a *binomial coefficient*. The binomial coefficient can also be interpreted as the number of ways in which i individuals can be chosen from a population of n, and it is therefore often read as "n choose i." The binomial distributions $b(n, p)$,

$$\text{Prob}(X = i) = \binom{n}{i} p^i(1-p)^{n-i}, \quad i = 0, 1, 2, \ldots, n, \qquad \text{(A.1)}$$

form a *conservative family* for a given p and $n = 1, 2, 3, \ldots$ because the sum of two independent binomially distributed stochastic variables $X \sim b(n, p)$ and

$Y \sim b(m, p)$ are binomially distributed $X + Y \sim b(n + m, p)$. The simplest binomial distribution is $b(1, p)$, and $b(n, p)$ can be seen as the distribution of the sum of n independent variables distributed $b(1, p)$.

A.1.1 Moments

Given the probabilities of the various values X can take, we may calculate its mean or *expected value*:

$$\mathrm{E} X = \sum_{i=0}^{n} i \operatorname{Prob}(X = i).$$

The mean is a simple linear property of stochastic variables. Consider, for instance, any two stochastic variables X and Y and a constant K; then

$$\mathrm{E}(X + Y) = \mathrm{E} X + \mathrm{E} Y \quad \text{and} \quad \mathrm{E}(KX) = K \mathrm{E} X. \tag{A.2}$$

For example, the expected value of the number of A- individuals in F_2 equals $\frac{3}{4}n$, and in general the expected value associated with the binomial distribution $b(n, p)$ is np. The variation in X is described by its *variance*, defined as

$$\operatorname{Var} X = \mathrm{E}(X - \mathrm{E} X)^2,$$

that is, the mean of the squared deviation of X from its expected value. The variance in the binomial distribution is $np(1 - p)$ and that of the number of A- individuals in F_2 equals $\frac{3}{4} \times \frac{1}{4} \times n$. The expected value and the variance are useful characteristics of a stochastic variable, as will become clear through their application in population genetics.

A convenient formula for the variance is

$$\operatorname{Var} X = \mathrm{E} X^2 - (\mathrm{E} X)^2. \tag{A.3}$$

This follows by expanding the squared deviation, that is, $\mathrm{E}(X - \mathrm{E} X)^2 = \mathrm{E}(X^2 - 2X\mathrm{E} X + (\mathrm{E} X)^2)$, and using the linearity (A.2) of the mean value.

The second of the formulae in A.2 has a similarly simple generalization to the variance as $\operatorname{Var}(KX) = K^2 \operatorname{Var} X$. The formula for $\operatorname{Var}(X + Y)$ is more elaborate because we have to evaluate the mean of

$$\big((X + Y) - (\mathrm{E} X + \mathrm{E} Y)\big)^2 = (X - \mathrm{E} X)^2 + (Y - \mathrm{E} Y)^2 + 2(X - \mathrm{E} X)(Y - \mathrm{E} Y).$$

Thus, we get

$$\operatorname{Var}(X + Y) = \operatorname{Var} X + \operatorname{Var} Y + 2\operatorname{Cov}(X, Y), \tag{A.4}$$

where

$$\operatorname{Cov}(X, Y) = \mathrm{E}(X - \mathrm{E} X)(Y - \mathrm{E} Y)$$

is the *covariance* of X and Y. The covariance is often used in the form of the normalized *correlation coefficient*

$$\operatorname{corr}(X, Y) = \frac{\operatorname{Cov}(X, Y)}{\sqrt{\operatorname{Var} X \operatorname{Var} Y}} \tag{A.5}$$

with the property $-1 \leq \mathrm{corr}(X, Y) \leq 1$.

Two stochastic variables X and Y taking the values $i = 1, 2, \ldots, n$ and $j = 1, 2, \ldots, m$ are *independent* when

$$\mathrm{Prob}(X = i \text{ and } Y = j) = \mathrm{Prob}(X = i)\mathrm{Prob}(Y = j)$$

for all $i = 1, 2, \ldots, n$ and $j = 1, 2, \ldots, m$. The covariance of two independent stochastic variables is zero, and thus $\mathrm{Var}(X + Y) = \mathrm{Var}\,X + \mathrm{Var}\,Y$ when X and Y are independent.

A.1.2   Conditional probabilities

The probability that pollen carries A given that the oocyte it fertilizes carries A is called a *conditional probability*. In a Mendelian cross this probability equals the probability that the pollen carries A, because the two events are independent. In general, the conditional probability of the event $\mathcal{A}$ given $\mathcal{B}$ is

$$\mathrm{Prob}(\mathcal{A}|\mathcal{B}) = \frac{\mathrm{Prob}(\mathcal{A} \text{ and } \mathcal{B})}{\mathrm{Prob}(\mathcal{B})},$$

provided, of course, that $\mathrm{Prob}(\mathcal{B}) > 0$. When the two events are independent, then $\mathrm{Prob}(\mathcal{A}|\mathcal{B}) = \mathrm{Prob}(\mathcal{A})$.

This concept is very applicable in genetics because we often have some knowledge of the parents and want to describe the probabilities of the various offspring types. The Mendelian analysis of the F_2 offspring, for example, is based on the conditional probability

$$\mathrm{Prob}(\mathsf{AA}|\mathsf{A}\text{-}) = \frac{\mathrm{Prob}(\mathsf{AA})}{\mathrm{Prob}(\mathsf{A}\text{-})} = \frac{\frac{1}{4}}{\frac{3}{4}} = \frac{1}{3}.$$

The random combination of male and female gametes may be expressed as

$$\mathrm{Prob}(\mathsf{A} \text{ in pollen}|\mathsf{A} \text{ in oocyte}) = \mathrm{Prob}(\mathsf{A} \text{ in pollen}).$$

Most of our applications of conditional probability are in terms of stochastic variables, and we therefore consider two discrete stochastic variables X and Y taking values $0, 1, 2, \ldots, n$ and $0, 1, 2, \ldots, m$, respectively. We may then consider the *conditional distribution* of Y given the event $X = i$, $\mathrm{Prob}(Y = j|X = i)$, $j = 0, 1, 2, \ldots, m$, in that

$$\sum_{j=0}^{m} \mathrm{Prob}(Y = j|X = i) = \frac{1}{\mathrm{Prob}(X = i)} \sum_{j=0}^{m} \mathrm{Prob}(Y = j \text{ and } X = i).$$

The events in the sum on the right side are disjoint (Y can only take one value), so the sum is the probability that $X = i$ and $Y = 0$ or 1 or ... or m, which obviously adds up to the probability that $X = i$.

We will need a couple of formulae to handle conditional probability. The first is *the law of total probability*, which states that

$$\text{Prob}(Y = j) = \sum_{i=0}^{n} \text{Prob}(Y = j | X = i) \, \text{Prob}(X = i). \qquad (A.6)$$

This is obtained by a straightforward calculation. First use the definition of conditional probability and get

$$\sum_{i=0}^{n} \text{Prob}(Y = j | X = i) \, \text{Prob}(X = i) = \sum_{i=0}^{n} \text{Prob}(Y = j \text{ and } X = i).$$

Then recognize the sum on the left side as probabilities of disjoint events adding up to the probability that $Y = j$. The second is *Bayes' formula*, which states that

$$\text{Prob}(X = i | Y = j) = \text{Prob}(Y = j | X = i) \times \frac{\text{Prob}(X = i)}{\text{Prob}(Y = j)}$$

when the conditional probabilities are defined, that is, $\text{Prob}(X = i) > 0$ and $\text{Prob}(Y = j) > 0$. This follows immediately from the definition of conditional probability.

The mean of the conditional distribution of Y given $X = i$ is written as $\text{E}(Y | X = i)$. This allows us to define a new stochastic variable $\text{E}(Y | X)$, which takes the value $\text{E}(Y | X = i)$ with probability $\text{Prob}(X = i)$. The stochastic variable $\text{E}(Y | X)$ is called *the conditional expectation of* Y *given* X. Just as in the definition of conditional distributions, we have to be careful with an event where $\text{Prob}(X = i) = 0$, but for these we can assign any value to $\text{E}(Y | X = i)$; it is not important because the event never occurs.

The mean of $\text{E}(Y | X)$ is particularly simple in that

$$\text{E}\big(\text{E}(Y | X)\big) = \text{E} \, Y. \qquad (A.7)$$

To see this, calculate the mean and get

$$
\begin{aligned}
\text{E}\big(\text{E}(Y | X)\big) &= \sum_{i=0}^{n} \text{E}(Y | X = i) \text{Prob}(X = i) \\
&= \sum_{i=0}^{n} \text{Prob}(X = i) \sum_{j=0}^{m} j \text{Prob}(Y = j | X = i) \\
&= \sum_{j=0}^{m} j \sum_{i=0}^{n} \text{Prob}(X = i) \text{Prob}(Y = j | X = i),
\end{aligned}
$$

and by using the law of total probability (A.6) we get

$$\text{E}\big(\text{E}(Y | X)\big) = \sum_{j=0}^{m} j \, \text{Prob}(Y = j),$$

Table A.1: Joint distribution of X and Y

X	0	1	2	$\cdots$	m	$\sum$
0	p_{00}	p_{01}	p_{02}	$\cdots$	p_{0m}	$p_{0\cdot}$
1	p_{10}	p_{11}	p_{12}	$\cdots$	p_{1m}	$p_{1\cdot}$
2	p_{20}	p_{21}	p_{22}	$\cdots$	p_{2m}	$p_{2\cdot}$
$\vdots$	$\vdots$	$\vdots$	$\vdots$		$\vdots$	$\vdots$
n	p_{n0}	p_{n1}	p_{n2}	$\cdots$	p_{nm}	$p_{n\cdot}$
$\sum$	$p_{\cdot 0}$	$p_{\cdot 1}$	$p_{\cdot 2}$	$\cdots$	$p_{\cdot m}$	1

which by definition is equal to $\mathrm{E}\,Y$.

A good illustration of a conditional distribution is found in Table A.1, which gives the probabilities $p_{ij} = \mathrm{Prob}(X = i, Y = j)$. The sum over the whole table is 1. The *marginal distribution* of X is $\mathrm{Prob}(X = i) = p_{i\cdot}$, $x = 0, 1, 2, \ldots, n$, given by the sums within rows, and the marginal distribution of Y is $\mathrm{Prob}(Y = j) = p_{\cdot j}$, $y = 0, 1, 2, \ldots, m$, given by the sums within columns. The conditional distribution of Y given $X = i$ is given by the probabilities in the ith row divided by $p_{i\cdot}$. In this sense the conditional expectation $\mathrm{E}(Y|X = i)$ is the average of Y within the ith row. In the same way we can talk about the variance $\mathrm{Var}(Y|X = i)$ of Y within the ith row. This conditional variance of Y given $X = i$ is simply the variance in the distribution $\mathrm{Prob}(Y = j|X = i) = p_{ij}/p_{\cdot i}$, $j = 0, 1, 2, \ldots, m$. These variances may then be viewed as realizations of a stochastic variable $\mathrm{Var}(Y|X)$ that attains the value $\mathrm{Var}(Y|X = i)$ with probability $p_{i\cdot}$. The average of the variable, $\mathrm{E}\,\mathrm{Var}(Y|X)$, is the average of the variance within columns. The variance between columns is the variance of the column means, that is, $\mathrm{Var}\big(\mathrm{E}(Y|X)\big)$, sometimes shortened to $\mathrm{Var}\,\mathrm{E}(Y|X)$.

These two variances may be viewed as components of the total variance in that we have

$$\mathrm{Var}\,Y = \mathrm{E}\,\mathrm{Var}(Y|X) + \mathrm{Var}\,\mathrm{E}(Y|X). \qquad (A.8)$$

This formula is a simple consequence of equation (A.7). Use formula (A.3):

$$\mathrm{Var}\,Y = \mathrm{E}\,Y^2 - (\mathrm{E}\,Y)^2$$

and evaluate the two means in terms of conditional expectations using equation (A.7) to get

$$\mathrm{Var}\,Y = \mathrm{E}\Big(\mathrm{E}(Y^2|X)\Big) - \Big(\mathrm{E}\big(\mathrm{E}(Y|X)\big)\Big)^2.$$

Now equation (A.3) for the conditional variance of Y given $X = i$ is

$$\mathrm{Var}(Y|X = i) = \mathrm{E}\big(Y^2|X = i\big) - \big(\mathrm{E}(Y|X = i)\big)^2,$$

and using this we get

$$\mathrm{Var}\,Y \;=\; \mathrm{E}\Big(\mathrm{Var}(Y|X) + \big(\mathrm{E}(Y|X)\big)^2\Big) - \Big(\mathrm{E}\big(\mathrm{E}(Y|X)\big)\Big)^2$$

$$\begin{aligned} &= \; \mathrm{E}\,\mathrm{Var}(Y|X) + \mathrm{E}\Big(\big(\mathrm{E}(Y|X)\big)^2 \Big) - \Big(\mathrm{E}\big(\mathrm{E}(Y|X)\big) \Big)^2 \\ &= \; \mathrm{E}\,\mathrm{Var}(Y|X) + \mathrm{Var}\,\mathrm{E}(Y|X) \end{aligned}$$

by using equation (A.3) for the variance of the stochastic variable $\mathrm{E}(Y|X)$.

A.1.3   The binomial distribution

The binomial distribution has already been introduced in equation (A.1). The mean and variance of $X \sim b(n,p)$ are $\mathrm{E}\,X = np$ and $\mathrm{Var}\,X = np(1-p)$. The mean is

$$\begin{aligned} \mathrm{E}\,X \; &= \; \sum_{i=0}^{n} i \, \binom{n}{i} p^i (1-p)^{n-i} \; = \; np \sum_{i=1}^{n} \binom{n-1}{i-1} p^{i-1}(1-p)^{n-i} \\ &= \; np \sum_{i=0}^{n-1} \binom{n-1}{i} p^i (1-p)^{(n-1)-i} \; = \; np \end{aligned}$$

because the last sum is of probabilities in the distribution $b(n-1,p)$. To find the variance we evaluate the expectation

$$\begin{aligned} \mathrm{E}\,X(X-1) \; &= \; \sum_{i=0}^{n} i(i-1) \binom{n}{i} p^i (1-p)^{n-i} \\ &= \; n(n-1)p^2 \sum_{i=2}^{n} \binom{n-2}{i-2} p^{i-2}(1-p)^{n-i} \; = \; n(n-1)p^2 \end{aligned}$$

because the last sum is of probabilities in the distribution $b(n-2,p)$. The variance formula in equation (A.3) has a useful counterpart, namely

$$\mathrm{Var}\,X = \mathrm{E}\,X(X-1) - \mathrm{E}\,X(\mathrm{E}\,X - 1),$$

that easily produces the binomial variance as $np(1-p)$.

The multinomial distribution corresponds to the binomial distribution, but instead of two classes it has many. With three genotypic classes the distribution is given by

$$m\big(n;p,q,1-p-q)\big) \sim \binom{n}{i \; j \; (n-i-j)} p^i q^j (1-p-q)^{n-i-j}, \qquad \text{(A.9)}$$

where the multinomial coefficient is given by

$$\binom{n}{i \; j \; (n-i-j)} = \frac{n!}{i!\,j!\,(n-i-j)!}.$$

The number of observations in any class is binomially distributed. For instance, if $(X,Y,Z) \sim m\big(n;p,q,1-p-q)$, then $X \sim b(n,p)$ or, equivalently, $X \sim m(n;p,1-p)$, and so the means and variances of the stochastic variables are the binomial means and variances corresponding to the class probability. The covariance of X and Y is found from the moment

$$
\mathrm{E}(XY) = \sum_{i=0}^{n} \sum_{j=0}^{n-i} ij \binom{n}{i \; j \; (n-i-j)} p^i q^j (1-p-q)^{n-i-j}
$$

$$
= n(n-1)pq \sum_{i=1}^{n} \sum_{j=1}^{n-i} \binom{n-2}{(i-1)\;(j-1)\;(n-i-j)} p^{i-1} q^{j-1} (1-p-q)^{n-i-j}
$$

and therefore $\mathrm{E}(XY) = n(n-1)pq$ because the sum is of probabilities in the distribution $m(n-2; p, q, 1-p-q)$. Thus,

$$
\mathrm{Cov}(X,Y) = \mathrm{E}(XY) - \mathrm{E}\,X\,\mathrm{E}\,Y = n(n-1)pq - np\,nq = -npq.
$$

Finally, the distribution of $X + Y$ is $b(n, p+q)$:

$$
\mathrm{Prob}(X+Y=i) = \sum_{j=0}^{i} \binom{n}{j \; i-j \; (n-i)} p^j q^{i-j} (1-p-q)^{n-i}
$$

$$
= \binom{n}{i}(1-p-q)^{n-i} \sum_{j=0}^{i} \binom{i}{j} p^j q^{i-j} = \binom{n}{i}(p+q)^i (1-p-q)^{n-i}.
$$

The multinomial distribution is defined for any number of classes, and it may be convenient to express that the vector (or matrix) $\boldsymbol{X}$ is multinomially distributed $\boldsymbol{X} \sim m(n; \boldsymbol{p})$, where $\boldsymbol{p}$ is a vector (or matrix) of probabilities that sum to one.

A.1.4   The geometric distribution

We will consider a stochastic variable T as geometrically distributed when

$$
\mathrm{Prob}(T=i) = \pi(1-\pi)^{i-1}, \quad i = 1, 2, \ldots
$$

where the parameter π is the event probability. These probabilities sum to one. First notice that

$$
\sum_{i=1}^{\infty} \pi(1-\pi)^{i-1} = \pi \sum_{i=0}^{\infty} (1-\pi)^i.
$$

The sum may be evaluated by using the formula

$$
\sum_{\ell=0}^{\infty} x^{\ell} = \frac{1}{1-x} \quad \text{for } x < 1. \tag{A.10}
$$

Thus, the sum over i equals $\frac{1}{\pi}$ and the desired result follows.

The mean of the geometric distribution is

$$
\mathrm{E}\,T = \sum_{i=1}^{\infty} i\pi(1-\pi)^{i-1} = \frac{\pi}{1-\pi} \sum_{i=1}^{\infty} \sum_{j=1}^{i} (1-\pi)^i = \frac{\pi}{1-\pi} \sum_{j=1}^{\infty} \sum_{i=j}^{\infty} (1-\pi)^i
$$

by reversing the order of summation. Now change the variable of the second sum to $\ell = i - j$ to get

$$\mathrm{E}\,T = \frac{\pi}{1 - \pi} \sum_{j=1}^{\infty} (1 - \pi)^j \sum_{\ell=0}^{\infty} (1 - \pi)^{\ell} = \sum_{j=1}^{\infty} (1 - \pi)^{j-1},$$

where the sum over ℓ was evaluated by using formula (A.10). Using this formula again on the sum over j gives the mean

$$\mathrm{E}\,T = \frac{1}{\pi}.$$

The variance of the geometric distribution is

$$\mathrm{Var}\,T = \frac{1 - \pi}{\pi^2},$$

which is shown using similar methods.

As shown in Box 6 on page 38, the geometric distribution has *no memory*, that is, $\mathrm{Prob}(T > j \,|\, T > i) = \mathrm{Prob}(T > j - i)$. This result is, of course, trivial if $j \leq i$ and we will therefore assume $j > i$. Here we deduce the result using the *distribution function* $F_{\pi}(i) = \mathrm{Prob}(T \leq i)$, $i = 1, 2, \ldots$:

$$F_{\pi}(i) = \sum_{j=1}^{i} \pi (1 - \pi)^{j-1} = \pi \sum_{j=0}^{i-1} (1 - \pi)^j = 1 - (1 - \pi)^i.$$

The conditional probability is then

$$\mathrm{Prob}(T > j \,|\, T > i) = \frac{\mathrm{Prob}(T > j)}{\mathrm{Prob}(T > i)} = \frac{1 - \mathrm{Prob}(T \leq j)}{1 - \mathrm{Prob}(T \leq i)} = \frac{1 - F_{\pi}(j)}{1 - F_{\pi}(i)},$$

and inserting the distribution function gives

$$\mathrm{Prob}(T > j \,|\, T > i) = (1 - \pi)^{(j-i)} = 1 - F_{\pi}(j - i) = \mathrm{Prob}(T > j - i).$$

We often need to compare geometrically distributed waiting times. Consider the two independent geometrically distributed stochastic variables X and Y that describe the waiting time to two different events. Their event probabilities are π_x and π_y. The question is which event occurs first, and what the waiting time is to the first event. Thus, we consider

$$\mathrm{Prob}(X \text{ event occurs first}) = \mathrm{Prob}(X < Y) = \sum_{i=1}^{\infty} \pi_x (1 - \pi_x)^{i-1} \sum_{j=i+1}^{\infty} \pi_y (1 - \pi_y)^{j-1}.$$

Summing over j yields

$$\mathrm{Prob}(X < Y) = \sum_{i=1}^{\infty} \pi_x (1 - \pi_x)^{i-1} (1 - \pi_y)^i = \frac{\pi_x (1 - \pi_y)}{1 - (1 - \pi_x)(1 - \pi_y)}$$

(change the summation into a summation over a geometric distribution with event probability $1 - (1 - \pi_x)(1 - \pi_y)$). In a similar way we get

$$\text{Prob}(X = Y) = \sum_{i=1}^{\infty} \pi_x(1 - \pi_x)^{i-1}\pi_y(1 - \pi_y)^{i-1} = \frac{\pi_x\pi_y}{1 - (1 - \pi_x)(1 - \pi_y)}.$$

The three possible events are therefore given in terms of relative probabilities:

$$\text{Prob}(X \text{ event occurs first}) = \frac{\pi_x(1 - \pi_y)}{\pi_x(1 - \pi_y) + \pi_y(1 - \pi_x) + \pi_x\pi_y},$$

$$\text{Prob}(Y \text{ event occurs first}) = \frac{\pi_y(1 - \pi_x)}{\pi_x(1 - \pi_y) + \pi_y(1 - \pi_x) + \pi_x\pi_y},$$

$$\text{Prob}(\text{events occur simultaneously}) = \frac{\pi_x\pi_y}{\pi_x(1 - \pi_y) + \pi_y(1 - \pi_x) + \pi_x\pi_y}.$$

This result generalizes to any number geometric waiting times.

The distribution of the waiting time until an event occurs can be found from

$$\text{Prob}(X > i \text{ and } Y > i) = \text{Prob}(X > i)\text{Prob}(Y > i)$$

$$= \sum_{j=i+1}^{\infty} \pi_x(1 - \pi_x)^j \sum_{j=i+1}^{\infty} \pi_y(1 - \pi_y)^j = \left((1 - \pi_x)(1 - \pi_y)\right)^{i+1},$$

where we used the independence of X and Y. The probability of an event at i is then

$$\text{Prob}(\min(X, Y) = i)$$
$$= \text{Prob}(X > i \text{ and } Y > i) - \text{Prob}(X > i - 1 \text{ and } Y > i - 1)$$
$$= \left(1 - (1 - \pi_x)(1 - \pi_y)\right)\left((1 - \pi_x)(1 - \pi_y)\right)^{i}.$$

The stochastic variable $\min(X, Y)$ thus has a geometric distribution with event probability equal to

$$1 - (1 - \pi_x)(1 - \pi_y) = \pi_x + \pi_y - \pi_x\pi_y.$$

In our applications of the geometric waiting time distribution we will consider rare events, that is, events where we can neglect the possibility of simultaneous occurrence when stochastic variables are compared.

Consider again two stochastic variables X and Y, now with small event probabilities, $\pi_x \ll 1$ and $\pi_y \ll 1$. The stochastic variable $\min(X, Y)$ then has a geometric distribution with event probability approximately equal to $\pi_x + \pi_y$. The average waiting time to the first event is therefore

$$\text{E}\left(\min(X, Y)\right) \approx \frac{1}{\pi_x + \pi_y},$$

and by neglecting products of event probabilities we get

$$\text{Prob}(X \text{ event occurs first}) \approx \frac{\pi_x}{\pi_x + \pi_y},$$

$$\text{Prob}(Y \text{ event occurs first}) \approx \frac{\pi_y}{\pi_x + \pi_y}.$$

Another geometric distribution

Frequently an alternative formulation of the geometric distribution is used, in that a stochastic variable S is said to be geometrically distributed when

$$\text{Prob}(S = i) = \pi(1 - \pi)^i, \quad i = 0, 1, 2, \dots.$$

Compared to our original definition, $S = T - 1$, and the mean and variance are therefore

$$\text{E}\, S = \text{E}(T - 1) = \frac{1}{\pi} - 1 \quad \text{and} \quad \text{Var}\, S = \text{Var}\, T.$$

A.1.5 The Poisson distribution

Mutations are rare events that occur independently over time. The probability of a mutation event in a gene during the time L is $\lambda = \mu L$, and the probability of i independent events is thus proportional to λ^i. A different event can be obtained by changing the order in which these i events occurred, so the probability of i mutations is proportional to $\lambda^i/i!$, where $i! = i \times (i - 1) \times \cdots \times 2 \times 1$. The probability of i mutations is therefore

$$\frac{\lambda^i}{i!}\, e^{-\lambda}, \quad i = 1, 2, \dots, \infty, \tag{A.11}$$

because

$$\sum_{i=0}^{\infty} \frac{\lambda^i}{i!} = e^{\lambda}.$$

The probabilities in equation (A.11) define the Poisson distribution with mean λ. The variance of this distribution is also λ.

The sum of two independent Poisson distributed variables X_1 and X_2 with means λ_1 and λ_2 are Poisson distributed with mean $\lambda_1 + \lambda_2$. Conversely, the distribution of X_1 given $X_1 + X_2 = n$ is binomial $b(p, n)$, where $p = \lambda_1/(\lambda_1 + \lambda_2)$. The binomial distribution and the Poisson distribution are thus closely related.

A.2 Markov Chains

The processes we consider in the Wright–Fisher model and its extensions are discrete time Markov processes, so-called Markov chains. Markov processes have the property that if we know the present state (in the Wright–Fisher process the number of A genes) then knowledge of past states of the process does not add any information about its future behavior. This is true if the present state is known exactly (in population genetics: the allele type of all genes in the population is known). If the present state is inferred from a sample of the population, earlier samples may add to our knowledge of the present state.

The state of a Wright–Fisher process is a number between zero and $2N$: it is a *finite state* Markov chain. These are simpler than general Markov chains, and the only ones we will consider here.

Markov chains are characterized by their transition probabilities $P^{t,\,t+1}_{i\to j}$ that the state i in generation t is followed by the state j in generation $t+1$. For the Wright–Fisher model the transition probabilities are given by equation (2.2), and for the Wright–Fisher process with mutation they are given by equation (4.4). In these Markov chains the transition probabilities depend only upon state, and not upon the time at which that state was reached. Such Markov chains are said to have *stationary transition probabilities*, and we will limit attention to these.

In a general finite state Markov chain we number the states as $i = 1, 2, \ldots, n$, where n is the number of states in the process. With stationary transition probabilities such a Markov chain may be described by the transition probabilities $P_{i\to j}$, where

$$P_{i\to j} \geq 0, \quad \sum_{k=1}^{n} P_{i\to k} = 1 \quad \text{for all } i \text{ and } j \in \{1, 2, \ldots, n\}. \tag{A.12}$$

Using this formulation, the equation carrying the distribution from generation 0 to generation 2 in the Wright–Fisher process (page 26) may now be written as

$$\text{Prob}(X_2 = k \,|\, X_0 = i) = \sum_{j=1}^{n} P_{i\to j} P_{j\to k}.$$

The probability of a given chain of events may be calculated from the definition of conditional probabilities because we get

$$\text{Prob}(X_0 = i_0, X_1 = i_1, \ldots, X_{t-1} = i_{t-1}, X_t = i_t) =$$
$$\text{Prob}(X_t = i_t \,|\, X_0 = i_0, X_1 = i_1, \ldots, X_{t-1} = i_{t-1})$$
$$\times \text{Prob}(X_0 = i_0, X_1 = i_1, \ldots, X_{t-1} = i_{t-1}).$$

In a Markov chain we have

$$\text{Prob}(X_t = i_t \,|\, X_0 = i_0, X_1 = i_1, \ldots, X_{t-1} = i_{t-1}) = \text{Prob}(X_t = i_t \,|\, X_{t-1} = i_{t-1})$$

and therefore

$$\text{Prob}(X_0 = i_0, X_1 = i_1, \ldots, X_{t-1} = i_{t-1}, X_t = i_t) =$$
$$\text{Prob}(X_0 = i_0, X_1 = i_1, \ldots, X_{t-1} = i_{t-1}) \times P_{i_{t-1}\to i_t}.$$

By repeated use of this formula we get

$$\text{Prob}(X_0 = i_0, X_1 = i_1, \ldots, X_{t-1} = i_{t-1}, X_t = i_t) =$$
$$\text{Prob}(X_0 = i_0) P_{i_0\to i_1} P_{i_1\to i_2} \cdots P_{i_{t-2}\to i_{t-1}} P_{i_{t-1}\to i_t}, \tag{A.13}$$

which in its simplest form connects two generations:

$$\text{Prob}(X_{t-1} = i_{t-1}, X_t = i_t) = \text{Prob}(X_{t-1} = i_{t-1}) P_{i_{t-1}\to i_t}. \tag{A.14}$$

A finite state Markov chain with stationary transition probabilities may thus be described by the matrix of transition probabilities:

$$\boldsymbol{P} = \left\{ \begin{array}{cccc} P_{1\to1} & P_{1\to2} & \cdots & P_{1\to n} \\ P_{2\to1} & P_{2\to2} & \cdots & P_{2\to n} \\ \vdots & \vdots & \ddots & \vdots \\ P_{n\to1} & P_{n\to2} & \cdots & P_{n\to n} \end{array} \right\}.$$

$\boldsymbol{P}$ is a stochastic matrix, that is, it satisfies the property (A.12), in particular

$$\left\{ \begin{array}{cccc} P_{1\to1} & P_{1\to2} & \cdots & P_{1\to n} \\ P_{2\to1} & P_{2\to2} & \cdots & P_{2\to n} \\ \vdots & \vdots & \ddots & \vdots \\ P_{n\to1} & P_{n\to2} & \cdots & P_{n\to n} \end{array} \right\} \left\{ \begin{array}{c} 1 \\ 1 \\ \vdots \\ 1 \end{array} \right\} = \left\{ \begin{array}{c} 1 \\ 1 \\ \vdots \\ 1 \end{array} \right\}, \qquad (A.15)$$

because from any state the process goes somewhere. The state at time t may be described by the distribution

$$\boldsymbol{p}_t = (p_{t1}, p_{t2}, \ldots, p_{tn}), \quad \text{where} \quad p_{ti} = \text{Prob}(X_t = i),$$

and so $p_{t1} + p_{t2} + \cdots + p_{tn} = 1$. Using equation (A.14) we have the recurrence equation for a finite state Markov chain with stationary transition probabilities as

$$\boldsymbol{p}_t = \boldsymbol{p}_{t-1}\boldsymbol{P} \qquad (A.16)$$

and the iteration of this equation provides a description of the state of the Markov chain at any time:

$$\boldsymbol{p}_t = \boldsymbol{p}_0\boldsymbol{P}^t. \qquad (A.17)$$

In Chapter 2 we saw that the Wright–Fisher process has two kinds of states, the states $1, 2, \ldots, 2N - 1$ from which all other states may be reached, and the states 0 and $2N$, which are *absorbing states* because the population cannot leave those states. The probability that the process ends up in state 0 or $2N$ depends on the initial conditions in the population. Due to conservation of the gene frequency the population ends up in 1 with probability p and in 0 with probability $q = 1 - p$.

All the states of the simple Wright–Fisher process with mutation can be reached from all the states in a finite number of generations (Chapter 4). Such Markov chains are called *recurrent*. Many Markov chains possess a unique stationary distribution $\boldsymbol{\pi}$ with the property

$$\boldsymbol{\pi} = \boldsymbol{\pi}\boldsymbol{P}, \qquad (A.18)$$

where $\pi_i \geq 0$, $i = 1, 2, \ldots, n$ and $\pi_1 + \pi_2 + \cdots + \pi_n = 1$. From any initial state the probability that the population is in state i, $i \in \{1, 2, \ldots, n\}$ converges to π_i as time goes to infinity. Thus, the population will after a long time be in state i with approximately the probability π_i. The proper condition for a unique stationary distribution to exist is that the finite state Markov chain is irreducible and aperiodic.

The Wright–Fisher processes are aperiodic, and we will therefore not consider periodic Markov chains; that is, processes like, for instance,

$$
P = \left\{ \begin{array}{cccc}
0 & 1 & 0 & 0 \\
0 & 0 & 1 & 0 \\
0 & 0 & 0 & 1 \\
1 & 0 & 0 & 0
\end{array} \right\}
$$

will develop as $1, 2, 3, 4, 1, 2, 3, 4, 1, 2, 3, 4, 1, \ldots$ if started in state 1. An example of a reducible Markov chain is

$$
P = \left\{ \begin{array}{ccccc}
\frac{3}{4} & \frac{1}{4} & 0 & 0 & 0 \\
\frac{1}{4} & \frac{3}{4} & 0 & 0 & 0 \\
\frac{1}{10} & \frac{1}{10} & \frac{3}{5} & \frac{1}{10} & \frac{1}{10} \\
0 & 0 & 0 & \frac{1}{2} & \frac{1}{2} \\
0 & 0 & 0 & \frac{1}{2} & \frac{1}{2}
\end{array} \right\}.
$$

Here the process ends up in either $\{1, 2\}$ or $\{4, 5\}$, but if the process starts in $\{1, 2\}$ (or $\{4, 5\}$) it stays there. Once the process is in $\{1, 2\}$ (or $\{4, 5\}$) it is recurrent within $\{1, 2\}$ (or $\{4, 5\}$), and the process will reach a unique stationary distribution (in this example, $\boldsymbol{\pi}_{\{12\}} = (\frac{1}{2}, \frac{1}{2})$). The Wright–Fisher process without mutation (Chapter 2) is reducible because a population starting in either state 0 or $2N$ will stay there ever after.

The existence of a unique solution to equation (A.18) may not at first seem surprising—it is a system of n linear equations in n variables: $\boldsymbol{\pi}(\boldsymbol{I} - \boldsymbol{P}) = 0$. It has, however, the solution $\pi_1 = \pi_2 = \cdots = \pi_n = 0$. From equation (A.15) we know that it also has the solution $\pi_1 = \pi_2 = \cdots = \pi_n = 1$ (meaning that the matrix $\boldsymbol{P}$ possesses an eigenvalue 1). Thus, the determinant of $\boldsymbol{I} - \boldsymbol{P}$ is therefore zero, $\det(\boldsymbol{I} - \boldsymbol{P}) = 0$, and equation (A.18) has a continuum of solutions, exactly one of which has

$$
\pi_i \geq 0 \quad \text{for all } i \in \{1, 2, \ldots, n\} \quad \text{and} \quad \sum_{j=0}^{n} \pi_j = 1.
$$

In addition, $\boldsymbol{p}_t \to \boldsymbol{\pi}$ for $t \to \infty$ no matter the initial state of the population ($p_{it} \to \pi_i$ for $t \to \infty$, $i = 1, 2, \ldots, n$). This result is known as the *basic limit theorem* for Markov chains.

The coalescent in a sample of n genes from a Wright–Fisher population or a Moran population is an example of a Markov chain on the states $1, 2, \ldots, n$. The state of the process is the number of ancestral genes. In each generation this number either stays the same or decreases by one. After a sufficiently long time the population ends up with one ancestral gene and no change happens after that. The waiting time to the next event is geometrically distributed, a property shared by all Markov chains. The continuous time coalescent processes considered in Section 2.6.2 and Box 14 are continuous time Markov processes, so-called birth-and-death processes. The general model for migration among d islands (Section 5.2 on page 118) describes immigration by the stochastic matrix $\boldsymbol{M}$. The geographical location of the ancestors of a gene is thus described by a Markov chain with transition matrix $\boldsymbol{M}$ (see, e.g., Section 5.2.3 on page 128).

The theory of Markov processes is well developed, and introductory textbooks are easy to find. A classic textbook geared towards applications, including genetic applications, is Karlin and Taylor (1975). The diffusion approximations we consider in Chapters 2 and 4 are, for instance, treated in Karlin and Taylor (1981).

A.3   Continuous Random Variables

The distribution of a continuous stochastic variable X is characterized by the distribution function

$$F(x) = \mathrm{Prob}(X < x) \quad \text{for} \quad -\infty < x < \infty$$

or the density

$$f(x) = \frac{d\,F}{d\,x}(x).$$

The density provides an approximation of the probability of observing X in a small interval, that is, $\mathrm{Prob}(x \le X < x + \Delta) \approx f(x)\Delta$ and

$$F(x) = \int_{-\infty}^{y} f(y)\,dy, \qquad \int_{-\infty}^{\infty} f(y)\,dy = 1.$$

The expected value and the variance are calculated from

$$\mathrm{E}\,X = \int_{-\infty}^{\infty} xf(x)\,dx \quad \text{and} \quad \mathrm{E}\,X^2 = \int_{-\infty}^{\infty} x^2 f(x)\,dx,$$

and in all the distributions that we apply, these integrals are well defined.

The conditional expectation of Y given X for a pair of continuous stochastic variables can be defined in a way similar to discrete variables. Assume that the joint density of Y and X is f and that f is positive only for positive numbers, $f(x,y) > 0, 0 < x < \infty, 0 < y < \infty$. The marginal densities of X and Y are f_X and f_Y, where

$$f_X(x) = \int_0^{\infty} f(x,y)\,dy \quad \text{and} \quad f_Y(y) = \int_0^{\infty} f(x,y)\,dx$$

(the arguments to follow are the same for $f(x,y) > 0$ in any interval of numbers). The conditional density of Y given X is then

$$f_{Y|X}(y|x) = \frac{f(x,y)}{f_X(x)}$$

and the conditional expectation is given by the expectations $\mathrm{E}(Y|X = x)$ of these distributions. The density of the conditional expectation $\mathrm{E}(Y|X)$ is f_X,

The Gaussian distributions form a conservative family because the sum of two independent normal distributed variables X_1 and X_2 with the same variance, that is, $X_1 \sim N(\mu_1, \sigma^2)$ and $X_2 \sim N(\mu_2, \sigma^2)$, has a normal distribution with the mean $\mu_1 + \mu_2$ and variance σ^2, or $X_1 + X_2 \sim N(\mu_1 + \mu_2, \sigma^2)$.

The Gaussian distribution provides a very useful approximation of the binomial distribution. The stochastic variable

$$\frac{X - np}{\sqrt{np(1-p)}}$$

where $X \sim b(p, n)$ is approximately standard normal distributed $N(0, 1)$ when $n \to \infty$. This result is known as the DeMoivre–Laplace theorem, and it is the first and simplest version of the central limit theorem, which states: For n independent identically distributed variables, $X_1, X_2, \ldots, X_n$ with expectation μ and variance σ^2 the distribution of the variable

$$\frac{1}{\sqrt{n\sigma^2}} \sum_{i=1}^{n} (X_i - \mu)$$

converges to the standard normal distribution $N(0, 1)$ when $n \to \infty$. The DeMoivre–Laplace theorem follows from this by viewing $X \sim b(p, n)$ as a sum of n independent variables with distribution $b(p, 1)$.

Statistics

The normal distribution is the default error distribution in statistics, and many standard statistical analyses of measurements refer to it. We will primarily use it in the description and analysis of quantitative characters. For these and many other applications it is often necessary to transform the data before the normal distribution becomes a good descriptor. In biological data a logarithmic transformation is commonly needed.

Statistical procedures for analyzing normal distributed data are well developed. They are based on specifications of linear structures, for instance in regression analyses (see, e.g., Blæsild and Granfeldt 2003).

Appendix B

Solutions to Exercises

Exercise 2.2.1 Use equation (A.4) and observe that

$$p_{A_1 0}(1 - p_{A_1 0}) + p_{A_2 0}(1 - p_{A_2 0}) - 2p_{A_1 0}p_{A_2 0} = (p_{A_1 0} + p_{A_2 0})(1 - (p_{A_1 0} + p_{A_2 0})).$$

The variance is thus obtained by pooling the two alleles A_1 and A_2.

Exercise 2.7.1 The event $i \to i+1$ occurs when an A breeds, probability $\frac{i}{2N}$, and an a dies, probability $\frac{2N-i}{2N}$, and vice versa for the event $i \to i-1$. The event $i \to i$ occurs when an A breeds and an A dies, or when an a breeds and an a dies.

$$\frac{i}{2N}\left(1 - \frac{i}{2N}\right) + \left(1 - \frac{i}{2N}\right)\frac{i}{2N} + \left(\frac{i}{2N}\right)^2 + \left(1 - \frac{i}{2N}\right)^2$$
$$= \frac{i}{2N} + \frac{i}{2N}\left(1 - \frac{i}{2N}\right) + \left(1 - \frac{i}{2N}\right) - \frac{i}{2N}\left(1 - \frac{i}{2N}\right) = 1.$$

Exercise 2.7.2

$E(X_t | X_{t-1} = i)$
$$= (i+1)\frac{i}{2N}\left(1 - \frac{i}{2N}\right) + (i-1)\left(1 - \frac{i}{2N}\right)\frac{i}{2N} + i\left(\frac{i}{2N}\right)^2 + i\left(1 - \frac{i}{2N}\right)^2$$
$$= i\frac{i}{2N} + \frac{i}{2N}\left(1 - \frac{i}{2N}\right) + i\left(1 - \frac{i}{2N}\right) - \frac{i}{2N}\left(1 - \frac{i}{2N}\right) = i.$$

Exercise 2.7.3

$E(X_t^2 | X_{t-1} = i)$
$$= (i+1)^2\frac{i}{2N}\left(1 - \frac{i}{2N}\right) + (i-1)^2\left(1 - \frac{i}{2N}\right)\frac{i}{2N} + i^2\left(\frac{i}{2N}\right)^2 + i^2\left(1 - \frac{i}{2N}\right)^2$$
$$= i^2 + 2i\frac{i}{2N}\left(1 - \frac{i}{2N}\right) - 2i\frac{i}{2N}\left(1 - \frac{i}{2N}\right) + 2\frac{i}{2N}\left(1 - \frac{i}{2N}\right),$$

because the i^2 terms sum as in Exercise 2.7.2. We thus get

$$\mathrm{Var}(X_t|X_{t-1}=i) = \mathrm{E}(X_t^2|X_{t-1}=i) - \Big(\mathrm{E}(X_t|X_{t-1}=i)\Big)^2 = \frac{i}{N}\Big(1-\frac{i}{2N}\Big).$$

Exercise 2.7.4   The proof of the formula proceeds as in Box 4 on page 29.

$$\begin{aligned}
\mathrm{Var}(X_t) &= \mathrm{E}\big(\mathrm{Var}(X_t|X_{t-1})\big) + \mathrm{Var}\big(\mathrm{E}(X_t|X_{t-1})\big) \\
&= \mathrm{E}\,\frac{X_{t-1}}{N}\Big(1-\frac{X_{t-1}}{2N}\Big) + \mathrm{Var}(X_{t-1}) \\
&= 2p - 2p^2 + \mathrm{Var}(X_{t-1})\Big(1-\frac{1}{2N^2}\Big),
\end{aligned}$$

where we used the formula $\mathrm{E}X^2 = \mathrm{Var}X + (\mathrm{E}X)^2$. The recurrence equation is therefore given by

$$\mathrm{Var}(X_t) = 2p(1-p) + \mathrm{Var}(X_{t-1})\Big(1-\frac{1}{2N^2}\Big).$$

The stationary variance is $4N^2 p(1-p)$ and the result quoted in the exercise follows.

Exercise 2.7.5   The main difference is the time unit. In the Wright–Fisher model the time unit is the generation time, whereas in the Moran model the time unit is the time between breeding of individuals. In the Moran model

$$\mathrm{Var}(p_t) = p(1-p)\left(1 - \Big(1-\frac{1}{2N^2}\Big)^t\right),$$

and for large N it becomes approximately

$$p(1-p)\exp\Big(\frac{t}{2N^2}\Big).$$

The two models therefore coincide if the time unit in the Moran model is that in which N breeding events occur.

Exercise 2.7.6   Assume at time t that all polymorphic classes are equally frequent, that is, the frequency f_{ti} is given by $f_{ti} = f_t$ for $i = 1, 2, \ldots, 2N-1$. For the classes $i = 2, 3, \ldots, 2N-2$ we thus have

$$\begin{aligned}
f_{t+1\,i} &= f_t\left(\frac{i-1}{2N}\Big(1-\frac{i-1}{2N}\Big) + \Big(1-\frac{i+1}{2N}\Big)\frac{i+1}{2N} + \Big(\frac{i}{2N}\Big)^2 + \Big(1-\frac{i}{2N}\Big)^2\right) \\
&= f_t\Big(1-\frac{1}{2N^2}\Big).
\end{aligned}$$

For the terminal classes

$$f_{t+1\,1} = f_t\left(\left(1 - \frac{2}{2N}\right)\frac{2}{2N} + \left(\frac{1}{2N}\right)^2 + \left(1 - \frac{1}{2N}\right)^2\right)$$

$$= f_t\left(1 - \frac{1}{2N^2}\right).$$

Thus, the distribution in Figure 2.15 is conserved and

$$f_{t+1} = f_t\left(1 - \frac{1}{2N^2}\right).$$

Exercise 3.0.1   The gene frequency in equation (3.2) is

$$p = p_♀p_♂ + \tfrac{1}{2}(p_♀q_♂ + q_♀p_♂) = \tfrac{1}{2}p_♀(p_♂ + q_♂) + \tfrac{1}{2}p_♂(p_♀ + q_♀) = \tfrac{1}{2}(p_♀ + p_♂).$$

The difference in **AA** frequency is

$$p_♀p_♂ - \left(\tfrac{1}{2}(p_♀ + p_♂)\right)^2 = p_♀p_♂ - \tfrac{1}{4}p_♀^2 - \tfrac{1}{2}p_♀p_♂ - \tfrac{1}{4}p_♂^2$$

$$= \tfrac{1}{4}p_♀(p_♂ - p_♀) + \tfrac{1}{4}p_♂(p_♀ - p_♂)$$

$$= -\tfrac{1}{4}(p_♀ - p_♂)^2.$$

The homozygote frequency is thus lower than the Hardy–Weinberg frequency that corresponds to the same frequency.

Exercise 3.1.1   From equation (3.12) we get

$$N_e \approx \left(\frac{1}{4000} + \frac{1}{4N_♂}\right)^{-1}.$$

For $N_♂ = 1000$ we get $N_e = 2000$, and

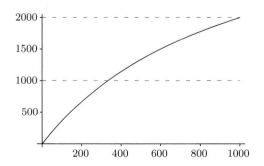

Exercise 3.1.2   The gene frequency of A is $p = z_{AA} + \frac{1}{2}z_{Aa}$ and the deviation from the corresponding Hardy–Weinberg frequency is

$$z_{AA} - \left(z_{AA} + \tfrac{1}{2}z_{Aa}\right)^2.$$

Now use that $z_{AA} + z_{Aa} + z_{aa} = 1$ to write the difference as

$$z_{AA} - p^2 = z_{AA} - \left(z_{AA} + z_{Aa} + z_{aa}\right) - \left(z_{AA} + \tfrac{1}{2}z_{Aa}\right)^2 = z_{AA}z_{aa} - \tfrac{1}{4}z_{Aa}^2.$$

(This trick transforms the difference into a homogeneous polynomial in the zs, i.e., all terms have the same power and this makes calculations a lot easier.) Thus, if we make the definition

$$\phi = \frac{z_{AA}z_{aa} - \tfrac{1}{4}z_{Aa}^2}{pq},$$

then we get $z_{AA} = p^2 + pq\phi$, and by symmetry $z_{aa} = q^2 + pq\phi$. Again using $z_{AA} + z_{Aa} + z_{aa} = 1$ gives the heterozygote frequency as $z_{AA} = 2pq - 2pq\phi$.

Exercise 3.2.1   Use the gamete pool model and let F_{2t-1} and F_{2t} be the consanguinity coefficients in the population in the two pools giving rise to the two generations in year t. Then we get

$$
\begin{aligned}
F_{2t-1} &= \frac{1}{2N_2} + \left(1 - \frac{1}{2N_2}\right)F_{2(t-1)},\\
F_{2t} &= \frac{1}{2N_1} + \left(1 - \frac{1}{2N_1}\right)F_{2t-1}.
\end{aligned}
$$

Combining the two equations provides

$$F_{2t} = \frac{1}{2N_1} + \left(1 - \frac{1}{2N_1}\right)\left(\frac{1}{2N_2} + \left(1 - \frac{1}{2N_2}\right)F_{2(t-1)}\right),$$

and collecting terms provides the recurrence equation:

$$F_{2t} = \left(\frac{1}{2N_1} + \left(1 - \frac{1}{2N_1}\right)\frac{1}{2N_2}\right) + \left(1 - \frac{1}{2N_1}\right)\left(1 - \frac{1}{2N_2}\right)F_{2(t-1)},$$

which describes the development of consanguinity in the population.

The effective population size is found from the equation

$$\left(1 - \frac{1}{2N_e}\right)^2 = \left(1 - \frac{1}{2N_1}\right)\left(1 - \frac{1}{2N_2}\right),$$

and by neglecting terms of the order $\left(\frac{1}{N}\right)^2$ we get

$$N_e \approx \tfrac{1}{2}\left(\frac{1}{N_1} + \frac{1}{N_2}\right)^{-1}.$$

Exercise 3.2.2 From equation (3.18) we get

$$N_e \approx \frac{1}{2} \left(\frac{1}{1000} + \frac{1}{N_2} \right)^{-1}.$$

For $N_1 = 1000$ we get $N_e = 1000$, and

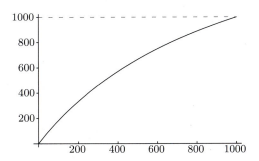

Exercise 4.2.1 At equilibrium $p = p' = \hat{p}$, so $\hat{p}$ solves the equation

$$\hat{p} = (1 - \mu_A)\hat{p} + \mu_a(1 - \hat{p}).$$

Exercise 4.2.2

$$
\begin{aligned}
p' - \hat{p} &= (1 - \mu_A)p + \mu_a q - \hat{p} \\
&= (1 - \mu_A)p + \mu_a q - \Big((1 - \mu_A)\hat{p} + \mu_a\hat{q} \Big) \\
&= (1 - \mu_A)(p - \hat{p}) + \mu_a(q - \hat{q}) \\
&= (1 - \mu_A - \mu_a)(p - \hat{p}).
\end{aligned}
$$

Exercise 4.5.1 The equilibrium homozygosity is

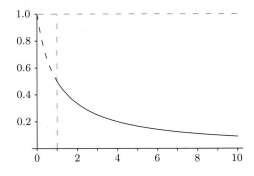

The heterozygosity is just $H = 1 - F$.

Exercise 4.5.2 For two alleles $H = 2p(1-p)$ and

$$\frac{dH}{dp} = 2(1-2p).$$

Thus, H has its maximum at $p = \frac{1}{2}$ and its value is $\frac{1}{2}$.

For three alleles $H = 2(p_1 p_2 + p_1 p_3 + p_2 p_3)$. Since $p_3 = 1 - p_1 - p_2$, we get

$$\frac{dH}{dp_1} = 2\Big(p_2 + p_3 - p_1 - p_2\Big).$$

The derivative is therefore zero when $p_1 = p_3$ and in the same way we get $p_2 = p_3$. Thus, the extreme value of H is at $p_1 = p_2 = p_3 = \frac{1}{3}$. This is clearly a maximum, and the value is at $\frac{2}{3}$.

The general result for k alleles requires a bit more bookkeeping. Again, let

$$p_k = 1 - \sum_{i=1}^{k-1} p_i.$$

The heterozygosity may be written as

$$H = 1 - \sum_{i=1}^{k-1} p_i^2 - p_k^2 \quad \text{with} \quad \frac{dH}{dp_j} = 2\Big(-p_j + p_k\Big).$$

Again, the derivative is zero when $p_j = p_k$ for $j = 1, 2, \ldots, k$, the extreme value of H is at $p_1 = p_2 = \cdots p_k = \frac{1}{k}$, and

$$\max H = 1 - (k-1)\left(\tfrac{1}{k}\right)^2 - \left(\tfrac{1}{k}\right)^2 = \tfrac{k-1}{k}.$$

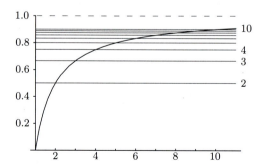

Exercise 5.2.1

$$c_i = \sum_{k=1}^{d} \tilde{m}_{ki}\tilde{c}_k = (1-m)\tilde{c}_i + m\tilde{c}_i \sum_{k=1}^{d} \tilde{c}_k = \tilde{c}_i.$$

The backward migration matrix is then given by

$$m_{ii} = \frac{\tilde{m}_{ii}\tilde{c}_i}{c_i} = \tilde{m}_{ii} = (1 - m) + m\tilde{c}_i = (1 - m) + mc_i,$$

$$m_{ij} = \frac{\tilde{m}_{ji}\tilde{c}_j}{c_i} = \frac{mc_i\tilde{c}_j}{\tilde{c}_i} = m\tilde{c}_j = mc_j.$$

Exercise 5.2.2 The migration matrix of Wright's two-island model is

$$\left\{ \begin{array}{cc} 1 - m + \frac{1}{2}m & \frac{1}{2}m \\ \frac{1}{2}m & 1 - m + \frac{1}{2}m \end{array} \right\} = \left\{ \begin{array}{cc} 1 - \frac{1}{2}m & \frac{1}{2}m \\ \frac{1}{2}m & 1 - \frac{1}{2}m \end{array} \right\}.$$

The migration matrix of the simple two-island model is

$$\left\{ \begin{array}{cc} 1 - m & m \\ m & 1 - m \end{array} \right\},$$

and its migration frequency parameter is therefore twice that of Wright's two-island model.

Exercise 5.2.3 If the population is well described by Wright's island model, we expect to observe variation that can be described by a coalescent process, if we may assume that each local population is sampled at most once.

If we have a good idea of the value of μ the estimate of θ would provide an estimate of N_e. This will, however, be considerably larger than expected based on knowledge of the effective population sizes of local populations. It will be inflated by the factor

$$\left(1 + \frac{1}{4Nm} \right).$$

Exercise 5.2.4 We need to find a solution to

$$(\pi_1, \pi_2) = (\pi_1, \pi_2) \left\{ \begin{array}{cc} 1 - \frac{1}{2}m & \frac{1}{2}m \\ m & 1 - m \end{array} \right\}.$$

This type of equation always simplifies as

$$(\pi_1, \pi_2) \left\{ \begin{array}{cc} -\frac{1}{2}m & \frac{1}{2}m \\ m & -m \end{array} \right\} = (0, 0).$$

The solution is $\pi_1 = 2\pi_2$, and as $\pi_1 + \pi_2 = 1$ we get the suggested stationary distribution. N_e is found from equation (5.14).

Exercise 5.2.5 We need to solve the equation $\hat{\pi}(\boldsymbol{M} - \boldsymbol{I}) = \boldsymbol{0}$, where $\boldsymbol{I}$ is the identity matrix. As $\boldsymbol{M}$ is conservative, the solution is $\pi_1 = \pi_2 = \pi_3 = \frac{1}{3}$, and therefore $N_e = 3N$.

Exercise 5.2.6 This $\boldsymbol{M}$ is not conservative. The stationary distribution is $\pi_1 = \pi_3 = \frac{1}{4}$, $\pi_2 = \frac{1}{2}$. Therefore $N_e = \frac{8}{3}N$.

Exercise 5.2.7 The strong migration limit provides N_e as the sum of the population sizes of the constituent parts. N_e in Figure 5.7 converges quickly to $2N = 100$. Equation (5.7) provides the coalescence time $\mathrm{E}\,T_{20}$ as the coalescence time in a population of size $2N$, and as does $\mathrm{E}\,T_{11}$ for Nm large.

Exercise 5.3.1 The frequency of AA is

$$\sum_{i=1}^{d} c_i p_i^2 = \mathrm{E}\,p^2 = \mathrm{Var}(p) + (\mathrm{E}\,p)^2 = \mathrm{Var}(p) + \bar{p}^2.$$

Exercise 5.3.2 The mean is $\frac{1}{2}(p + \delta) + \frac{1}{2}(p - \delta) = p$. The variance is $\frac{1}{2}(p + \delta)^2 + \frac{1}{2}(p - \delta)^2 = \delta^2$. For $p = \frac{1}{2}$ the maximum excess of homozygotes is $\frac{1}{4}$, and in general it is $\min(p^2, q^2)$ because the gene frequencies cannot be negative.

Exercise 5.3.3 The gene frequency in the inner Danish waters is 0.610, and the similarity between the observed genotypic counts and those expected with Hardy–Weinberg proportions is excellent (Table B.1)—may be tested by a χ^2 goodness-of-fit-test with one degree of freedom. The gene frequency in the Baltic

Table B.1: Danish and Baltic samples

	A_1A_1	A_1A_2	A_2A_2	Total
Denmark	258	324	106	688
expected HW	256.4	327.2	104.3	688
Baltic	0	10	230	240
expected HW	0.1	9.8	230.1	240

is 0.021, and again correspondence to Hardy–Weinberg expectations is excellent (the χ^2 goodness-of-fit-test cannot be used because of the low expectation in the A_1A_1 class).

In the western Baltic sample 1 the gene frequency is 0.474 and in sample 2 it is 0.342. In both cases the deviation from Hardy–Weinberg proportions is positive (Table B.2): $\hat{F}_1 = 0.110$ and $\hat{F}_2 = 0.351$. The deviation in sample 1 is high,

Table B.2: Samples in the western Baltic Sea

Sample	A_1A_1	A_1A_2	A_2A_2	Total
1	67	118	81	266
expected HW	59.7	132.6	73.7	266
2	49	73	128	250
expected HW	29.2	112.5	108.2	250

but not statistically significant. That in sample 2 is highly significant. Further samples close to the island of Bornholm also showed an excess of homozygotes.

Given the large genetic differentiation between the inner Danish waters and the Baltic Sea, we can propose that the observations around Bornholm show population mixing. This is consistent with the observed genetic heterogeneity in contrast to the homogeneity in the main areas of investigation.

Let a sample consist of the fraction c from inner Danish waters and $1 - c$ from the Baltic. Then $\bar{p} = cp_1 + (1 - c)p_2$, and therefore $c = (\bar{p} - p_2)/(p_1 - p_2)$. Doing this, we get $c_1 = 0.77$ and $c_2 = 0.55$, and calculating the theoretical Fs from these produces $F_1 = 0.10$ and $F_2 = 0.22$. These values do not seem to contradict the conclusion of the analysis.

Exercise 6.1.1 The A genes segregate in the first division, the B genes in the second.

Exercise 6.1.2 The easy solution is to count recombinants: the number of observed recombinants in the left sector is $r_1 n$, and in the right it is $r_2 n$. In both of these numbers we counted double crossovers, but they do not appear as recombinants between the outer loci. We thus have $r_1 + r_2 - 2\iota r_1 r_2$ as the frequency of recombination.

Suppose the two alleles at each of the three loci are named 0 and 1 and we consider gametes from the heterozygote 000/111. The frequency of double crossover gametes, 010 or 101, is $\iota r_1 r_2$ from the definition of the coefficient of coincidence. The total frequency of the gametes 011, 100, 010, and 101 is r_1 and the gamete frequencies are therefore

$$
\begin{aligned}
&\text{000 or 111: } 1 - r_1 - r_2 + \iota r_1 r_2 \\
&\text{001 or 110: } r_2 - \iota r_1 r_2 \\
&\text{011 or 100: } r_1 - \iota r_1 r_2 \\
&\text{010 or 101: } \iota r_1 r_2
\end{aligned}
$$

The frequency of recombination between the outer loci is therefore

$$(r_2 - \iota r_1 r_2) + (r_1 - \iota r_1 r_2) = r_1 + r_2 - 2\iota r_1 r_2.$$

Exercise 6.2.1 The observed linkage disequilibrium value needs to be related to its allowed range before it can be evaluated.

For $p_A = p_B = 0.5$ we get $-0.25 < D < 0.25$, for $p_A = p_B = 0.2$ we get $-0.04 < D < 0.16$, and for $p_A = 0.4$ and $p_B = 0.2$ we get $-0.1 < D < 0.1$. In the third case D is at its maximum value, and in the other cases D is large, but not very large.

Exercise 6.2.2 We use the definition $D = x_1 - p_A p_B$ and want to get rid of $p_A = x_1 + x_2$ and $p_B = x_1 + x_3$:

$$
\begin{aligned}
D &= x_1 - (x_1 + x_2)(x_1 + x_3) \\
&= x_1(x_1 + x_2 + x_3 + x_4) - (x_1 + x_2)(x_1 + x_3) \\
&= x_1 x_4 - x_2 x_3.
\end{aligned}
$$

Note the trick of multiplying x_1 by 1 to ease calculations.

Exercise 6.2.3   The covariance is $x_4 - q_A q_B$, and the variances at the two loci are $p_A q_A$ and $p_B q_B$.

Exercise 6.4.1   No! It corresponds to the type of recombination event shown in equation (6.11) on page 168.

Exercise 7.2.1   From the definitions of the additive and the dominance variance we get

$$
\begin{aligned}
V_A + V_D &= \sum_{i=1}^{k}\sum_{j=1}^{k}(a_i + a_j)^2 p_i p_j + \sum_{i=1}^{k_\ell}\sum_{j=1}^{\ell}(g_{ij} - (a_i + a_j))^2 p_i p_j \\
&= \sum_{i=1}^{k}\sum_{j=1}^{k}(a_i^2 + a_j^2 + 2a_i a_j) p_i p_j + \sum_{i=1}^{k}\sum_{j=1}^{k} g_{ij}^2 p_i p_j \\
&\quad - 2\sum_{i=1}^{k} a_i^2 p_i - 2\sum_{j=1}^{k} a_j^2 p_j
\end{aligned}
$$

by using the definition of additive effects. From equation (7.9) the first and the last term are both equal to twice the additive variance, and the result follows.

Exercise 7.2.2   The variance is $V_A = 2a^2 p_A q_a$ from equation (7.18), and in Exercise 4.5.2 we showed that $p_A q_a$ is at maximum for even gene frequencies.

Exercise 7.2.3   The probabilities that a pair of half-sibs share none, one, or two identical genes at a given locus are $\frac{1}{2}$, $\frac{1}{2}$, and 0, respectively. Thus, we have $C_{\mathcal{HS}} = \frac{1}{2} \times \frac{1}{2} V_A$.

Exercise 8.2.1   We get

$$
W_A = p + wq, \quad W_a = wp + w^2 q = w(p + wq), \quad W = (p + wq)^2,
$$

and therefore

$$
p' = \frac{p}{p + wq} \quad \text{and} \quad q' = \frac{wq}{p + wq}.
$$

The selection model thus corresponds a haploid model with the viabilities $w_A = 1$ and $w_a = w$.

Exercise 8.2.2   Let A be the *carbonaria* allele and a the *typica* allele, that is, $w_{Aa} = w_{AA}$. Thus, $W = (p^2 + 2pq)w_{Aa} + q^2 w_{AA}$, $W_A = w_{AA}$, and equation (8.1) becomes

$$
p' = \frac{p w_{AA}}{(p^2 + 2pq)w_{AA} + q^2 w_{aa}}.
$$

Dividing both the numerator and the denominator by w_{AA} shows that only the ratio of the survival probabilities matters, so let $w_{AA} = 1$ and $w_{aa} = 1 - s$. The average fitness is then $W = 1 - sq^2$ and the recurrence equation (8.1) becomes

$$p' = \frac{p}{1 - sq^2} \, .$$

In polluted areas with $p \approx 0$ we resort to a Taylor expansion of the right side of the recurrence equation:

$$p' = \frac{d}{dp} \left. \frac{p}{1 - s(1 - p)^2} \right|_{p=0} \times p + \frac{d^2}{dp^2} \left. \frac{p}{1 - s(1 - p)^2} \right|_{p=0} \times \tfrac{1}{2} p^2 + \cdots .$$

The derivative is

$$\frac{d}{dp} \frac{p}{1 - sq^2} = \frac{1}{1 - sq^2} + \frac{-2spq}{(1 - sq^2)^2}$$

and at $p = 0$ this is $\frac{1}{1-s}$. The recurrence equation is then approximately

$$p' \approx \frac{p}{1 - s} \quad \text{and therefore} \quad p^{(t)} \approx \frac{p}{(1 - s)^t} \, ,$$

where we neglected terms of the order of $\left(p^{(t)} \right)^2$ when t is not too large. Thus, the frequency of *carbonaria* allele initially shows exponential increase by the factor $\frac{1}{1-s}$. Exponential growth implies that the time it takes for the gene frequency to double is constant, and in this case it will double about every

$$-\frac{\log 2}{\log(1 - s)} \approx \frac{\log 2}{s} \quad \text{generations.}$$

In pristine areas with $q \approx 0$ we might again use a Taylor expansion of the recurrence equation around $p = 1$, or rather $q = 0$,

$$q' = \frac{q(1 + sq)}{1 + sq^2} \, ,$$

where we assumed $w_{aa} = 1 + s$. The first derivative is evidently equal to one. To express that directional selection favors the *typica* allele, the second term of the Taylor expansion therefore needs to be included, that is $q' \approx q + sq^2$, which is hard to iterate. Numerical calculation of the doubling time is, however, more informative, and below is an example with $w_{aa} = 1.5$ with initial gene frequency $q = 0.001$ and doubling times indicated. The figure below shows decreasing doubling times (from the first to the second doubling the time decreases by about a factor 2) and the increase of the gene frequency is therefore not exponential. In fact, as the initial gene frequency becomes smaller and smaller the doubling time increases without bounds. Exercise 8.2.3 discusses this phenomenon more thoroughly.

The final elimination of the *typica* allele is slow in polluted areas. The halving time increases as the gene frequency decreases (see Exercise 8.2.3), and the increase of the *carbonaria* allele becomes slow when it is common. In pristine areas elimination of the *carbonaria* allele is fast, approximately by exponential decrease by the factor $\frac{1}{1+s}$.

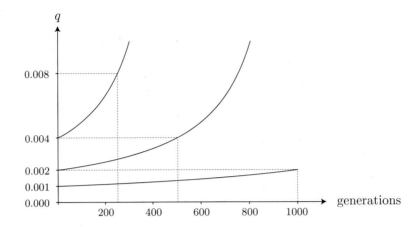

Exercise 8.2.3 The recurrence equation for a recessive allele was found in Exercise 8.2.2, and with $s = 1$ we get:

$$q' = \frac{q(1-q)}{1-q^2},$$

which immediately simplifies to

$$q' = \frac{q}{1+q}.$$

The frequency of a therefore decreases, but the rate of decrease slows as the recessive allele becomes rarer. The recurrence equation is so simple that it may be iterated, for instance,

$$q'' = \frac{q'}{1+q'} = \frac{q}{(1+q)+q} = \frac{q}{1+2q},$$

which suggests that after t generations we have

$$q^{(t)} = \frac{q}{1+tq}.$$

To verify this we find $q^{(t+1)}$ as

$$q^{(t+1)} = \frac{q^{(t)}}{1+q^{(t)}} = \frac{q}{(1+tq)+q} = \frac{q}{1+(t+1)q}.$$

When is $q^{(t)} = \frac{1}{2}q$? When t solves the equation

$$\tfrac{1}{2}q = \frac{q}{1+tq}!$$

That happens when $t = q^{-1}$. The time to reach half the frequency of a lethal allele thus increases strongly as the gene frequency decreases. For example, a

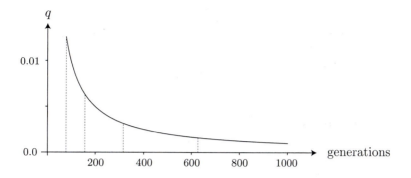

population with $q = 0.01$ reaches $q = 0.005$ in 100 generations, whereas the subsequent journey to $q = 0.0025$ takes 500 generations. Below is an example with halving times indicated.

Selection happens by elimination of the aa homozygotes. The frequency of a alleles that reside in homozygotes is q, and as the allele becomes rarer, the fraction exposed as the recessive phenotype becomes smaller.

This argument relies heavily on the assumption of Hardy–Weinberg proportions. With inbreeding, for instance, the frequency of homozygotes is close to Fq for a rare recessive, where F is the population inbreeding coefficient. The rate of elimination therefore becomes much higher, resembling the situation of a rare dominant allele.

Exercise 8.2.4 The equilibrium equation is $W_A = W_a$, namely

$$pw_{AA} + (1-p)w_{Aa} = pw_{Aa} + (1-p)w_{aa}.$$

Collecting terms with and without p immediately gives the solution (8.3), except when

$$2w_{Aa} = w_{AA} + w_{aa}.$$

The heterozygote fitness is here exactly in the middle between the homozygote fitnesses (it is the additive alleles model for the quantitative character fitness).

The equilibrium gene frequencies are

$$\hat{p} = \frac{w_{Aa} - w_{aa}}{2w_{Aa} - w_{AA} - w_{aa}}$$

$$\hat{q} = \frac{w_{Aa} - w_{AA}}{2w_{Aa} - w_{AA} - w_{aa}}.$$

The solution provides valid gene frequencies when $\hat{p}$ and $\hat{q}$ are both positive. The two numerators must then have the same sign as the denominator, which is their sum, and they therefore both have to be either positive or negative. That is, w_{Aa} either has to be larger than w_{AA} and w_{aa} or smaller than both.

Exercise 8.2.5 The correspondence between the direction of change in the gene frequency and the behavior of the average fitness may be written into the recurrence equation in the gene frequencies. The difference equation corresponding to equation (8.1) is

$$p' - p = \frac{p(W_A - W)}{W}.$$

The average fitness is $W = pW_A + qW_a$, and therefore $W_A - W = q(W_A - W_a)$. Using this we get

$$\frac{dW}{dp} = 2(W_A - W_a) = 2\frac{W_A - W}{q}.$$

Inserting this result into the difference equation produces equation (8.4).

Exercise 8.2.6 The observed population could be of mixed origin, but given the amount of variation in the studied area this explanation this seems unlikely (Christiansen et al. 1977). The studied population is situated in the westernmost of the small inlets just north of the gene frequency of 0.36 close to the center of the map in Figure 5.10 on page 132.

Exercise 8.3.1 The one-to-one correspondence between genotypic values and parameters is shown by a straightforward calculation of averages. Remember, however, to subtract the population mean

$$E\,G = 2(\varepsilon_A p + \varepsilon_a q) + (p - q)^2 \eta$$

to get genotypic effects. The additive effect of A is

$$
\begin{aligned}
\alpha_A &= (2\varepsilon_A + \eta)p + (\varepsilon_A + \varepsilon_a - \eta)q - \left(2(\varepsilon_A p + \varepsilon_a q) + (p - q)^2\eta\right) \\
&= q(\varepsilon_A - \varepsilon_a) + 2q(p - q)\eta.
\end{aligned}
$$

We thus get $a = (\varepsilon_A - \varepsilon_a) + 2(p - q)\eta$, and the additive variance is given by equation (7.18). The dominance effect of AA is

$$
\begin{aligned}
d_{Aa} &= (\varepsilon_A + \varepsilon_a - \eta) - \left(2(\varepsilon_A p + \varepsilon_a q) + (p - q)^2\eta\right) \\
&\quad - \left(\left(q(\varepsilon_A - \varepsilon_a) + 2q(p - q)\eta\right) + \left(-p(\varepsilon_A - \varepsilon_a) - 2p(p - q)\eta\right)\right) \\
&= -(\varepsilon_A - \varepsilon_a)(p - q) - \left(1 + (p - q)^2\right)\eta \\
&\quad - \left(\left(q(\varepsilon_A - \varepsilon_a) + 2q(p - q)\eta\right) + \left(-p(\varepsilon_A - \varepsilon_a) - 2p(p - q)\eta\right)\right) \\
&= -4pq\eta.
\end{aligned}
$$

We thus get $d = 4\eta$, and the dominance variance is given by equation (7.18).

Exercise 8.3.2 Using Table 7.3 we find

$$\frac{w_{AA}}{w_{Aa}} = \frac{1 + j2q_a\alpha}{1 - j(p_A - q_a)\alpha} \approx 1 + j2q_a\alpha + j(p_A - q_a)\alpha = 1 + j\alpha,$$

where $j = i/\sqrt{V_E}$. Similarly, $w_{aa}/w_{Aa} = 1 - j\alpha$.

Exercise 8.3.3 The allele fitnesses are $W_{\mathsf{A}} = pw + q$, $W_{\mathsf{a}} = p + qw$ and the average fitness is $W = (p^2 + q^2)w + 2pq$. At equilibrium $W_{\mathsf{A}} = W_{\mathsf{a}}$, and therefore $pw + q = p + qw$. This is true only when $p = q$ and therefore $\hat{p} = \frac{1}{2}$.

The additive effect of A is

$$W_{\mathsf{A}} - W = pw + q - (p^2 + q^2)w - 2pq = pqw - q^2w - (p-q)q = q(p-q)(w-1),$$

and therefore $\alpha = (p-q)(w-1)$ and the additive variance is

$$V_A = 2pq(p-q)^2(1-w)^2$$

from equations (7.18).

The genotypic deviation of the heterozygote is

$$1 - (p^2 + q^2)w - 2pq = (p^2 + q^2)(1-w).$$

In terms of the parameters in Table 7.3 this genotypic deviation is

$$(p-q)^2(1-w) - pq\delta.$$

Equating these gives $\delta = 2(1-w)$, and from equations (7.18) we get the dominance variance as

$$V_D = \big(2pq(1-w)\big)^2.$$

The variance in fitness among the individuals in the population is

$$
\begin{aligned}
V_G &= (p^2 + q^2)w^2 + 2pq - \big((p^2 + q^2)w + 2pq\big)^2 \\
&= (1 - 2pq)w^2 + 2pq - \big((1 - 2pq)w + 2pq\big)^2 \\
&= 2pq(1 - 2pq)w^2 - 4pq(1 - 2pq)w + 2pq(1 - 2pq) \\
&= 2pq(1 - 2pq)(1 - w)^2.
\end{aligned}
$$

The sum of the variance components is

$$
\begin{aligned}
V_A + V_D &= 2pq(p-q)^2(1-w)^2 + \big(2pq(1-w)\big)^2 \\
&= 2pq(1-w)^2\big((p-q)^2 + 2pq\big) = V_G.
\end{aligned}
$$

The variance components are shown in Figure 8.11, bottom left, and the scale is given by $\hat{V}_G = \frac{1}{4}(1-w)^2$ for $p = \frac{1}{2}$.

Exercise 8.3.4 The covariance is

$$2W_{\mathsf{A}}w_{\mathsf{AA}}p^2 + (W_{\mathsf{A}} + W_{\mathsf{a}})w_{\mathsf{Aa}}2pq + 2W_{\mathsf{a}}w_{\mathsf{aa}}q^2 - W^2$$

$$
\begin{aligned}
&= 2\Big(pW_{\mathsf{A}}(w_{\mathsf{AA}}p + w_{\mathsf{Aa}}q) + qW_{\mathsf{a}}(w_{\mathsf{Aa}}p + w_{\mathsf{aa}}q) - W^2\Big) \\
&= 2\Big(pW_{\mathsf{A}}^2 + qW_{\mathsf{a}}^2 - W^2\Big) = V_A.
\end{aligned}
$$

This result is more general in that using the specification in equation (7.13) we get

$$\sum_{i=1}^{k}\sum_{j=1}^{k} g_{ij}(a_i + a_j)p_i p_j = \sum_{i=1}^{k}\sum_{j=1}^{k}(a_i + a_j + d_{ij})(a_i + a_j)p_i p_j = \sum_{i=1}^{k}\sum_{j=1}^{k}(a_i + a_j)^2 p_i p_j.$$

Exercise 8.3.5   In Exercise 8.2.5 we showed that

$$p' - p = \frac{p(W_\mathsf{A} - W)}{W}$$

and $\alpha_w = W_\mathsf{A} - W_\mathsf{a}$.

The additive effect has the same sign for all gene frequencies with directional selection, and fitness is not heritable when it is zero.

Exercise 8.3.6

$$
\begin{aligned}
\mathrm{Cov}(w, \Gamma) &= \mathrm{E}(w, \Gamma) - (\mathrm{E}\,w)(\mathrm{E}\,\Gamma) = 2pqw_\mathsf{Aa} + 2q^2 w_\mathsf{aa} - 2qW \\
&= 2q(W_\mathsf{a} - W) = -2pq\,(W_\mathsf{A} - W_\mathsf{a})\,.
\end{aligned}
$$

Exercise 8.3.7   In Table 8.7 use the genotypic proportions in equation (5.18) on page 135 instead of the Hardy–Weinberg proportions. The average fitness is

$$W_F = (p^2 + pqF)w_\mathsf{AA} + 2pq(1 - F)w_\mathsf{Aa} + (q^2 + pqF)w_\mathsf{aa}.$$

The average fitnesses of the two alleles are

$$
\begin{aligned}
W_{\mathsf{A}F} &= (p + qF)w_\mathsf{AA} + q(1 - F)w_\mathsf{Aa}, \\
W_{\mathsf{a}F} &= p(1 - F)w_\mathsf{Aa} + (q + pF)w_\mathsf{aa},
\end{aligned}
$$

and $W_F = pW_{\mathsf{A}F} + qW_{\mathsf{a}F}$. The recurrence equation becomes

$$p' = \frac{pW_{\mathsf{A}F}}{W_F} \quad \text{or} \quad p' - p = \frac{pq(W_{\mathsf{A}F} - W_{\mathsf{a}F})}{W_F}\,.$$

As in Exercise 8.3.6, we get

$$
\begin{aligned}
\mathrm{Cov}(w, \Gamma) &= 2pq(1 - F)w_\mathsf{Aa} + 2(q^2 + pqF)w_\mathsf{aa} - 2qW_F \\
&= 2q(W_{\mathsf{a}F} - W_F) = -2pq\,(W_{\mathsf{A}F} - W_{\mathsf{a}F})\,.
\end{aligned}
$$

Thus, the change in gene frequency is again given by Price's formula (8.14).

Exercise 8.4.1   From equation (8.16) we have

$$\hat{q} = \sqrt{\frac{\mu}{s}}$$

where we have written $w_\mathsf{aa} = 1 - s$ relative to $w_\mathsf{AA} = w_\mathsf{Aa} = 1$. The equilibrium viability of allele A is therefore $\hat{W}_\mathsf{A} = 1$. The equilibrium viability of allele a is $\hat{W}_\mathsf{a} = 1 - s\hat{q}$, which is $\hat{W}_\mathsf{a} = 1 - \sqrt{s\mu}$.

Exercise 8.4.2   The fitness of dwarfs relative to that of normal sibs is

$$w_{Aa} = \frac{27}{108} \times \frac{457}{582} = 0.20.$$

The equilibrium frequency is

$$\frac{10}{188,000} = 0.0004 = \frac{\mu}{1 - w_{Aa}}.$$

A mutation rate of

$$\frac{7}{188,000} = 4 \times 10^{-5}$$

then corresponds to

$$w_{Aa} = 1 - \frac{7}{10} = 0.30.$$

This is in reasonable agreement with the directly measured value, which would correspond exactly to 8 out of 10 newborn dwarfs having normal parents.

Exercise 8.4.3

$$\hat{q} = \frac{(1 + \mu_A)}{2(1 - 2w_{Aa} + w_{aa})} \left((1 - w_{Aa}) - \sqrt{(1 - w_{Aa})^2 - \frac{4\mu_A(1 - 2w_{Aa} + w_{aa})}{(1 + \mu_A)^2}} \right)$$

for $1 - 2w_{Aa} + w_{aa} \neq 0$, and when the effect of the alleles on the fitness is additive we get

$$\hat{q} = \frac{2\mu_A}{(1 + \mu_A)(1 - w_{aa})} \quad \text{for } w_{Aa} = \tfrac{1}{2}(1 + w_{aa}).$$

Exercise 8.5.1   The influx of mutations is $2N\mu$ per generation, so the added frequency is μ per generation.

Exercise 8.6.1   For the last two columns in Table 8.9 the pairs AG, AA, CA, and CG are observed. In the last column and in each of the two closest to the electrophoretic allele site we have CG, CA, TA, and TG.

Exercise 9.1.1   The equilibrium is given by

$$\hat{x}_1 = \hat{x}_4 = \tfrac{1}{4}\left(1 + \sqrt{1 - \frac{4r}{\beta + \gamma - \alpha}} \right),$$

$$\hat{x}_2 = \hat{x}_3 = \tfrac{1}{4}\left(1 - \sqrt{1 - \frac{4r}{\beta + \gamma - \alpha}} \right),$$

or vice versa. Let S designate the square root, then

$$D = \hat{x}_1\hat{x}_4 - \hat{x}_2\hat{x}_3 = \tfrac{1}{16}\left((1 + S)^2 - (1 - S)^2 \right) = \tfrac{1}{4}S.$$

As r increases from 0 to $\tfrac{1}{4}(\beta + \gamma - \alpha)$, S decreases from 1 to 0.

Exercise 9.1.2 Refer to Table 9.7 on page 278. The first two conditions are the conditions that the homozygotes have a higher fitness than the single heterozygotes. The condition that the complementary gamete cannot increase is $w_{11} > (1 - r)w_{14}$, which in the symmetric viability model is $1 - \alpha > (1 - r)$, or equivalently, $r > \alpha$.

Exercise 9.4.1 Use the approximations $\hat{q}(F) \approx \hat{q}(1)/F$ and $\hat{p}(F)\hat{q}(F) \approx \hat{q}(F)$ and note that the dominance is given by $-w_{\mathsf{AA}} + 2w_{\mathsf{Aa}} - w_{\mathsf{aa}} = 1 - w_{\mathsf{aa}}$.

Exercise 9.5.1 Limits to selection due to random genetic drift give a lower response per generation. Random genetic drift causes loss of variation and the buildup of stronger associate fitness effects—both effects give a lower total response to selection. The stronger associate fitness effects give a quicker and deeper decrease in the character mean when selection ceases.

Exercise 9.5.2 We see still tighter limits to selection and stronger associate fitness effects. Large blocks of linked genes are produced by selection and drift, and the breakup of such a block may produce the renewed response after an intermittent equilibrium. The renewed response is even accelerated compared to the initial response in accordance with the expectation that the responses need not be related.

Exercise 10.1.1 Directional selection for allele a corresponds to $s > 0$. The allele fitnesses are $W_{2\mathsf{A}} = 1 - s - sp_2$, $W_{2\mathsf{a}} = 1 - sp_2$, and $W_2 = 1 - 2sp_2$. The balance equilibrium (10.4) is not exact in this case, but

$$\hat{p}_2 = \frac{m_2}{s(1 + m_2)}.$$

This equilibrium is valid and stable for $m_2 < s/(1-s)$. Otherwise the frequency of allele a decreases to zero. It corresponds well to the balance equilibrium (10.4) when m_2 is small. With polymorphism in population 1 we obtain a picture much like that in Figure 10.3.

Exercise 10.1.2 The contrast between the larger and smaller islands may have several explanations, and the following is only a short and incomplete sketch.

The simplest explanation is random genetic drift within the populations and migration between them. The urge of a butterfly to flutter out to sea is probably small, so migration probably happens as a result of accidental dispersals. A butterfly over the sea is doomed unless it happens to pass over an island. We can thus suppose that immigration is proportional to the size of the island. Considerably more than one immigrant per generation in the large islands and a lot fewer in the small islands could then explain the observed contrast between large and small islands (Wright's island model with migration). Placing Cornwall into the equation requires that the peninsula produce a sufficient number of immigrants to at least one of the larger islands. Alternatively, the Scilly Isles harbor a large population founded by a high number of individuals originating from the population that also founded the Cornwall population.

Inclusion of selection may provide what seems an easier explanation of the homogeneity among the four large populations. The environments are supposedly similar, and the spot configuration may be under selection—sexual selection by mate preferences, for instance. Then each of the small islands may be described by the classical island model with little immigration.

Finally, an explanation not referring to random genetic drift could posit different selection in different environments, with the large islands and the peninsula having similar distributions of these. The small islands with fewer and less variable habitats should then be more diverse.

The drift explanation is the most parsimonious, and to refute it we either have to obtain information on dispersal, population sizes, and the history of the populations (difficult requirements), or try a comparison to variation patterns of molecular genetic polymorphisms.

Bibliography

Aguadé, M., Miyashita, N., and Langley, C. H. (1989), "Reduced variation in the *yellow-achaete-scute* region in natural populations of *Drosophila melanogaster*," *Genetics* **122**, 607–615.

Akin, E. (1979), *The Geometry of Population Genetics*, vol. 31 of *Lecture Notes in Biomathematics*, Springer Verlag, Berlin, Heidelberg, New York.

Allison, A. C. (1964), "Polymorphism and natural selection in human populations," *Cold Spring Harbor Symp. Quant. Biol.* **29**, 137–149.

Angelini, D. R., and Kaufman, T. C. (2005), "Comparative developmental genetics and the evolution of arthropod body plans," *Annu. Rev. Genet.* **39**, 95–119.

Antony, J. M., van Marle, G., Opii, W., Butterfield, D. A., Mallet, F., Yong, V. W., Wallace, J. L., Deacon, R. M., Warren, K., and Power, C. (2004), "Human endogenous retrovirus glycoprotein-mediated induction of redox reactants causes oligodendrocyte death and demyelination," *Nat. Neurosci.* **7**, 1088–1095.

Ayala, F. J., Tracey, M. L., Barr, L. G., MacDonald, J. F., and Perez-Salas, S. (1974), "Genetic variation in natural populations of five *Drosophila* species and the hypothesis of the selective neutrality of protein polymorphisms," *Genetics* **77**, 343–384.

Bahn, E. (1968), "Crossing over in the chromosomal region determining amylase isoenzymes in *Drosophila melanogaster*," *Hereditas* **58**, 1–12.

Bannert, N., and Kurth, R. (2006), "The evolutionary dynamics of human endogenous retroviral families," *Annu. Rev. Genomics Hum. Genet.* **7**, 149–173.

Barker, J. S. F. (1977), "Population genetics of a sex-linked locus in *Drosophila melanogaster*. I. Linkage disequilibrium and associative overdominance," *Hereditas* **85**, 169–198.

Barrett, S. C. H., ed. (1992), *Evolution and Function of Heterostyly*, Springer Verlag, Berlin, Heidelberg.

Barton, N. H., and Bengtsson, B. O. (1986), "The barrier to genetic exchange between hybridising populations," *Heredity* **56**, 357–376.

Barton, N. H., and Keightley, P. D. (2002), "Understanding quantitative genetic variation," *Nat. Rev. Genet.* **3**, 11–21.

Bateson, W., and Saunders, E. R. (1902), "Experimental studies in the physiology of heredity," *Rep. Evol. Comm. R. Soc.* **1**, 1–160.

Bateson, W., Saunders, E. R., and Punnett, R. C. (1905), "Experimental studies in the physiology of heredity," *Rep. Evol. Comm. R. Soc.* **2**, 1–55 and 80–99.

Beja-Pereira, A., Luikart, G., England, P. R., Bradley, D. G., Jann, O. C., Bertorelle, G., Chamberlain, A. T., Nunes, T. P., Metodiev, S., Ferrand, N., and Erhardt, G. (2003), "Gene-culture coevolution between cattle milk protein genes and human lactase genes," *Nat. Genet.* **35**, 311–313.

Bengtsson, B. O. (1978), "Avoiding inbreeding – at what cost," *J. Theor. Biol.* **73**, 439–444.

Bengtsson, B. O. (1985), The flow of genes through a genetic barrier, *in* J. J. Greenwood, P. H. Harvey, and M. Slatkin, eds., *Evolution essays in Honour of John Maynard Smith*, Cambridge Univ. Press, Cambridge, New York, pp. 31–42.

Bengtsson, B. O., and Bodmer, W. F. (1976), "On the increase of chromosome mutations under random mating," *Theor. Popul. Biol.* **9**, 260–281.

Bengtsson, B. O., Weibull, P., and Ghatnekar, L. (1995), "The loss of alleles by sampling: A study of the common outbreeding grass *Festuca ovina* over three geographic scales," *Hereditas* **122**, 221–238.

Bennett, J. H. (1954), "On the theory of random mating," *Ann. Eugenics* **18**, 311–317.

Berlin, S., and Ellegren, H. (2006), "Fast accumulation of nonsynonymous mutations on the female-specific W chromosome in birds," *J. Mol. Evol.* **62**, 66–72.

Berry, A., and Kreitman, M. (1993), "Molecular analysis of an allozyme cline: Alcohol dehydrogenase in *Drosophila melanogaster* on the east coast of North America," *Genetics* **134**, 869–893.

Blæsild, P., and Granfeldt (2003), *Statistics with Applications in Biology and Geology*, Chapman & Hall/CRC, London.

Blond, J. L., Beseme, F., Duret, L., Bouton, O., Bedin, F., Perron, H., Mandrand, B., and Mallet, F. (1999), "Molecular characterization and placental expression of HERV-W, a new human endogenous retrovirus family," *J. Virol.* **73**(2), 1175–1185.

Bodmer, W. F., and Cavalli-Sforza, L. L. (1968), "A migration matrix model for the study of random genetic drift," *Genetics* **59**, 565–592.

Bodmer, W. F., and Felsenstein, J. (1967), "Linkage and selection: Theoretical analysis of the deterministic two locus random mating model," *Genetics* **57**, 237–265.

Bodmer, W. F., and Parsons, P. A. (1962), "Linkage and recombination in evolution," *Advan. Genet.* **11**, 1–100.

Bodmer, W. F., Trowsdale, J., Yong, J., and Bodmer, J. (1986), "Gene clusters and the evolution of the major histocompatibility system," *Phil. Trans. Roy. Soc. Lond. B Biol. Sci.* **312**, 303–315.

Bridges, C. B. (1936), "The Bar 'gene' a duplication," *Science* **83**, 210–211.

Bridges, C. B., and Olbrycht, T. M. (1926), "The multiple stock "Xple" and its use," *Genetics* **11**, 41–56.

Britton-Davidian, J., Nadeau, J. H., Croset, H., and Thaler, L. (1989), "Genic differentiation and origin of Robertsonian populations of the house mouse (*Mus musculus domesticus* Rutty)," *Genet. Res. (Camb.)* **53**, 29–44.

Broman, K. W., Murray, J. C., Sheffield, V. C., White, R. L., and Weber, J. L. (1998), "Comprehensive human genetic maps: Individual and sex-specific variation in recombination," *Amer. J. Hum. Genet.* **63**, 861–869. Data reside at http://www.marshmed.org/genetics/.

Brown, A. H. D. (1990), Genetic characterization of plant mating systems, *in* A. H. D. Brown, M. T. Clegg, A. L. Kahler, and B. S. Weir, eds., *Plant Population Genetics, Breeding and Genetic Resources*, Sinauer, Sunderland, MA, pp. 145–162.

Bulmer, M. G. (1971), "The effect of selection on genetic variability," *Amer. Natur.* **105**, 201–211.

Bürger, R. (2000), *The Mathematical Theory of Selection, Recombination, and Mutation*, John Wiley & Sons, Chichester.

Cain, A. J., and Sheppard, P. M. (1954), "Natural selection in *Cepaea*," *Genetics* **39**, 89–116.

Cavalli-Sforza, L. L., and Feldman, M. W. (1971), *Cultural Transmission and Evolution: A Quantitative Approach*, Princeton Univ. Press, Princeton, NJ.

Cavalli-Sforza, L. L., Menozzi, P., and Piazza, A. (1994), *The History and Geography of Human Genes*, Princeton Univ. Press, Princeton, NJ.

Chanter, D. O., and Owen, D. F. (1972), "The inheritance and population genetics of sex ratio in the butterfly *Acraea encedon*," *J. Zool.* **166**, 363–383.

Charlesworth, B. (1996), "Background selection and patterns of genetic diversity in *Drosophila melanogaster*," *Genet. Res. (Camb.)* **68**, 131–149.

Charlesworth, B., and Charlesworth, D. (1998), "The effect of recombination on background selection," *Genetica* **102/103**, 3–19.

Charlesworth, B., and Charlesworth, D. (1999), "The genetic basis of inbreeding depression," *Genet. Res. (Camb.)* **74**, 329–340.

Charlesworth, B., Morgan, M. T., and Charlesworth, D. (1993a), "The effect of deleterious mutations on neutral molecular variation," *Genetics* **134**, 1289–1303.

Charlesworth, D. (1991), "The apparent selection on neutral marker loci in partially inbreeding populations," *Genet. Res. (Camb.)* **57**, 159–175.

Charlesworth, D., Morgan, M. T., and Charlesworth, B. (1993b), "Mutation accumulation in finite outbreeding and inbreeding populations," *Genet. Res. (Camb.)* **61**, 39–56.

Chow, J. C., Yen, Z., Ziesche, S. M., and Brown, C. J. (2005), "Silencing of the mammalian x chromosome," *Annu. Rev. Genomics Hum. Genet.* **6**, 69–92.

Christensen, T., Schierup, M. H., and Hein, J. (2006), "Coalescent: Recombination and gene conversion – finite sites," http://www.coalescent.dk.

Christiansen, F. B. (1974), "Sufficient conditions for protected polymorphism in a subdivided population," *Amer. Natur.* **108**, 157–166.

Christiansen, F. B. (1975), "Hard and soft selection in a subdivided population," *Amer. Natur.* **109**, 11–16.

Christiansen, F. B. (1985), "Selection and population regulation with habitat variation," *Amer. Natur.* **126**, 418–429.

Christiansen, F. B. (1988), "Epistasis in the multiple locus symmetric viability model," *J. Math. Biol.* **26**, 595–618.

Christiansen, F. B. (1990a), Genetic comparisons of life stages in natural populations of *Zoarces viviparus*, *in* D. L. J. Adams, A. Hermalin, D. Lam, and P. Smouse, eds., *Convergent Questions in Genetics and Demography*, Oxford Univ. Press, Oxford, New York, pp. 287–305.

Christiansen, F. B. (1990b), Population consequences of genetic design in sexually reproducing organisms, *in* H. Mooney, and G. Bernardi, eds., *Introduction of Genetically Modified Organisms into the Environment*, John Wiley & Sons, London, New York, Sidney, Toronto, pp. 43–55. Available at http://www.icsu-scope.org/downloadpubs/scope44/chapter06.html.

Christiansen, F. B. (2000), *Population Genetics of Multiple Loci*, John Wiley & Sons, Chichester.

Christiansen, F. B. (2004), Density dependent selection, *in* R. S. Singh, S. Jain, and M. Uyenoyama, eds., *The Evolution of Population Biology: Modern Synthesis*, Cambridge Univ. Press, New York, pp. 139–155.

Christiansen, F. B., and Feldman, M. W. (1975), "Subdivided populations: A review of the one- and two-locus deterministic theory," *Theor. Popul. Biol.* **7**, 13–38.

Christiansen, F. B., and Feldman, M. W. (1986), *Population Genetics*, Blackwell Sci. Publ., Palo Alto, CA.

Christiansen, F. B., and Fenchel, T. M. (1977), *Theories of Populations in Biological Communities*, Springer Verlag, Berlin, Heidelberg, New York.

Christiansen, F. B., and Frydenberg, O. (1973), "Selection component analysis of natural polymorphisms using population samples including mother-offspring combinations," *Theor. Popul. Biol.* **4**, 425–445.

Christiansen, F. B., and Frydenberg, O. (1974), "Geographical patterns of four polymorphisms in *Zoarces viviparus* as evidence of selection," *Genetics* **77**, 765–770.

Christiansen, F. B., and Frydenberg, O. (1976), Selection component analysis of natural polymorphisms using mother–offspring samples of successive cohorts, *in* S. Karlin, and E. Nevo, eds., *Population Genetics and Ecology*, Academic Press, New York, San Francisco, London, pp. 277–301.

Christiansen, F. B., and Frydenberg, O. (1977), "Selection–mutation balance for two nonallelic recessives producing an inferior double homozygote," *Amer. J. Hum. Genet.* **29**, 195–207.

Christiansen, F. B., Frydenberg, O., and Simonsen, V. (1977), "Genetics of *Zoarces* populations X. selection component analysis of the *EstIII* polymorphism using samples of successive cohorts," *Hereditas* **87**, 129–150.

Christiansen, F. B., Nielsen, V. H., and Simonsen, V. (1988), "Genetics of *Zoarces* populations XV. Genetic and morphological variation in Mariager Fjord," *Hereditas* **109**, 99–112.

Christiansen, F. B., and Prout, T. (1999), Aspects of fitness, *in* R. S. Singh, and C. B. Krimbas, eds., *Evolutionary Genetics from Molecules to Morphology*, Cambridge Univ. Press, New York.

Clarke, A. G. (1998), "Mutation-selection balance with multiple alleles," *Genetica* **102/103**, 41–47.

Cook, L. M. (2003), "The rise and fall of the *carbonaria* form of the peppered moth," *Q. Rev. Biol.* **78**, 399–417.

Cooper, G. M. (1995), *Oncogenes*, 2nd edn., Jones & Bartlett Publishers Inc., Sudbury, MA.

Corpet, F. (1988), "Multiple sequence alignment with hierarchical clustering," *Nucl. Acid Res.* **16**, 10881–10890.

Cowie, R. H., and Jones, J. S. (1983), "Climatic selection on body color in *Cepaea*," *Heredity* **55**, 261–267.

Crow, J. F. (1954), Breeding structure of populations. I. effective population number, *in* O. Kempthorne, T. A. Bancroft, J. W. Gowen, and J. L. Lush, eds., *Statistics and Mathematics in Biology*, Iowa State Univ. Press, Ames, Iowa, pp. 543–556.

Crow, J. F., and Kimura, M. (1970), *An Introduction to Population Genetics Theory*, Harper & Row, New York.

Dahlgaard, J., and Hoffmann, A. A. (2000), "Stress resistance and environmental dependency of inbreeding depression in *Drosophila melanogaster*," *Conserv. Biol.* **14**, 1187–1192.

Deakin, M. A. B. (1966), "Sufficient conditions for genetic polymorphism," *Amer. Natur.* **100**, 690–692.

Dempster, E. R. (1955), "Maintenance of genetic heterogeneity," *Cold Spring Harbor Symp. Quant. Biol.* **20**, 140–143.

Diekmann, O., Christiansen, F. B., and Law, R., eds. (1996), *Evolutionary Dynamics*, vol. 34(5/6) of *J. Math. Biol.*, Springer Verlag, Berlin, Heidelberg.

Dobson-Stone, C., Gatt, J. M., Kuan, S. A., Grieve, S. M., Gordon, E., Williams, L. M., and Schofield, P. R. (2007), "Investigation of MCPH1 G37995C and ASPM A44871G polymorphisms and brain size in a healthy cohort," *Neuroimage* **37**, 394–400.

Dobzhansky, T. (1950), "Mendelian populations and their evolution," *Amer. Natur.* **84**, 401–418.

Dobzhansky, T. (1970), *Genetics and the Evolutionary Process*, Columbia Univ. Press, New York.

Dobzhansky, T., Hunter, A. S., Pavlovsky, O., Spassky, B., and Wallace, B. (1963), "Genetics of natural populations. XXXI. Genetics of an isolated marginal population of *Drosophila pseudoobscura*," *Genetics* **48**, 91–103.

Donnelly, P. (1986), "Partition structures, Polya urns, the Ewens sampling formula, and the age of alleles," *Theor. Popul. Biol.* **30**, 271–288.

Dunning Hotopp, J. C., Clark, M. E., Oliveira, D. C. S. G., Foster, J. M.,
Fischer, P., Torres, M. C. M., Giebel, J. D., Kumar, N., Ishmael, N.,
Wang, S., Ingram, J., Nene, R. V., Shepard, J., Tomkins, J., Richards, S.,
Spiro, D. J., Ghedin, E., Slatko, B. E., Tettelin, H., and Werren, J. H.
(2007), "Widespread lateral gene transfer from intracellular bacteria to
multicellular eukaryotes," *Sciencexpress* (DOI: 10.1126/science.1142490).

Edwards, A. W. F. (1977), *Foundations of Mathematical Genetics*, Cambridge
Univ. Press, Cambridge.

Ehrlich, P. R., and Raven, P. H. (1969), "Differentiation of populations," *Science*
165, 1228–1232.

Eshel, I. (1991), Game theory and population dynamics in complex genetical
systems: The role of sex in short-term and in long-term evolution, *in Game
Equilibrium Models. vol. I Evolution and Game Dynamics*, Springer-Verlag,
Berlin, pp. 6–28.

Eshel, I. (1996), "On the changing concept of evolutionary population stability
as a reflection of a changing point of view in the quantitative theory of
evolution," *J. Math. Biol.* **34**, 485–510.

Eshel, I., and Feldman, M. W. (1970), "On the evolutionary effect of recombination," *Theor. Popul. Biol.* **1**, 88–100.

Evans, P. D., Anderson, J. R., Vallender, E. J., Choi, S. S., and Lahn, B. T.
(2004), "Reconstructing the evolutionary history of microcephalin, a gene
controlling human brain size," *Hum. Mol. Genet.* **13**, 1139–1145.

Evans, P. D., Gilbert, S. L., Mekel-Bobrov, N., Vallender, E. J., Anderson, J. R.,
Vaez-Azizi, L. M., Tishkoff, S. A., Hudson, R. R., and Lahn, B. T. (2005),
"Microcephalin, a gene regulating brain size, continues to evolve adaptively
in humans," *Science* **309**, 1717–1720.

Evans, P. D., Mekel-Bobrov, N., Vallender, E. J., Hudson, R. R., and Lahn,
B. T. (2006), "Evidence that the adaptive allele of the brain size gene mi-
crocephalin introgressed into Homo sapiens from an archaic Homo lineage,"
Proc. Natl. Acad. Sci. USA **103**, 18178–18183.

Ewens, W. J. (1968), "A genetic model having complex linkage behavior," *Theoret. Appl. Genet.* **38**, 140–143.

Ewens, W. J. (1972), "The sampling theory of selectively neutral alleles," *Theor.
Popul. Biol.* **3**, 87–112.

Ewens, W. J. (1976), "Remarks on the evolutionary effect of natural selection,"
Genetics **83**, 601–607.

Ewens, W. J. (1979), *Mathematical Population Genetics*, vol. 9 of *Biomathematics*, Springer Verlag, Berlin, Heidelberg, New York.

Ewens, W. J. (2004), *Mathematical Population Genetics. I Theoretical Intro-duction*, vol. 27 of *Interdisciplinary Applied Mathematics*, Springer Verlag, Berlin, Heidelberg, New York.

Ewens, W. J., and Thomson, G. (1977), "Properties of equilibria in multi-locus systems," *Genetics* **87**, 807–819.

Falconer, D. S., and Mackay, T. (1996), *Introduction to Quantitative Genetics*, 4th edn., Longman, Essex, England.

Fearnhead, P., and Donnelly, P. (2001), "Estimating recombination rates from population genetic data," *Genetics* **159**, 1299–1318.

Feldman, M. W., and Cavalli-Sforza, L. L. (1989), On the theory of evolution under genetic and cultural transmission with application to the lactose absorption problem, *in* M. W. Feldman, ed., *Mathematical Evolutionary Theory*, Princeton Univ. Press, pp. 145–173.

Feldman, M. W., and Christiansen, F. B. (1975), "The effect of population subdivision on two loci without selection," *Genet. Res. (Camb.)* **24**, 151–162.

Feldman, M. W., and Lewontin, R. C. (1975), "The heritability hangup," *Science* **190**, 1163–1168.

Feldman, M. W., Otto, S. P., and Christiansen, F. B. (2000), Genes, culture and inequality, *in* K. Arrow, S. Bowles, and S. Durlauf, eds., *Meritocracy and Economic Inequality*, Princeton Univ. Press, Princeton, NJ, pp. 61–85.

Felsenstein, J. (1965), "The effect of linkage on directional selection," *Genetics* **52**, 349–363.

Felsenstein, J. (1971), "The rate of loss of multiple alleles in finite haploid populations," *Theor. Popul. Biol.* **2**, 391–403.

Felsenstein, J. (2004), *Inferring Phylogenies*, Sinauer Associates, Sunderland, MA.

Fenchel, T. M. (2001), *Det Første Liv—livets oprindelse og tidlige udvikling*, Gad, Copenhagen.

Fenchel, T. M. (2002), *Origin and Early Evolution of Life*, Oxford Univ. Press, Oxford, New York.

Fenchel, T. M., and Kolding, S. (1979), "Habitat selection and distribution patterns of five species of the amphipod genus *Gammarus*," *Oikos* **33**, 316–322.

Fire, A., Xu, S., Montgomery, M. K., Kostas, S. A., Driver, S. E., and Mello, C. C. (1998), "Potent and specific genetic interference by double-stranded RNA in Caenorhabditis elegans," *Nature (London)* **391**, 806–811.

Fisher, R. A. (1918), "The correlation between relatives on the supposition of Mendelian inheritance," *Trans. Roy. Soc. Edinburgh* **52**, 399–433.

Fisher, R. A. (1922), "On the dominance ratio," *Proc. Roy. Soc. Edinburgh* **42**, 321–431.

Fisher, R. A. (1930*a*), "The distribution of gene ratios for rare mutations," *Proc. Roy. Soc. Edinburgh* **50**, 205–220.

Fisher, R. A. (1930*b*), *The Genetical Theory of Natural Selection*, Clarendon Press, Oxford.

Fisher, R. A. (1935), "The sheltering of lethals," *Amer. Natur.* **69**, 446–455.

Fitch, W. M. (1970), "Distinguishing homologous from analogous proteins," *Syst. Zool.* **19**, 99–113.

Ford, E. B. (1971), *Ecological Genetics*, Chapman & Hall, London.

Frankham, R., Jones, L. P., and Barker, J. S. F. (1968*a*), "The effects of population size and selection intensity in selection for a quantitative character in *Drosophila*. I. Short-term response to selection," *Genet. Res. (Camb.)* **12**, 237–248.

Frankham, R., Jones, L. P., and Barker, J. S. F. (1968*b*), "The effects of population size and selection intensity in selection for a quantitative character in *Drosophila*. III. Analyses of the lines," *Genet. Res. (Camb.)* **12**, 267–283.

Frydenberg, O. (1963), "Population studies of a lethal mutant in *Drosophila melanogaster* I. Behaviour in populations with discrete generations," *Hereditas* **50**, 89–116.

Frydenberg, O. (1964), "Long-term instability of an *ebony* polymorphism in artificial populations of *Drosophila melanogaster*," *Hereditas* **51**, 198–206.

Frydenberg, O., Gyldenholm, A. O., Hjorth, J. P., and Simonsen, V. (1973), "Genetics of *Zoarces* populations III. Geographic variation in the esterase polymorphism *EstIII*," *Hereditas* **73**, 233–238.

Fullerton, S., Bond, J., Schneider, J., Hamilton, B., Harding, R., Boyce, A., and Clegg, J. (2000), "Polymorphism and divergence in the beta-globin replication origin initiation region," *Mol. Biol. Evol.* **17**, 179–188.

Geiringer, H. (1944), "On the probability theory of linkage in Mendelian heredity," *Ann. Math. Statist.* **15**, 25–57.

GenBank (2006), "The NIH genetic sequence database," available at the general resource http://www.ncbi.nlm.nih.gov.

Ghatnekar, L., Jaarola, M., and Bengtsson, B. O. (2006), "The introgression of a functional nuclear gene from *Poa* to *Festuca ovina*," *Phil. Trans. Roy. Soc. London B* **273**, 395–399.

Gibbs, R. A., Belmont, J. W., Hardenbol, P., Willis, T. D., Yu, F. L., Yang,
H. M., Ch'ang, L. Y., Huang, W., Liu, B., Shen, Y., Tam, P. K. H., Tsui,
L. C., Waye, M. M. Y., Wong, J. T. F., Zeng, C. Q., Zhang, Q. R., Chee,
M. S., Galver, L. M., Kruglyak, S., Murray, S. S., Oliphant, A. R., Mont-
petit, A., Hudson, T. J., Chagnon, F., Ferretti, V., Leboeuf, M., Phillips,
M. S., Verner, A., Kwok, P. Y., Duan, S. H., Lind, D. L., Miller, R. D., Rice,
J. P., Saccone, N. L., Taillon-Miller, P., Xiao, M., Nakamura, Y., Sekine,
A., Sorimachi, K., Tanaka, T., Tanaka, Y., Tsunoda, T., Yoshino, E., Bent-
ley, D. R., Deloukas, P., Hunt, S., Powell, D., Altshuler, D., Gabriel, S. B.,
Qiu, R. Z., Ken, A., Dunston, G. M., Kato, K., Niikawa, N., Knoppers,
B. M., Foster, M. W., Clayton, E. W., Wang, V. O., Watkin, J., Gibbs,
R. A., Belmont, J. W., Sodergren, E., Weinstock, G. M., Wilson, R. K.,
Fulton, L. L., Rogers, J., Birren, B. W., Han, H., Wang, H. G., God-
bout, M., Wallenburg, J. C., L'Archeveque, P., Bellemare, G., Todani, K.,
Fujita, T., Tanaka, S., Holden, A. L., Lai, E. H., Collins, F. S., Brooks,
L. D., McEwen, J. E., Guyer, M. S., Jordan, E., Peterson, J. L., Spiegel,
J., Sung, L. M., Zacharia, L. F., Kennedy, K., Dunn, M. G., Seabrook, R.,
Shillito, M., Skene, B., Stewart, J. G., Valle, D. L., Clayton, E. W., Jorde,
L. B., Belmont, J. W., Chakravarti, A., Cho, M. K., Duster, T., Foster,
M. W., Jasperse, M., Knoppers, B. M., Kwok, P. Y., Licinio, J., Long,
J. C., Marshall, P. A., Ossorio, P. N., Wang, V. O., Rotimi, C. N., Royal,
C. D. M., Spallone, P., Terry, S. F., Lander, E. S., Lai, E. H., Nickerson,
D. A., Abecasis, G. R., Altshuler, D., Bentley, D. R., Boehnke, M., Cardon,
L. R., Daly, M. J., Deloukas, P., Douglas, J. A., Gabriel, S. B., Hudson,
R. R., Hudson, T. J., Kruglyak, L., Kwok, P. Y., Nakamura, Y., Nussbaum,
R. L., Royal, C. D. M., Schaffner, S. F., Sherry, S. T., Stein, L. D., and
Tanaka, T. (2003), "The international hapmap project," *Nature (London)*
426, 789–796. The HapMap resource resides at http://www.hapmap.org.

Gilligan, D. M., Briscoe, D. A., and Frankham, R. (2005), "Comparative losses
of quantitative and molecular genetic variation in finite populations of
drosophila melanogaster," *Genet. Res. (Camb.)* **85**, 47–55.

Goldman, N., and Yang, Z. (1994), "A codon-based model of nucleotide substitution for protein-coding DNA sequences," *Mol. Biol. Evol.* **11**, 725–736.

Goodman, M., and Moore, G. W. (1973), "Phylogeny of hemoglobin," *Syst.
Zool.* **22**, 508–532.

Gould, S. J., and Lewontin, R. C. (1979), "The spandrels of San Marco and the
Panglossian paradigm. A critique of the adaptionist programme," *Proc.
Roy. Soc. London B* **205**, 581–598.

Grant, V. (1963), *The Origin of Adaptations*, Columbia Univ. Press, New York.

Green, R. E., Krause, J., Ptak, S. E., Briggs, A. W., Ronan, M. T., Simons,
J. F., Du, L., Egholm, M., Rothberg, J. M., Paunovic, M., and Pääbo, S.

(2006), "Analysis of one million base pairs of Neanderthal DNA," *Nature (London)* **444**, 330–336.

Griffiths, R. C., and Marjoram, P. (1997), An ancestral recombination graph, *in* P. Donnelly, and S. Tavaré, eds., *Progress in Population Genetics and Human Evolution*, vol. 87 of *The IMA Volumes in Mathematics and Its Applications*, Springer Verlag, Berlin, Heidelberg, New York, pp. 257–270.

Haldane, J. B. S. (1924), "A mathematical theory of natural and artificial selection," *Trans. Camb. Phil. Soc.* **23**, 19–41.

Haldane, J. B. S. (1927), "A mathematical theory of natural and artificial selection. IV. Selection and mutation," *Proc. Camb. Phil. Soc.* **23**, 838–844.

Haldane, J. B. S. (1930), "A mathematical theory of natural and artificial selection. VI. Isolation," *Proc. Camb. Phil. Soc.* **26**, 220–230.

Haldane, J. B. S. (1932), "A mathematical theory of natural and artificial selection. IX. Rapid selection," *Proc. Camb. Phil. Soc.* **28**, 244–248.

Haldane, J. B. S. (1937), "The effect of variation on fitness," *Amer. Natur.* **71**, 337–349.

Hall, J. M., Lee, M. K., Newman, B., Morrow, J. E., Anderson, L. A., Huey, B., and King, M. C. (1990), "Linkage of early-onset familial breast cancer to chromosome 17q21," *Science* **4988**, 1684–1689.

Hanks, M. C., Loomis, C. A., Harris, E., Tong, C. X., Anson-Cartwright, L., Auerbach, A., and Joyner, A. (1998), "*Drosophila engrailed* can substitute for mouse *Engrailed1* function in mid-hindbrain, but not limb development," *Development* **125**, 4521–4530.

Hanski, I., and Gilpin, M. E., eds. (1997), *Metapopulation Biology: Ecology, Genetics and Evolution*, Academic Press, London.

Harris, H. (1966), "Enzyme polymorphisms in man," *Proc. Roy. Soc. London B* **164**, 298–310.

Hasegawa, M., Kishino, H., and Yano, T. (1985), "Dating the human-ape splitting by a molecular clock of mitochondrial DNA," *J. Mol. Evol.* **22**, 160–174.

Hastings, A. (1981), "Stable cycling in discrete-time genetic models," *Proc. Natl. Acad. Sci. USA* **78**, 7224–7225.

Hastings, A. (1982), "Unexpected behavior in two locus genetic systems: An analysis of marginal underdominance at a stable equilibrium," *Genetics* **102**, 129–138.

He, L., and Hannon, G. J. (2004), "MicroRNAs: Small RNAs with a big role in gene regulation," *Nat. Rev. Genet.* **5**, 522–531.

Hedrick, P. W., Klitz, W., Robinson, W. P., Kuhner, M. K., and Thomson, G. (1991), Population genetics of HLA, *in* R. K. Selander, A. G. Clarke, and T. S. Whittam, eds., *Evolution at The Molecular Level*, Sinauer Assoc, Sunderland, Mass., pp. 248–271.

Hein, J., Schierup, M. H., and Wiuf, C. (2005), *Gene Genealogies, Variation and Evolution*, Oxford University Press, Oxford.

Hellmann, I., Ebersberger, I., Ptak, S. E., Paabo, S., and Przeworski, M. (2003), "A neutral explanation for the correlation of diversity with recombination rates in humans," *Amer. J. Hum. Genet.* **72**, 1527–1535.

Heuch, I. (1978), "Maintenance of butterfly populations with all-female broods under recurrent extinction and recolonization," *J. Theor. Biol.* **75**, 115–122.

Hill, W. G., and Robertson, A. (1966), "The effect of linkage on limits to artificial selection," *Genet. Res. (Camb.)* **8**, 269–294.

Hill, W. G., and Robertson, A. (1968), "Linkage disequilibrium in finite populations," *Theoret. Appl. Genet.* **38**, 226–231.

Hjorth, J. P., and Simonsen, V. (1975), "Genetics of *Zoarces* populations VIII. Geographic variation common to the polymorphic loci *HbI* and *EstIII*," *Hereditas* **81**, 173–184.

Hobolth, A., Christensen, O. F., Mailund, T., and Schierup, M. H. (2007), "Genomic relationships and speciation times of human, chimpanzee, and gorilla inferred from a coalescent hidden Markov model," *PLoS Genetics* **3**, 294–304.

Holsinger, K. (2006), "Population biology simulations," http://darwin.eeb.uconn.edu/simulations/simulations.html.

Houle, D. (1992), "Comparing evolvability and variability of quantitative traits," *Genetics* **130**, 195–204.

Hubby, J. L., and Lewontin, R. C. (1966), "A molecular approach to the study of genic heterozygosity in natural populations. I. The number of alleles at different loci in *Drosophila pseudoobscura*," *Genetics* **57**, 577–594.

Hudson, R. R. (1983), "Properties of the neutral allele model with intragenic recombination," *Theor. Popul. Biol.* **23**, 183–201.

Hudson, R. R. (1991), "Gene genealogies and the coalescent process," *Oxford Surveys in Evolutionary Biology* **7**, 1–49.

Hudson, R. R., and Kaplan, N. L. (1985), "Statistical properties of the number of recombination events in the history of a sample of DNA sequences," *Genetics* **111**, 147–164.

Hudson, R. R., and Kaplan, N. L. (1995), "Deleterious background selection with recombination," *Genetics* **141**, 1605–1617.

Hueber, S. D., Bezdan, D., Henz, S. R., Blank, M., Wu, H. J., and Lohmann, I. (2007), "Comparative analysis of Hox downstream genes in *Drosophila*," *Development* **134**(2), 381–392.

Hughes, A. L., and Nei, M. (1988), "Pattern of nucleotide substitution at major histocompatibility complex class I loci reveals overdominant selection," *Nature (London)* **335**, 167–170.

International HapMap Consortium (2006), "International HapMap Project," http://www.hapmap.org.

Jackson, A. P., Eastwood, H., Bell, S. M., Adu, J., Toomes, C., Carr, I. M., Roberts, E., Hampshire, D. J., Crow, Y. J., Mighell, A. J., Karbani, G., Jafri, H., Rashid, Y., F, M. R., Markham, A. F., and Woods, C. G. (2002), "Identification of microcephalin, a protein implicated in determining the size of the human brain," *Am. J. Hum. Genet.* **71**, 136–142.

Jeffreys, A. J., Kauppi, L., and Neumann, R. (2001), "Intensely punctate meiotic recombination in the class II region of the major histocompatibility complex," *Nat. Genet.* **29**, 217–222.

Jiggins, F. (2003), "Male-killing Wolbachia and mitochondrial DNA: Selective sweeps, hybrid introgression and parasite population dynamics," *Genetics* **164**, 5–12.

Jiggins, F. M., Hurst, G. D. D., Dolman, C. E., and Majerus, M. E. N. (2003), "High prevalence male-killing *Wolbachia* in the butterfly *Acraea encedon*," *J. Evol. Biol.* **13**, 495–501.

Jiggins, F. M., Hurst, G. D. D., and Majerus, M. E. N. (1998), "Sex ratio distortion in *Acraea encedon* (Lepidoptera: Nymphalidae) is caused by a male-killing bacterium," *Heredity* **81**, 87–91.

Johannsen, W. W. (1903), *Über Erblichkeit in Populationen und in reinen Linien*, Gustav Fisher, Jena.

Johannsen, W. W. (1905), *Arvelighedslærens elementer*, Gyldendalske Boghandel, Nordiske Forlag, Copenhagen.

Johannsen, W. W. (1909), *Elemente der exakten Erblichkeitslehre*, Gustav Fisher, Jena.

Jones, J. S., Leith, B. H., and Rawlings, P. (1977), "Polymorphism in Cepaea: a problem with too many solutions?," *Annu. Rev. Ecol. Syst.* **8**, 109–143.

Jones, L. P., Frankham, R., and Barker, J. S. F. (1968), "The effects of population size and selection intensity in selection for a quantitative character in *Drosophila*. II. Long-term response to selection," *Genet. Res. (Camb.)* **12**, 249–266.

Jukes, T. H., and Cantor, C. R. (1969), Evolution of protein molecules, *in* H. N. Munro, ed., *Mammalian Protein Metabolism*, Academic Press, New York, pp. 21–132.

Karlin, S., and Feldman, M. W. (1970), "Linkage and selection: Two-locus symmetric viability model," *Theor. Popul. Biol.* **1**, 39–71.

Karlin, S., and McGregor, J. (1972), "Polymorphisms for genetic and ecological systems with weak coupling," *Theor. Popul. Biol.* **3**, 210–238.

Karlin, S., and Taylor, H. M. (1975), *A First Course in Stochastic Processes*, 2nd edn., Academic Press, New York, San Francisco, London.

Karlin, S., and Taylor, H. M. (1981), *A Second Course in Stochastic Processes*, Academic Press, New York, San Francisco, London.

Kassen, R., and Bataillon, T. (2006), "The distribution of fitness effects among beneficial mutations prior to selection in experimental populations of bacteria," *Nat. Genet.* **38**, 484–488.

Kettlewell, B. (1973), *The Evolution of Melanism*, Clarendon Press, Oxford.

Kimura, M. (1955*a*), "Solution of a process of random genetic drift with a continuous model," *Proc. Natl. Acad. Sci. USA* **41**, 144–150.

Kimura, M. (1955*b*), "Stochastic processes and distribution of gene frequencies under natural selection," *Cold Spring Harbor Symp. Quant. Biol.* **20**, 33–53.

Kimura, M. (1965), "Attainment of quasi linkage equilibrium when gene frequencies are changing by natural selection," *Genetics* **52**, 875–890.

Kimura, M. (1968), "Evolutionary rate at the molecular level," *Nature (London)* **217**, 624–626.

Kimura, M. (1969*a*), "The number of heterozygous nucleotide sites maintained in a finite population due to steady flux of mutations," *Genetics* **61**, 893–903.

Kimura, M. (1969*b*), "The rate of molecular evolution considered from the standpoint of population genetics," *Proc. Natl. Acad. Sci. USA* **63**, 1181–1188.

Kimura, M. (1980), "A simple method for estimating evolutionary rates of base substitutions through comparative studies of nucleotide sequences," *J. Mol. Evol.* **16**, 111–120.

Kimura, M. (1983*a*), *The Neutral Theory of Molecular Evolution*, Cambridge Univ. Press, Cambridge.

Kimura, M. (1983*b*), The neutral theory of molecular evolution, *in* M. Nei, and R. K. Koehn, eds., *Evolution of Genes and Proteins*, Sinauer, Sunderland, MA, pp. 208–233.

Kimura, M., and Crow, J. (1964), "The number of alleles that can be maintained in a finite population," *Genetics* **49**, 725–738.

Kimura, M., and Ohta, T. (1973), "The age of a neutral mutant persisting in a finite population," *Genetics* **75**, 199–212.

Kimura, M., and Weiss, G. H. (1964), "The stepping stone model of population structure and the decrease of genetic correlation with distance," *Genetics* **49**, 561–576.

King, J. L., and Jukes, T. H. (1969), "Non-Darwinian evolution," *Science* **164**, 788–789.

Kingman, J. F. C. (1961), "A mathematical problem in population genetics," *Proc. Camb. Phil. Soc.* **57**, 574–582.

Kingman, J. F. C. (1982*a*), "The coalescent," *Stoch. Process. Appl.* **13**, 235–248.

Kingman, J. F. C. (1982*b*), Exchangeability and the evolution of large populations, *in* G. Koch, and F. Spizzichino, eds., *Exchangeability in Probability and Statistic*, North-Holland, pp. 97–112.

Kingman, J. F. C. (1982*c*), "On the genealogy of large populations," *J. Appl. Prob.* **19A**, 27–43.

Kingman, J. F. C. (2000), "Origins of the coalescent: 1974–1982," *Genetics* **156**, 1461–1468.

Kirkpatrick, M., and Barton, N. (2006), "Chromosome inversions, local adaptation and speciation," *Genetics* **173**, 419–434.

Koehn, R. K., and Eanes, W. F. (1978), "Molecular structure and protein variation within and among populations," *Evol. Biol.* **11**, 39–100.

Kolmogorov, A. N. (1935), "Deviations from Hardy's formula in partial isolation," *C. R. Acad. Sci. URSS* **3**, 129–132.

Koonin, E. V. (2005), "Orthologs, paralogs, and evolutionary genomics," *Annu. Rev. Genet.* **39**, 309–338.

Kraft, T., Säll, T., Magnusson-Rading, I., Nilsson, N.-O., and Halldén, C. (1998), "Positive correlation between recombination rates and levels of genetic variation in natural populations of sea beet *Beta vulgaris* subsp. *maritima*," *Genetics* **150**, 631–633.

Kreitman, M. (1983), "Nucleotide polymorphism at the alcohol dehydrogenase locus of *Drosophila melanogaster*," *Nature (London)* **304**, 412–417.

Kristensen, T. N., Sørensen, A. C., Sorensen, D., Pedersen, K. S., Sørensen, J. G., and Loeschcke, V. (2005), "A test of quantitative genetic theory using drosophila - effects of inbreeding and rate of inbreeding on heritabilities and variance components," *J. Evol. Biol.* **18**, 763–770.

Krone, S. M., and Neuhauser, C. (1997), "Ancestral processes with selection," *Theor. Popul. Biol.* **51**, 210–237.

Lander, E. S., Linton, L. M., Birren, B., Nusbaum, C., Zody, M. C., Baldwin, J., Devon, K., Dewar, K., Doyle, M., FitzHugh, W., Funke, R., Gage, D., Harris, K., Heaford, A., Howland, J., Kann, L., Lehoczky, J., LeVine, R., McEwan, P., McKernan, K., Meldrim, J., Mesirov, J. P., Miranda, C., Morris, W., Naylor, J., Raymond, C., Rosetti, M., Santos, R., Sheridan, A., Sougnez, C., Stange-Thomann, N., Stojanovic, N., Subramanian, A., Wyman, D., Rogers, J., Sulston, J., Ainscough, R., Beck, S., Bentley, D., Burton, J., Clee, C., Carter, N., Coulson, A., Deadma, R., Deloukas, P., Dunham, A., Dunham, I., Durbin, R., French, L., Grafham, D., Gregory, S., Hubbard, T., Humphray, S., Hunt, A., Jones, M., Lloyd, C., McMurray, A., Matthews, L., Mercer, S., Milne, S., Mullikin, J. C., Mungall, A., Plumb, R., Ross, M., Shownkeen, R., Sims, S., Waterston, R. H., Wilson, R. K., Hillier, L. W., McPherson, J. D., Marra, M. A., Mardis, E. R., Fulton, L. A., Chinwalla, A. T., Pepin, K. H., Gish, W. R., Chissoe, S. L., Wendl, M. C., Delehaunty, K. D., Miner, T. L., Delehaunty, A., Kramer, J. B., Cook, L. L., Fulton, R. S., Johnson, D. L., Minx, P. J., Clifton, S. W., Hawkins, T., Branscomb, E., Predki, P., Richardson, P., Wenning, S., Slezak, T., Doggett, N., Cheng, J. F., Olsen, A., Lucas, S., Elkin, C., Uberbacher, E., Frazier, M., Gibbs, R. A., Muzny, D. M., Scherer, S. E., Bouck, J. B., Sodergren, E. J., Worley, K. C., Rives, C. M., Gorrell, J. H., Metzker, M. L., Naylor, S. L., Kucherlapati, R. S., Nelson, D. L., Weinstock, G. M., Sakaki, Y., Fujiyama, A., Hattori, M., Yada, T., Toyoda, A., Itoh, T., Kawagoe, C., Watanabe, H., Totoki, Y., Taylor, T., Weissenbach, J., Heilig, R., Saurin, W., Artiguenave, F., Brottier, P., Bruls, T., Pelletier, E., Robert, C., Wincker, P., Smith, D. R., Doucette-Stamm, L., Rubenfield, M., Weinstock, K., Lee, H. M., Dubois, J., Rosenthal, A., Platzer, M., Nyakatura, G., Taudien, S., Rump, A., Yang, H., Yu, J., Wang, J., Huang, G., Gu, J., Hood, L., Rowen, L., Madan, A., Qin, S., Davis, R. W., Federspiel, N. A., Abola, A. P., Proctor, M. J., Myers, R. M., Schmutz, J., Dickson, M., Grimwood, J., Cox, D. R., Olson, M. V., Kaul, R., Raymond, C., Shimizu, N., Kawasaki, K., Minoshima, S., Evans, G. A., Athanasiou, M., Schultz, R., Roe, B. A., Chen, F., Pan, H., Ramser, J., Lehrach, H., Reinhardt, R., McCombie, W. R., de la Bastide, M., Dedhia, N., Blocker, H., Hornischer, K., Nordsiek, G., Agarwala, R., Aravind, L., Bailey, J. A., Bateman, A., Batzoglou, S., Birney, E., Bork, P., Brown, D. G., Burge, C. B., Cerutti, L., Chen, H. C., Church, D., Clamp, M., Copley, R. R., Doerks, T., Eddy, S. R., Eichler, E. E., Furey, T. S., Galagan, J., Gilbert, J. G., Harmon, C., Hayashizaki, Y., Haussler, D., Hermjakob, H., Hokamp, K., Jang, W., Johnson, L. S., Jones, T. A., Kasif, S., Kaspryzk, A., Kennedy, S., Kent, W. J., Kitts, P., Koonin, E. V., Korf, I., Kulp, D., Lancet, D., Lowe, T. M., McLysaght, A., Mikkelsen, T., Moran, J. V., Mulder, N., Pollara, V. J., Ponting, C. P., Schuler, G., Schultz, J., Slater, G., Smit, A. F., Stupka, E., Szustakowski, J., Thierry-Mieg, D., Thierry-Mieg, J., Wagner, L., Wallis, J., Wheeler, R., Williams, A., Wolf, Y. I., Wolfe, K. H., Yang, S. P., Yeh, R. F., Collins, F., Guyer, M. S., Peterson, J., Felsenfeld, A., Wetterstrand, K. A., Patrinos, A., Morgan, M. J., de Jong, P., Catanese, J. J., Osoegawa, K., Shizuya, H., Choi, S., and Chen, Y. J. (2001), "Initial sequencing and analysis of the human genome," *Nature (London)* **409**, 860–892.

Langley, C. H., and Aquadro, C. F. (1987), "Restriction-map variation in natural populations of *Drosophila melanogaster*: white-locus region," *Mol. Biol. Evol.* **4**, 651–663.

Latter, B. D. H. (1973), "Island model of population differentiation – general solution," *Genetics* **73**, 147–157.

Latter, B. D. H. (1998), "Mutant alleles of small effect are primarily responsible for the loss of fitness with slow inbreeding in *Drosophila melanogaster*," *Genetics* **148**, 1143–1158.

Lemons, D., and McGinnis, W. (2006), "Genomic evolution of Hox gene clusters," *Science* **313**, 1918–1922.

Levene, H. (1953), "Genetic equilibrium when more than one niche is available," *Amer. Natur.* **87**, 331–333.

Levin, S. A., and Paine, R. T. (1974), "Disturbance, patch formation, and community structure," *Proc. Natl. Acad. Sci. USA* **71**, 2744–2747.

Levins, R. (1970), "Extinction," *Lecture Notes in Mathematics* **2**, 75–107.

Levy, S., Sutton, G., Ng, P. C., Feuk, L., Halpern, A. L., Walenz, B. P., Axelrod, N., Huang, J., Kirkness, E. F., Denisov, G., Lin, Y., MacDonald, J. R., Pang, A. W. C., Shago, M., Stockwell, T. B., Tsiamouri, A., Bafna, V., Bansal, V., Kravitz, S. A., Busam, D. A., Beeson, K. Y., McIntosh, T. C., Remington, K. A., Abril, J. F., Gill, J., Borman, J., Rogers, Y.-H., Frazier, M. E., Scherer, S. W., Strausberg, R. L., and Venter, J. C. (2007), "The diploid genome sequence of an individual human," *PLoS Biol.* **5**(doi:10.1371/journal.pbio.0050254).

Lewis, E. B. (1978), "A gene complex controlling segmentation in *Drosophila*," *Nature (London)* **276**, 565–570.

Lewontin, R. C. (1964), "The interaction of selection and linkage. II. Optimum models," *Genetics* **50**, 757–782.

Lewontin, R. C. (1974), *The Genetic Basis of Evolutionary Change*, Columbia Univ. Press, New York.

Lewontin, R. C., and Hubby, J. L. (1966), "A molecular approach to the study of genic heterozygosity in natural populations. II. Amount of variation and degree of heterozygosity in natural populations of *Drosophila pseudoobscura*," *Genetics* **57**, 577–594.

Lewontin, R. C., and Kojima, K. (1960), "The evolutionary dynamics of complex polymorphism," *Evolution* **14**, 458–472.

Li, W.-H. (1997), *Molecular Evolution*, Sinauer Assoc., Sunderland, MA.

Lindow, M., and Krogh, A. (2005), "Computational evidence for hundreds of non-conserved plant microRNAs," *BMC Genomics* **6**, 119.

Luria, S. E., and Delbrück, M. (1943), "Mutations of bacteria from virus sensitivity to virus resistance," *Genetics* **28**, 491–511.

Lynch, M., and Walsh, B. (1998), *Genetics and Analysis of Quantitative Traits*, Sinauer Assoc., Sunderland, MA.

MacLean, J. A., Chen, M. A., Wayne, C. M., Bruce, S. R., Rao, M., Meistrich, M. L., Macleod, C., and Wilkinson, M. F. (2005), "*Rhox*: a new homeobox gene cluster," *Cell* **120**, 369–382.

MacLean, J. A., Lorenzetti, D., Hu, Z. Y., Salerno, W. J., Miller, J., and Wilkinson, M. F. (2006), "*Rhox* homeobox gene cluster: Recent duplication of three family members," *Genesis* **44**, 122–129.

Maconochie, M., Nonchev, S., Morrison, A., and Krumlauf, R. (1996), "Paralogous *Hox* genes: function and regulation," *Annu. Rev. Genet.* **30**, 529–556.

Malécot, G. (1948), *Les Mathématiques de l'Hérédité*, Masson et Cie, Paris. See Malécot (1969).

Malécot, G. (1969), *The Mathematics of Heredity*, W. H. Freeman, San Francisco.

Mallet, F., Bouton, O., Prudhomme, S., Cheynet, V., Oriol, G., Bonnaud, B., Lucotte, G., Duret, L., and Mandrand, B. (2004), "The endogenous retroviral locus ERVWE1 is a bona fide gene involved in hominoid placental physiology," *Proc. Natl. Acad. Sci. USA* **101**, 1731–1736.

Martin, E. R., Lai, E. H., Gilbert, J. R., Rogala, A. R., Afshari, A. J., Riley, J., Finch, K. L., Stevens, J. F., Livak, K. J., Slotterbeck, B. D., Slifer, S. H., Warren, L. L., Conneally, P. M., Schmechel, D. E., Purvis, I., Pericak-Vance, M. A., Roses, A. D., and Vance, J. M. (2000), "SNPing away at complex diseases: analysis of single-nucleotide polymorphisms around APOE in Alzheimer disease," *Amer. J. Hum. Genet.* **67**, 383–394.

Matsen, E., and Wakeley, J. (2006), "Convergence to the island-model coalescent process in populations with restricted migration," *Genetics* **172**, 701–708.

Maynard Smith, J. (1970), "Sufficient conditions for multiple niche polymorphism," *Amer. Natur.* **104**, 487–490.

Maynard Smith, J. (1999), "The detection and measurement of recombination from sequence data," *Genetics* **153**, 1021–1027.

Maynard Smith, J., and Haig, J. (1974), "The hitch-hiking effect of a favourable gene," *Genet. Res. (Camb.)* **23**, 23–35.

McKusick, V. A. (2006), "A 60-year tale of spots, maps, and genes," *Annu. Rev. Genomics Hum. Genet.* **7**, 1–27.

McVean, G. A. T., Myers, S. R., Hunt, S., Deloukas, P., Bentley, D. R., and Donnelly, P. (2004), "The fine-scale structure of recombination rate variation in the human genome," *Science* **304**, 581–584.

Mellars, P. A. (1992), "Archaeology and the population-dispersal hypothesis of modern human origins in Europe," *Phil. Trans. Roy. Soc. Lond. B Biol. Sci.* **337**, 225–234.

Mendel, G. (1866), "Versuche über Pflanzen-Hybriden," *Verhandlungen des naturforschenden Vereines in Brünn* **4**, 3–47. English translation in Stern and Sherwood (1966).

Merriwether, D. A., Clark, A. G., Ballinger, S. W., Schurr, T. G., Soodyall, H., Jenkins, T., Sherry, S. T., and Wallace, D. C. (1991), "The structure of human mitochondrial-DNA variation," *J. Mol. Evol.* **33**, 543–555.

Mi, S., Lee, X., Li, X., Veldman, G. M., Finnerty, H., Racie, L., LaVallie, E., Tang, X. Y., Edouard, P., Howes, S., Keith, J. C. J., and McCoy, J. M. (2000), "Syncytin is a captive retroviral envelope protein involved in human placental morphogenesis," *Nature (London)* **403**, 785–789.

Miki, Y., Swensen, J., Shattuck-Eidens, D., Futreal, P. A., Harshman, K., Tavtigian, S., Liu, Q., Cochran, C., Bennett, L. M., and Ding, W. (1994), "A strong candidate for the breast and ovarian cancer susceptibility gene BRCA1," *Science* **266**, 66–71.

Mikkelsen, A. M., Schierup, M. H., and Hein, J. (2006), "Wright-Fisher animator," http://www.coalescent.dk.

Miller, G. F. (1962), "The evaluation of eigenvalues of a differential equation arising in a problem in genetics," *Proc. Camb. Phil. Soc.* **58**, 588–593.

Mitton, J. B., Koehn, R. K., and Prout, T. (1973), "Population genetics of marine pelecypods. III. Epistasis between functionally related isoenzymes in *Mytilus edulis*," *Genetics* **73**, 487–496.

Moran, P. A. P. (1958), "Random processes in genetics," *Proc. Camb. Phil. Soc.* **54**, 69–71.

Moran, P. A. P. (1968), "On the theory of selection dependent on two loci," *Ann. Hum. Genet.* **32**, 574–582.

Mørch, E. T. (1941), *Chondrodystrophic Dwarfism in Denmark*, Munksgaard, Copenhagen.

Mukai, T., Chigusa, S. I., Mettler, L. E., and Crow, J. F. (1972), "Mutation rate and dominance of genes affecting viability in *Drosophila melanogaster.*," *Genetics* **72**, 335–355.

Muona, O. (1990), Population genetics in forest tree improvement, *in* A. H. D. Brown, M. T. Clegg, A. L. Kahler, and B. S. Weir, eds., *Plant Population Genetics, Breeding and Genetic Resources*, Sinauer, Sunderland, MA, pp. 145–162.

Myers, S., Bottolo, L., Freeman, C., McVean, G., and Donnelly, P. (2005), "A fine-scale map of recombination rates and hotspots across the human genome," *Science* **310**, 321–324.

Nachman, M. W., Bauer, V. L., Crowell, S. L., and Aquadro, C. F. (1998), "DNA variability and recombination rates at X-linked loci in humans," *Genetics* **150**, 1133–1141.

Nadeau, J. H., Britton-Davidian, J., Bonhomme, F., and Thaler, L. (1988), "H-2 polymorphisms are more uniformly distributed than allozyme polymorphisms in natural populations of house mice," *Genetics* **118**(1), 131–140.

Nagylaki, T. (1980), "The strong-migration limit in geographically structured populations," *J. Math. Biol.* **9**, 101–114.

Nagylaki, T., Hofbauer, J., and Brunovski, P. (1999), "Convergence of multilocus systems under weak epistasis or weak selection," *J. Math. Biol.* **38**, 103–133.

Nasrallah, J. B. (2000), "Cell-cell signaling in the self-incompatibility response," *Curr. Opin. Plant Biol.* **3**, 368–373.

Nath, H. B., and Griffiths, R. C. (1993), "The coalescent in two colonies with symmetric migration," *J. Math. Biol.* **31**, 841–851.

Nei, M. (1977), "*F*-statistics and the analysis of gene diversity in subdivided populations," *Ann. Hum. Genet.* **41**, 225–233.

Nei, M. (1987), *Molecular Evolutionary Genetics*, Columbia Univ. Press, New York.

Nei, M., and Feldman, M. W. (1972), "Identity of genes by descent within and between populations under mutation and migration pressures," *Theor. Popul. Biol.* **3**, 460–465.

Nei, M., and Hughes, A. L. (1991), Polymorphism and the evolution of the major histocompatibility complex loci in mammals, *in* R. K. Selander, A. G. Clark, and T. S. Whittam, eds., *Evolution at The Molecular Level*, Sinauer Assoc., Sunderland, Mass., pp. 222–247.

Nei, M., and Li, W.-H. (1973), "Linkage disequilibrium in subdivided populations," *Genetics* **75**, 213–219.

Neuhauser, C., and Krone, S. M. (1997), "The genealogy of samples in models with selection," *Genetics* **145**, 519–534.

Nielsen, J. T. (1977), "Variation in the number of genes coding for salivary amylase in the bank vole, *Clethrionomys glareola*," *Genetics* **85**, 155–169.

Nielsen, R., Williamson, S., Kim, Y., Hubisz, M. J., Clark, A. G., and Bustamante, C. (2005), "Genomic scans for selective sweeps using SNP data," *Genome Res.* **15**, 1566–1575.

Nielsen, V. H., and Andersen, S. (1987), "Selection for growth on normal and reduced protein diets in mice. I. Direct and correlated responses for growth," *Genet. Res. (Camb.)* **50**, 7–15.

Nilsson-Ehle, H. (1909), "Kreuzungsuntersuchungen an Hafer und Weizen," *Acta Univ. Lund* **5**(2), 1–122.

Nordborg, M., Charlesworth, B., and Charlesworth, D. (1996), "The effect of recombination on background selection," *Genet. Res. (Camb.)* **67**, 159–174.

Nordborg, M., Franklin, I. R., and Feldman, M. W. (1995), "Effects of *cis-trans* viability selection on some two-locus models," *Theor. Popul. Biol.* **47**, 365–392.

Nusslein-Volhard, C., and Wieschaus, E. (1980), "Mutations affecting segment number and polarity in *Drosophila*," *Nature (London)* **287**, 795–801.

Oakeshott, J. G., Gibson, J. B., Anderson, P. R., and Chambers, R. K. (1982), "Alcohol dehydrogenase and glycerol-3-phosphate dehydrogenase clines in *Drosophila melanogaster* on different continents," *Evolution* **36**, 86–96.

Ochman, H., Jones, J. S., and Selander, R. K. (1983), "Molecular area effects in *Cepaea*," *Proc. Natl. Acad. Sci. USA* **80**, 4189–4193.

Ohno, S. (1970), *Evolution by Gene Duplication*, Allen and Unwin, London.

Ohta, T. (1973), "Slightly deleterious mutant substitutions in evolution," *Nature (London)* **246**, 96–98.

Ohta, T., and Kimura, M. (1969), "Linkage disequilibrium due to random genetic drift," *Genet. Res. (Camb.)* **13**, 47–55.

OMIM (2006), "Online Mendelian Inheritance in Man," human genetic database at http://www.ncbi.nlm.nih.gov/entrez/query.fcgi?db=OMIM.

Oostra, B. A., and Chiurazzi, P. (2001), "The fragile X gene and its function," *Clin. Genet.* **60**, 399–408.

Painter, T. S. (1934), "A new method for the study of chromosome aberrations and the plotting of chromosome maps in *Drosophila melanogaster*," *Genetics* **19**, 175–188.

Patterson, N. J. (2005), "How old is the most recent ancestor of two copies of an allele?," *Genetics* **169**, 1093–1104.

Pedersen, J. S., and Hein, J. (2003), "Gene finding with a hidden Markov model of genome structure and evolution," *Bioinformatics* **19**, 219–227.

Price, G. R. (1972), "Fisher's Fundamental Theorem made clear," *Ann. Hum. Genet.* **36**, 129–140.

Price, M. V., and Waser, N. M. (1979), "Pollen dispersal and optimal outcrossing in *Delphinium nelsonii*," *Nature (London)* **277**, 294–297.

Prout, T. (1968), "Sufficient conditions for multiple niche polymorphism," *Amer. Natur.* **102**, 493–496.

Provine, W. B. (1971), *The Origins of Theoretical Population Genetics*, Univ. Chicago Press, Chicago.

Pylkov, K. V., Zhivotovsky, L. A., and Christiansen, F. B. (1999), "The strength of the selection barrier between populations," *Genet. Res. (Camb.)* **76**, 179–185.

Rasmuson, M. (1952), "Variation in bristle number of *Drosophila melanogaster*," *Acta Zoolog., Stockholm* **51**, 210–237.

Redon, R., Ishikawa, S., Fitch, K. R., Feuk, L., Perry, G. H., Andrews, T. D., Fiegler, H., Shapero, M. H., Carson, A. R., Chen, W., Cho, E. K., Dallaire, S., Freeman, J. L., Gonzalez, J. R., Gratacos, M., Huang, J., Kalaitzopoulos, D., Komura, D., MacDonald, J. R., Marshall, C. R., Mei, R., Montgomery, L., Nishimura, K., Okamura, K., Shen, F., Somerville, M. J., Tchinda, J., Valsesia, A., Woodwark, C., Yang, F., Zhang, J., Zerjal, T., Zhang, J., Armengol, L., Conrad, D. F., Estivill, X., Tyler-Smith, C., Carter, N. P., Aburatani, H., Lee, C., Jones, K. W., Scherer, S. W., and Hurles, M. E. (2006), "Global variation in copy number in the human genome," *Nature (London)* **444**, 444–454.

Reiersøl, O. (1962), "Genetic algebras studied recursively and by means of differential operators," *Math. Scand.* **10**, 25–44.

Richman, A. D., Uyenoyama, M. K., and Kohn, J. R. (1996), "Allelic diversity and gene genealogy at the self-incompatibility locus in the Solanaceae," *Science* **273**, 1212–1214.

Rigoutsos, I., Huynh, T., Miranda, K., Tsirigos, A., McHardy, A., and Platt, D. (2006), "Short blocks from the noncoding parts of the human genome have instances within nearly all known genes and relate to biological processes," *Proc. Natl. Acad. Sci. USA* **103**, 6605–6610.

Robbins, R. B. (1918), "Some applications of mathematics to breeding problems, III," *Genetics* **3**, 375–389.

Robertson, A. (1962), "Selection for heterozygotes in small populations," *Genetics* **47**, 1291–1300.

Robertson, A. (1966), "Artificial selection in plants and animals," *Proc. Roy. Soc. Lond. B* **164**, 341–349.

Robertson, A. (1970), A theory of limits in artificial selection with many linked loci, *in* K. Kojima, ed., *Mathematical Topics in Population Genetics*, vol. 1 of *Biomathematics*, Springer Verlag, Berlin, Heidelberg, New York, pp. 246–288.

Roelofs, W. L., Glover, T. J., Tang, X.-H., Sreng, I., Robbins, P. S., Eckenrode, E. E., Löfstedt, C., Hansson, B. S., and Bengtsson, B. O. (1987), "Sex pheromone production and perception in European corn borer moths is determined by both autosomal and sex-linked genes," *Proc. Natl. Acad. Sci. USA* **84**, 7585–7589.

Rosenblatt, A., Brinkman, R. R., Liang, K. Y., Almqvist, E. W., Margolis, R. L., Huang, C. Y., Sherr, M., Franz, M. L., Abbott, M. H., Hayden, M. R., and Ross, C. A. (2001), "Familial influence on age of onset among siblings with Huntington disease," *Am. J. Med. Genet.* **105**, 399–403.

Sabeti, P. C., Schaffner, S. F., Fry, B., Lohmueller, J., Varilly, P., Shamovsky, O., Palma, A., Mikkelsen, T. S., Altshuler, D., and Lander, E. S. (2006), "Positive natural selection in the human lineage," *Science* **312**, 1614–1620.

Sachidanandam, R., Weissman, D., Schmidt, S. C., Kakol, J. M., Stein, L. D., Marth, G., Sherry, S., Mullikin, J. C., Mortimore, B. J., Willey, D. L., Hunt, S. E., Cole, C. G., Coggill, P. C., Rice, C. M., Ning, Z., Rogers, J., Bentley, D. R., Kwok, P. Y., Mardis, E. R., Yeh, R. T., Schultz, B., Cook, L., Davenport, R., Dante, M., Fulton, L., Hillier, L., Waterston, R. H., McPherson, J. D., Gilman, B., Schaffner, S., Van Etten, W. J., Reich, D., Higgins, J., Daly, M. J., Blumenstiel, B., Baldwin, J., Stange-Thomann, N., Zody, M. C., Linton, L., Lander, E. S., and Altshuler, D. (2001), "A map of human genome sequence variation containing 1.42 million single nucleotide polymorphisms," *Nature (London)* **409**, 928–933.

Sagitov, S., and Jagers, P. (2005), "The coalescent effective size of age-structured populations," *Ann. Appl. Prob.* **15**, 1778–1797.

Sanger, F., Nicklen, S., and Coulson, A. R. (1977), "DNA sequencing with chain-terminating inhibitors.," *Proc. Natl. Acad. Sci. USA* **74**, 5463–5467.

Sax, K. (1923), "The association of size differences with seed coat pattern and pigmentation in *Phaseolus vulgaris*," *Genetics* **8**, 552–560.

Scharloo, W., van Dijken, F. R., Hoorn, A. J. W., de Jong, G., and Thörig, G. E. W. (1977), Functional aspects of genetic variation, *in* F. B. Christiansen, and T. Fenchel, eds., *Measuring Selection in Natural Populations*, vol. 19 of *Lecture Notes in Biomathematics*, Springer-Verlag, Berlin, Heidelberg, New York, pp. 131–147.

Scheel, J. J. (1968), *Rivulins of the Old World*, T. F. H. Publ., Neptune City, NJ.

Schierup, M. H., and Christiansen, F. B. (1996), "Inbreeding depression and outbreeding depression in plants," *Heredity* **77**, 461–468.

Schierup, M. H., and Hein, J. (2000*a*), "Consequences of recombination on traditional phylogenetic analysis," *Genetics* **156**, 879–891.

Schierup, M. H., and Hein, J. (2000*b*), "Recombination and the molecular clock," *Mol. Biol. Evol.* **17**, 1578–1579.

Schierup, M. H., Vekemans, X., and Christiansen, F. B. (1997), "Evolutionary dynamics of sporophytic self-incompatibility alleles in plants," *Genetics* **147**, 835–846.

Schierup, M. H., Vekemans, X., and Christiansen, F. B. (1998), "Allelic genealogies in sporophytic self-incompatibility systems in plants," *Genetics* **150**, 1187–1198.

Schmidt, J. (1917), "Racial investigations. I. *Zoarces viviparus* L. and local races of the same," *C. R. Trav. Lab. Carlsberg* **13**, 277–397.

Schmidt, J. (1918), "Racial studies in fishes. I. Statistical investigations with *Zoarces viviparus* L," *J. Genet.* **7**, 105–118.

Seielstad, M. T., Minch, E., and Cavalli-Sforza, L. L. (1998), "Genetic evidence for a higher female migration rate in humans," *Nat. Genet.* **20**, 278–280.

Sheridan, A. K., Frankham, R., Jones, L. P., Rathie, K. A., and Barker, J. S. F. (1968), "Partitioning of the variance and estimation of genetic parameters for various bristle number characters of *Drosophila melanogaster*.," *Theoret. Appl. Genet.* **38**, 179–187.

Sick, K. (1965), "Heamoglobin polymorphism of cod in the Baltic and the Danish Belt Sea," *Hereditas* **54**, 19–48.

Sinnock, P., and Sing, C. F. (1972), "Analysis of multilocus genetic systems in Tecumseh Michigan. II. Consideration of the correlation between non-alleles in gametes," *Amer. J. Hum. Genet.* **24**, 393–415.

Slade, P. F., and Wakeley, J. (2005), "The structured ancestral selection graph and the many-demes limit," *Genetics* **169**, 1117–1131.

Slatkin, M., and Hudson, R. R. (1991), "Pairwise comparisons of mitochondrial DNA sequences in stable and exponentially growing populations," *Genetics* **129**, 555–562.

Smith, K. (1921), "Racial investigations. VI. Statistical investigations on inheritance in *Zoarces viviparus* L.," *C. R. Trav. Lab. Carlsberg* **14**(11), 1–60.

Stebbins, G. L. (1950), *Variation and Evolution in Plants*, Columbia Univ. Press, New York.

Stern, C., and Sherwood, E. R., eds. (1966), *The Origin of Genetics. A Mendel Source Book*, Freeman, San Francisco.

Stoneking, M. (1998), "Genetic evidence for a higher female migration rate in humans," *Nat. Genet.* **20**, 219–220.

Strobeck, C. (1987), "Average number of nucleotide differences in a sample from a single subpopulation: A test for population subdivision," *Genetics* **117**, 149–153.

Sturm, R. A., and Frudakis, T. N. (2004), "Eye colour: portals into pigmentation genes and ancestry," *Trends Genet.* **20**, 327–332.

Sturtevant, A. H. (1925), "The effects of unequal crossing over at the Bar locus in *Drosophila*," *Genetics* **10**, 117–147.

Sturtevant, A. H., and Beadle, G. W. (1939), *An Introduction to Genetics*, W. B. Saunders, New York.

Sturtevant, A. H., and Morgan, T. H. (1923), "Reverse mutation of the Bar gene correlated with crossing over," *Science* **57**, 746–747.

Sved, J. (1971), "Linkage disequilibrium and homozygosity of chromosome segments in finite populations," *Theor. Popul. Biol.* **2**, 125–141.

Szymura, J. M., and Barton, N. H. (1991), "The genetic structure of the hybrid zone between the fire-bellied toads *Bombina bombina* and *B. variegata*: Comparisons between transects and between loci," *Evolution* **45**, 237–261.

Tajima, F. (1983), "Evolutionary relationship of DNA sequences in finite populations," *Genetics* **105**, 437–460.

Tajima, F. (1989a), "Statistical-method for testing the neutral mutation hypothesis by DNA polymorphism," *Genetics* **123**, 585–595.

Tajima, F. (1989*b*), "The effect of change in population-size on DNA polymorphism," *Genetics* **123**, 597–601.

Takahata, N. (1988), "The coalescent in two partially isolated diffusion populations," *Genet. Res. (Camb.)* **52**, 213–222.

Takahata, N. (1990), "A simple genealogical structure of strongly balanced allelic lines and *trans*-species evolution of polymorphism," *Proc. Natl. Acad. Sci. USA* **87**, 2419–2423.

Tamura, K., and Nei, M. (1993), "Estimation of the number of nucleotide substitutions in the control region of mitochondrial DNA in humans and chimpanzees," *Mol. Biol. Evol.* **10**, 512–526.

Tang, H., Siegmund, D. O., Shen, P., Oefner, P. J., and Feldman, M. W. (2002), "Frequentist estimation of coalescence times from nucleotide sequence data using a tree-based partition," *Genetics* **161**, 447–459.

Tavaré, S. (1984), "Line-of-descent and genealogical processes, and their applications in population genetics models," *Theor. Popul. Biol.* **26**, 119–184.

Teshima, K. M., Coop, G., and Przeworski, M. (2006), "How reliable are empirical genomic scans for selective sweeps?," *Genome Res.* **16**, 702–712.

Theissen, G., Becker, A., Di Rosa, A., Kanno, A., Kim, J. T., Munster, T., Winter, K. U., and Saedler, H. (2000), "A short history of MADS-box genes in plants," *Plant Mol. Biol.* **42**, 115–149.

Thoday, J. M. (1955), "Balance, heterozygosity and developmental stability," *Cold Spring Harbor Symp. Quant. Biol.* **20**, 318–326.

Thomson, G. (1977), "The effect of a selected locus on linked neutral loci," *Genetics* **85**, 753–788.

Thorisson, G. A., Smith, A. V., Krishnan, L., and Stein, L. D. (2005), "The international hapmap project web site," *Genome Res.* **15**, 1592–1593. The HapMap resource resides at http://www.hapmap.org.

Tishkoff, S. A., Reed, F. A., Ranciaro, A., Voight, B. F., Babbitt, C. C., Silverman, J. S., Powell, K., Mortensen, H. M., Hirbo, J. B., Osman, M., Ibrahim, M., Omar, S., Lema, G., Nyambo, T. B., Ghori, J., Bumpstead, S., Pritchard, J. K., Wray, G. A., and Deloukas, P. (2007), "Convergent adaptation of human lactase persistence in Africa and Europe," *Nat. Genet.* **39**, 31–40.

True, J. R. (2003), "Insect melanism: the molecules matter," *Trends Ecol. Evol.* **18**, 640–647.

Tuzun, E., Sharp, A. J., Bailey, J. A., Kaul, R., Morrison, V. A., Pertz, L. M., Haugen, E., Hayden, H., Albertson, D., Pinkel, D., Olson, M. V., and Eichler, E. E. (2005), "Fine-scale structural variation of the human genome," *Nature Genetics* **37**, 727–732.

Vekemans, X., and Slatkin, M. (1994), "Gene and allelic genealogies at a gametophytic self-incompatibility locus," *Genetics* **137**, 1157–1165.

Venter, J. C., Adams, M. D., Myers, E. W., Li, P. W., Mural, R. J., Sutton, G. G., Smith, H. O., Yandell, M., Evans, C. A., Holt, R. A., Gocayne, J. D., Amanatides, P., Ballew, R. M., Huson, D. H., Wortman, J. R., Zhang, Q., Kodira, C. D., Zheng, X. H., Chen, L., Skupski, M., Subramanian, G., Thomas, P. D., Zhang, J., Gabor Miklos, G. L., Nelson, C., Broder, S., Clark, A. G., Nadeau, J., McKusick, V. A., Zinder, N., Levine, A. J., Roberts, R. J., Simon, M., Slayman, C., Hunkapiller, M., Bolanos, R., Delcher, A., Dew, I., Fasulo, D., Flanigan, M., Florea, L., Halpern, A., Hannenhalli, S., Kravitz, S., Levy, S., Mobarry, C., Reinert, K., Remington, K., Abu-Threideh, J., Beasley, E., Biddick, K., Bonazzi, V., Brandon, R., Cargill, M., Chandramouliswaran, I., Charlab, R., Chaturvedi, K., Deng, Z., Di Francesco, V., Dunn, P., Eilbeck, K., Evangelista, C., Gabrielian, A. E., Gan, W., Ge, W., Gong, F., Gu, Z., Guan, P., Heiman, T. J., Higgins, M. E., Ji, R. R., Ke, Z., Ketchum, K. A., Lai, Z., Lei, Y., Li, Z., Li, J., Liang, Y., Lin, X., Lu, F., Merkulov, G. V., Milshina, N., Moore, H. M., Naik, A. K., Narayan, V. A., Neelam, B., Nusskern, D., Rusch, D. B., Salzberg, S., Shao, W., Shue, B., Sun, J., Wang, Z., Wang, A., Wang, X., Wang, J., Wei, M., Wides, R., Xiao, C., Yan, C., Yao, A., Ye, J., Zhan, M., Zhang, W., Zhang, H., Zhao, Q., Zheng, L., Zhong, F., Zhong, W., Zhu, S., Zhao, S., Gilbert, D., Baumhueter, S., Spier, G., Carter, C., Cravchik, A., Woodage, T., Ali, F., An, H., Awe, A., Baldwin, D., Baden, H., Barnstead, M., Barrow, I., Beeson, K., Busam, D., Carver, A., Center, A., Cheng, M. L., Curry, L., Danaher, S., Davenport, L., Desilets, R., Dietz, S., Dodson, K., Doup, L., Ferriera, S., Garg, N., Gluecksmann, A., Hart, B., Haynes, J., Haynes, C., Heiner, C., Hladun, S., Hostin, D., Houck, J., Howland, T., Ibegwam, C., Johnson, J., Kalush, F., Kline, L., Koduru, S., Love, A., Mann, F., May, D., McCawley, S., McIntosh, T., McMullen, I., Moy, M., Moy, L., Murphy, B., Nelson, K., Pfannkoch, C., Pratts, E., Puri, V., Qureshi, H., Reardon, M., Rodriguez, R., Rogers, Y. H., Romblad, D., Ruhfel, B., Scott, R., Sitter, C., Smallwood, M., Stewart, E., Strong, R., Suh, E., Thomas, R., Tint, N. N., Tse, S., Vech, C., Wang, G., Wetter, J., Williams, S., Williams, M., Windsor, S., Winn-Deen, E., Wolfe, K., Zaveri, J., Zaveri, K., Abril, J. F., Guigo, R., Campbell, M. J., Sjolander, K. V., Karlak, B., Kejariwal, A., Mi, H., Lazareva, B., Hatton, T., Narechania, A., Diemer, K., Muruganujan, A., Guo, N., Sato, S., Bafna, V., Istrail, S., Lippert, R., Schwartz, R., Walenz, B., Yooseph, S., Allen, D., Basu, A., Baxendale, J., Blick, L., Caminha, M., Carnes-Stine, J., Caulk, P., Chiang, Y. H., Coyne, M., Dahlke, C., Mays, A., Dombroski, M., Donnelly, M., Ely, D., Esparham, S., Fosler, C., Gire, H., Glanowski, S., Glasser,

K., Glodek, A., Gorokhov, M., Graham, K., Gropman, B., Harris, M., Heil, J., Henderson, S., Hoover, J., Jennings, D., Jordan, C., Jordan, J., Kasha, J., Kagan, L., Kraft, C., Levitsky, A., Lewis, M., Liu, X., Lopez, J., Ma, D., Majoros, W., McDaniel, J., Murphy, S., Newman, M., Nguyen, T., Nguyen, N., Nodell, M., Pan, S., Peck, J., Peterson, M., Rowe, W., Sanders, R., Scott, J., Simpson, M., Smith, T., Sprague, A., Stockwell, T., Turner, R., Venter, E., Wang, M., Wen, M., Wu, D., Wu, M., Xia, A., Zandieh, A., and Zhu, X. (2001), "The sequence of the human genome," *Science* **291**, 1304–1351.

Villesen, P., Aagaard, L., Wiuf, C., and Pedersen, F. S. (2004), "Identification of endogenous retroviral reading frames in the human genome," *Retrovirology* **1**(1), 32.

Wahlund, S. (1928), "Zusammensetzung von Populationen und Korrelationserscheinungen von Standpunkt der Vererbungslehre aus betrachtet," *Hereditas* **11**, 65–106.

Wakeley, J. (1998), "Segregating sites in Wright's island model," *Theor. Popul. Biol.* **53**, 166–174.

Wakeley, J. (1999), "Nonequilibrium migration in human history," *Genetics* **153**, 1863–1871.

Walsh, J. B. (1984), "Hard lessons for soft selection," *Amer. Natur.* **124**, 518–526.

Wang, R.-L., Stec, A., Hey, J., Lukens, L., and Doebley, J. (1999), "The limits of selection during maize domestication," *Nature* **398**, 236–239.

Waser, N. M., and Price, M. V. (1994), "Crossing-distance effects in *Delphinium nelsonii:* outbreeding and inbreeding depression in progeny fitness," *Evolution* **48**, 842–852.

Watson, J. D., and Crick, F. H. C. (1953), "Genetical implications of the structure of deoxyribose nucleic acid," *Nature (London)* **171**, 964–967.

Watterson, G. A. (1962), "Some theoretical aspects of diffusion theory in population genetics," *Ann. Math. Stat.* **33**, 939–957.

Watterson, G. A. (1975), "On the number of segregating sites in genetical models without recombination," *Theor. Popul. Biol.* **7**, 256–276.

Watterson, G. A. (1976), "Reversibility of the age of an allele. I. Morgan's infinitely many alleles model," *Theor. Popul. Biol.* **10**, 239–253.

Watterson, G. A. (1978), "The homozygosity test of neutrality," *Genetics* **88**, 405–417.

Weir, B. S. (1996), *Genetic Data Analysis II*, Sinauer Assoc., Sunderland, MA.

White, M. J. D. (1978), *Modes of Speciation*, W. H. Freeman and Co., San Francisco.

Williams, E., and Hurst, L. D. (2000), "The proteins of linked genes evolve at similar rates," *Nature (London)* **407**, 900–903.

Williams, W. R., and Anderson, D. E. (1984), "Genetic epidemiology of breast cancer: Segregation analysis of 200 Danish pedigrees," *Genet. Epidemiol.* **1**, 7–20.

Wiuf, C. (2000), "A coalescence approach to gene conversion," *Theor. Popul. Biol.* **57**, 357–367.

Wiuf, C., and Hein, J. (1997), "On the number of ancestors to a DNA sequence," *Genetics* **147**, 1459–1468.

Wiuf, C., and Hein, J. (2000), "The coalescent with gene conversion," *Genetics* **155**, 451–462.

Wooster, R., Neuhausen, S. L., Mangion, J., Quirk, Y., Ford, D., Collins, N., Nguyen, K., Seal, S., Tran, T., and Averill, D. (1994), "Localization of a breast cancer susceptibility gene, BRCA2, to chromosome 13q12–13," *Science* **265**, 2088–2090.

Wright, S. (1931), "Evolution in Mendelian populations," *Genetics* **16**, 97–159.

Wright, S. (1933), "Inbreeding and recombination," *Proc. Natl. Acad. Sci. USA* **19**, 420–433.

Wright, S. (1935), "Evolution in populations in approximate equilibrium," *J. Genet.* **30**, 257–266.

Wright, S. (1937), "The distribution of gene frequencies in populations," *Proc. Natl. Acad. Sci. USA* **87**, 430–431.

Wright, S. (1938), "Size of population and breeding structure in relation to evolution," *Science* **87**, 430–431.

Wright, S. (1939), "The distribution of self-sterility alleles in populations," *Genetics* **24**, 538–552.

Wright, S. (1943), "Isolation by distance," *Genetics* **16**, 114–138.

Wright, S. (1949), Adaptation and selection, *in* G. L. Jepson, G. G. Simpson, and E. Mayr, eds., *Genetics, Paleontology and Evolution*, Princeton Univ. Press, Princeton, N. J., pp. 365–89.

Wright, S. I., Foxe, J. P., DeRose-Wilson, L., Kawabe, A., Looseley, M., Gaut, B. S., and Charlesworth, D. (2006), "Testing for effects of recombination rate on nucleotide diversity in natural populations of *Arabidopsis lyrata*," *Genetics* **174**, 1421–1430.

Yoo, B. H. (1980*a*), "Long-term selection for a quantitative character in large replicate populations of *Drosophila melanogaster*. I. Response to selection," *Genet. Res. (Camb.)* **35**, 1–17.

Yoo, B. H. (1980*b*), "Long-term selection for a quantitative character in large replicate populations of *Drosophila melanogaster*. II. Lethals and visible mutants with large effects," *Genet. Res. (Camb.)* **35**, 19–31.

Yu, F., Hill, R. S., Schaffner, S. F., Sabeti, P. C., Wang, E. T., Mignault, A. A., Ferland, R. J., Moyzis, R. K., Walsh, C. A., and Reich, D. (2007), "Comment on "ongoing adaptive evolution of ASPM, a brain size determinant in *Homo sapiens*"," *Science* **316**, 370–372.

Zhivotovsky, L. A., and Christiansen, F. B. (1995), "The selection barrier between populations," *Evolution* **49**, 490–501.

Zhu, G., Evans, D. M., Duffy, D. L., Montgomery, G. W., Medland, S. E., Gillespie, N. A., Ewen, K. R., Jewell, M., Liew, Y. W., Hayward, N. K., Sturm, R. A., Trent, J. M., and Martin, N. G. (2004), "A genome scan for eye color in 502 twin families: Most variation is due to a QTL on chromosome 15q," *Twin Res.* **7**, 197–210.

Zuckerkandl, E., and Pauling, L. (1962), Molecular disease, evolution, and genetic heterogeneity, *in* M. Kasha, and B. Pullman, eds., *Horizons in Biochemistry*, Academic Press, New York, pp. 189–225.

Index